LONDON MATHEMATICAL SOCIETY LECTURE NOTE SERIES

Managing Editor: Professor I.M.James,
Mathematical Institute, 24-29 St Giles, Oxford

1. General cohomology theory and K-theory, P.HILTON
4. Algebraic topology: a student's guide, J.F.ADAMS
5. Commutative algebra, J.T.KNIGHT
8. Integration and harmonic analysis on compact groups, R.E.EDWARDS
9. Elliptic functions and elliptic curves, P.DU VAL
10. Numerical ranges II, F.F.BONSALL & J.DUNCAN
11. New developments in topology, G.SEGAL (ed.)
12. Symposium on complex analysis, Canterbury, 1973, J.CLUNIE & W.K.HAYMAN (eds.)
13. Combinatorics: Proceedings of the British combinatorial conference 1973, T.P.McDONOUGH & V.C.MAVRON (eds.)
14. Analytic theory of abelian varieties, H.P.F.SWINNERTON-DYER
15. An introduction to topological groups, P.J.HIGGINS
16. Topics in finite groups, T.M.GAGEN
17. Differentiable germs and catastrophes, Th.BROCKER & L.LANDER
18. A geometric approach to homology theory, S.BUONCRISTIANO, C.P.ROURKE & B.J.SANDERSON
20. Sheaf theory, B.R.TENNISON
21. Automatic continuity of linear operators, A.M.SINCLAIR
23. Parallelisms of complete designs, P.J.CAMERON
24. The topology of Stiefel manifolds, I.M.JAMES
25. Lie groups and compact groups, J.F.PRICE
26. Transformation groups: Proceedings of the conference in the University of Newcastle upon Tyne, August 1976, C.KOSNIOWSKI
27. Skew field constructions, P.M.COHN
28. Brownian motion, Hardy spaces and bounded mean oscillation, K.E.PETERSEN
29. Pontryagin duality and the structure of locally compact abelian groups, S.A.MORRIS
30. Interaction models, N.L.BIGGS
31. Continuous crossed products and type III von Neumann algebras, A.VAN DAELE
32. Uniform algebras and Jensen measures, T.W.GAMELIN
33. Permutation groups and combinatorial structures, N.L.BIGGS & A.T.WHITE
34. Representation theory of Lie groups, M.F.ATIYAH et al.
35. Trace ideals and their applications, B.SIMON
36. Homological group theory, C.T.C.WALL (ed.)
37. Partially ordered rings and semi-algebraic geometry, G.W.BRUMFIEL
38. Surveys in combinatorics, B.BOLLOBAS (ed.)
39. Affine sets and affine groups, D.G.NORTHCOTT
40. Introduction to H_p spaces, P.J.KOOSIS
41. Theory and applications of Hopf bifurcation, B.D.HASSARD, N.D.KAZARINOFF & Y-H.WAN
42. Topics in the theory of group presentations, D.L.JOHNSON
43. Graphs, codes and designs, P.J.CAMERON & J.H.VAN LINT
44. Z/2-homotopy theory, M.C.CRABB
45. Recursion theory: its generalisations and applications, F.R.DRAKE & S.S.WAINER (eds.)
46. p-adic analysis: a short course on recent work, N.KOBLITZ
47. Coding the Universe, A. BELLER, R. JENSEN & P. WELCH
48. Low-dimensional topology, R. BROWN & T.L. THICKSTUN (eds.)
49. Finite geometries and designs, P. CAMERON, J.W.P. HIRSCHFELD & D.R. HUGHES (eds.)
50. Commutator Calculus and groups of homotopy classes, H.J. BAUES
51. Synthetic differential geometry, A. KOCK
52. Combinatorics, H.N.V. TEMPERLEY (ed.)
53. Singularity theory, V.I. ARNOLD
54. Markov processes and related problems of analysis, E.B. DYNKIN
55. Ordered permutation groups, A.M.W. GLASS
56. Journees arithmetiques 1980, J.V. ARMITAGE (ed.)
57. Techniques of geometric topology, R.A. FENN
58. Singularities of differentiable functions, J. MARTINET
59. Applicable differential geometry, F.A.E. PIRANI and M. CRAMPIN
60. Integrable systems, S.P. NOVIKOV et al.

61. The core model, A. DODD
62. Economics for mathematicians, J.W.S. CASSELS
63. Continuous semigroups in Banach algebras, A.M. SINCLAIR
64. Basic concepts of enriched category theory, G.M. KELLY

London Mathematical Society Lecture Note Series. 61

The Core Model

A. DODD
Junior Research Fellow at Merton College, Oxford

CAMBRIDGE UNIVERSITY PRESS

CAMBRIDGE

LONDON NEW YORK NEW ROCHELLE

MELBOURNE SYDNEY

CAMBRIDGE UNIVERSITY PRESS
Cambridge, New York, Melbourne, Madrid, Cape Town, Singapore, São Paulo

Cambridge University Press
The Edinburgh Building, Cambridge CB2 8RU, UK

Published in the United States of America by Cambridge University Press, New York

www.cambridge.org
Information on this title: www.cambridge.org/9780521285308

First published 1982
Re-issued in this digitally printed version 2007

A catalogue record for this publication is available from the British Library

Library of Congress Catalogue Card Number: 81–17989

ISBN 978-0-521-28530-8 paperback

PREFACE

This monograph is intended to give a self-contained presentation of the core model, adding details to the brief account that Ronald Jensen and I presented in [10] and [11]. By "self-contained" I mean that any result not proved should be easy to find in a textbook: I have used Jech's book ([25]) for references wherever possible. I have included just about everything I know about K, sometimes without proofs, and a few details about larger core models.

I am conscious, looking over the text, of many places where the explanations could be clearer, of important results that never quite surface as lemmas and of clumsiness in some of the proofs. I hope that proofs of everything that needs proving can be extracted from the text by a careful reader: I shall be most grateful for comments from anyone who finds the presentation incomprehensible (or erroneous). Two particular comments may help: firstly the exercises range from the very easy to the unsolved. Sometimes they are essential to later results in the text; I hope these ones are not too difficult! Secondly there is no good reason why appendix II is not inserted at appropriate points in the text, as, apart from one result in descriptive set theory, it is self-contained. Were I rewriting the text there would be more to be said about models of ZFC and less about R; but the reader will have to extract general proofs from those given in chapter 12.

Numbers in square brackets thus [1] refer to the bibliography. I have added to this a set of notes on the history of the results in the text, and used this as an excuse not to give attributions of results elsewhere. But I should add the traditiional disclaimer about not being a historian of the subject; it will be apparent that I have relied heavily on the notes in Jech's book. There is one historical point on which I can offer no help: people occasionally ask why mice are so-called. I am afraid that neither Jensen nor I can remember why, but plausible explanations would be welcomed.

Many thanks are due to those who have helped by their own explanations or by their criticism of mine. I should mention particularly Keith Devlin, Bill Mitchell and Philip Welch; and among the discoverers of errors in previous editions Peter Koepke and Lee Stanley. Robin Gandy has been a constant inspirer of improvements from the time when as my supervisor he cast a critical eye over the almost incomprehensible first draft of my thesis up to a recent seminar talk on rudimentary functions which compelled me completely to rewrite part one.

It is appropriate that I should dedicate this work to Ronald Jensen, since so much of it is his work. How much is not apparent from the references: one must also take into account his patient explanations in answer to my endless questionning. I hope the references do make it clear that, although I alone am responsible for the exposition, the major results are our joint work and many of the others are his alone.

I was supported while writing by Junior Research Fellowships first at New College and then from Merton College. To have been able to work in two such delightful communities has meant a great deal to me and I owe more than I can express to the friendship and generosity of the fellows of each.

Professor Ioan James first encouraged me to write this monograph, and to him, as well as to David Tranah and the Cambridge University Press, who gave much advice and help, and waited very patiently for the result, I express my gratitude.

Tony Dodd.
Merton College,
Oxford.

CONTENTS

PRELIMINARIES

My intention in writing this book was that it should be accessible to anyone who had a reasonable background in axiomatic set theory up to Gödel's consistency results; and, if this failed, at least that readers should not be expected to hunt through piles of old journals in search of alleged folklore. The appearance of Jech's book ([25]), and more recently of that of Kunen, have made it easier to find these results and I have not hesitated to give references for results that lie away from the main development of the fine structural theory. Various odd facts have, no doubt, slipped in along the way unproved, but consultation of one of the textbooks mentioned should help. Chapter 0 of Devlin [6] also contains many of the things I feel I ought to have said here. A few notes on particular points are necessary.

LANGUAGE

The language L is the usual first order language for set theory. On official occasions its primitives are $-, \wedge, \exists, (,)$ with binary relation symbols $=$ and $\in$ and variables v_i. Formulae are also treated as objects (definition 1.16). A bounded quantifier is one of the form $(\exists v_i \in v_j)$, where $(\exists v_i \in v_j)\phi$ abbreviates $(\exists v_i)(v_i \in v_j \wedge \phi)$; a formula all of whose quantifiers are bounded is called restricted or Σ_0. In induction on Σ_0 structure, therefore, $(\exists v_i \in v_j)\phi$ is more complex than $v_i \in v_j \wedge \phi$. Generally if variables are displayed after a formula then all, but not necessarily only, free variables are displayed. But this is not a hard and fast rule, and the absence of a variable list is certainly not an indication that we are dealing with a sentence.

The language L_N is L together with N additional unary predicates, usually called $A_1 \ldots A_N$. If they are to be given other names, or we are to add binary relations or constants or anything else, or if we want to rename the $\in$ predicate then the complete list of non-logical symbols is displayed thus: $L_{R,f,c}$.

AXIOMS

A sequence of weak theories is introduced in part one. For reference the axioms of ZF are the universal closures of the following:

(1) $x=y \leftrightarrow \forall z(z\in x \leftrightarrow z\in y)$ (extensionality)

(2) $x\neq\phi \rightarrow (\exists y\in x)(y\cap x=\phi)$ (foundation)

(3) $\exists z(z=\{x,y\})$ (pairing)

(4) $\exists z(z=\bigcup x)$ (union)

$(5)_\phi$ $\exists y\forall z(z\in y \leftrightarrow z\in x\wedge\phi)$ (separation)

$(6)_\phi$ $\forall x\exists y\phi(x,y,p) \rightarrow \forall u\exists v(\forall x\in u)(\exists y\in v)\phi(x,y,p)$ (replacement)

(7) $\exists y(y=P(x))$ (power set)

(8) $\exists y(y=\omega)$ (infinity).

The axiom of choice is

(9) $\forall x(\forall z,z'(z,z'\in x \rightarrow z=z'\vee z\cap z'=\phi) \rightarrow \exists y\forall z\in x\exists! w\, w\in y\cap z)$.

(1)-(9) is called ZFC. (1)-(6),(8) is called ZF^-. The theory KP consists of axioms (1)-(4) together with $(5)_\phi$ and $(6)_\phi$ for restricted ϕ. Its models are called admissible. Admissible sets are too narrow a collection for the fine-structural theory used in part one - the theory R which we use is weaker than KP - but similar considerations motivate the removal of, for example, full replacement from the theory.

FUNCTIONS

The ordered pair $\langle x,y\rangle$ is $\{\{x,y\},\{x\}\}$. A relation is a collection of ordered pairs. A relation R is a function provided $Rxy\wedge Rxz \rightarrow y=z$; then instead of Rxy we may write $R(x)=y$. This rather trivial point needs attention: usually the literature of fine structure has followed Gödel in identifying the function f with $\{\langle f(x),x\rangle : x\in dom(f)\}$. I have used the different notation here because it seems to have become the more commonly used one. It is natural, given this usage, to define the ordered n+1-tuple so that $\langle x_1\ldots x_{n+1}\rangle=\langle\langle x_1\ldots x_n\rangle,x_{n+1}\rangle$; thus n+1-ary functions are functions. Parts of the theory - lemma 1.5 for example - are sensitive to the choice of convention - although they would not be if we were to redefine $\langle x,y\rangle=\{\{x,y\};\{y\}\}$.

I have tried to avoid reliance on the convention by adding unnecessary $\langle\ \rangle$ in places. Sometimes an inverse defined by $x=\langle(x)_0\ldots(x)_{n-1}\rangle$ is used, but care is needed as this is ambiguous. $\vec{x}$ represents a list and not a sequence: so $\vec{x}\in X$ means $x_1\ldots x_n\in X$ and not $\langle x_1\ldots x_n\rangle\in X$. Exceptions occur if there is no danger of confusion.

id is the function $id(x)=x$.

ORDINALS AND CARDINALS

$\overline{\overline{X}}$ denotes the cardinality of X: this is always identified with the least ordinal in one-one correspondence with X (in all our weak systems other than R this can be shown to exist). The order-type of

$\langle X,<\rangle$ is written $otp(\langle X,<\rangle)$; $<$ may be omitted when it is $\in$. $\aleph_\alpha$ and ω_α are used interchangeably.

If κ is regular and uncountable and $C\subseteq\kappa$ then C is closed in κ if $X\subseteq C \Rightarrow \sup X\in C\cup\{\kappa\}$. It is unbounded if $\sup C=\kappa$.

$S\subseteq\kappa$ is stationary provided $S\cap C\neq\phi$ whenever C is closed and unbounded in κ. If C,C' are closed unbounded in κ then so is $C\cap C'$. Indeed, if $\gamma<\kappa$ and $\delta<\gamma \Rightarrow C_\delta$ closed unbounded in κ then $\bigcap_{\delta<\gamma} C_\delta$ is closed unbounded in κ. Fodor's theorem states that if S is stationary in κ and $f:S\to\kappa$ is regressive (i.e. for $x\in S$ $f(x)<x$) then there is $\beta\in\kappa$ and stationary $S'\subseteq S$ such that $f''S'=\{\beta\}$.

STRUCTURES

A structure for L is a pair $\langle M,E\rangle$. We distinguish the structure $\langle M,E\rangle$ from the set M by underlining: $\underline{M}$ denotes $\langle M,E\rangle$. There are some exceptions (other than carelessness) to this rule, though. One is discussed in chapter 4; another will be introduced in a moment. But we have tried to adhere strictly to the rule that underlining should not be used for any other purpose: so $\underline{M}$ and M will never be used simply as different variables.

V denotes the universe, even if the theory is weaker than ZF; this looks bizarre until you get used to it. When $V\models ZFC$, we call a structure $\underline{M}$ an inner model of a theory T provided $\underline{M}\models T$ and M is a transitive class containing On. It is usually safe to ignore the underlining convention with inner models, and in particular with V itself (as we did at the start of the previous sentence).

Another abuse of notation is the writing of $\langle M,A\rangle$ to denote $\langle M,A\cap M\rangle$. This never causes confusion. If $\underline{M}=\langle M,E,A_1\ldots A_N\rangle$ then $\underline{M}|X$ denotes $\langle M\cap X,E\cap X^2,A_1\cap X,\ldots,A_N\cap X\rangle$. In listing structures E may be omitted if it is clear what it should be. If t is a term then $t^M(\vec{y})$ denotes that x such that $\underline{M}\models x=t(\vec{y})$. If convenient a subscript rather than a superscript may be used.

The Mostowski collapsing lemma (a theorem of ZF) says that if $\langle X,E\rangle\models$axiom (1) and E is well-founded then there is a unique transitive set M and a unique isomorphism π such that $\pi:\langle X,E\rangle\cong\langle M,\in\rangle$. If X is a proper class it must also be specified that for all $x\in X$ $\{y:yEx\}$ is a set.

THE LEVY HIERARCHY

Σ_0 has already been defined. Suppose Σ_n is defined: then Π_n is the collection of negations of formulae in Σ_n. Σ_{n+1} is the set of formulae of the form $\exists y\phi$ where $\phi\in\Pi_n$. If Γ is a set of formulae then $\Gamma^T=\{\phi:T\vdash\phi\leftrightarrow\psi$ for some ψ in $\Gamma\}$. T is always omitted; it is taken to be whatever set theory we are working in.

$\pi:\underline{N}\to_{\Sigma_m}\underline{M}$ means that for all Σ_m ϕ and $x_1\ldots x_k\in N$

$\underline{N}\models\phi(x_1\ldots x_k) \Leftrightarrow \underline{M}\models\phi(\pi(x_1)\ldots\pi(x_k))$.

$X\prec_{\Sigma_m}\underline{M}$ (where $X\subseteq M$) means $id|X:\underline{M}|X\to_{\Sigma_m}\underline{M}$. A relation R is $\Sigma_m(\underline{M})$ with parameter p provided there is a Σ_m formula ϕ such that

$R(y) \Leftrightarrow \underline{M}\models\phi(y,p)$.

$\Sigma_m(\underline{M})$ denotes the set of relations which are $\Sigma_m(\underline{M})$ in some parameter p. Really we should use a bold-face Σ for this, but instead we specify explicitly if there is a restriction on parameters.

Note that if $\underline{M},\underline{N}$ are models of ZF^-+AC and $\pi:\underline{M}\to_{\Sigma_1}\underline{N}$ maps On_M cofinally into On_N then $\pi:\underline{M}\to_{\Sigma_m}\underline{N}$ for all m. Suppose this proved for $n\leq m$ and let ϕ be $\exists y\psi$ with ψ Π_m. Say $\underline{N}\models\psi(y,\pi(x_1)\ldots\pi(x_k))$. So if $y\in V^M_{\pi(\beta)}$ $\underline{N}\models(\exists y\in V_{\pi(\beta)})\psi(y,\pi(x_1)\ldots\pi(x_k))$. Since $\pi:\underline{M}\to_{\Sigma_1}\underline{N}$ $\pi(V^M_\beta)=V^N_{\pi(\beta)}$; and $(\exists y\in z)\psi(y,\vec{x})$ is Π_m (see [25] lemma 14.2(ii): it is here that we need the full force of ZF^-) so $\underline{M}\models(\exists y\in V_\beta)\psi(y,x_1\ldots x_k)$, i.e. $\underline{M}\models\phi(x_1\ldots x_k)$. $\pi:\underline{M}\to_{\Sigma_m}\underline{N}$ for all m is abbreviated $\pi:\underline{M}\to_e\underline{N}$ (e for elementary). $id|X:\underline{M}|X\to_e\underline{M}$ is written $X\prec\underline{M}$. The usual terminology, $\underline{N}\prec\underline{M}$ may also be used.

$\Delta_n=\Sigma_n\cap\Pi_n$. Δ_1 formulae are absolute; that is, if ϕ is Δ_1 and $X\subseteq M$, $\vec{x}\in X$ then $\underline{M}\models\phi(\vec{x}) \Leftrightarrow \underline{M}|X\models\phi(\vec{x})$. Δ^T_1 are absolute between models of T; for example, if $\underline{M}$ is an inner model of ZF and $R\in M$ is a partial order then R is well-founded if and only if $\underline{M}\models$R is well-founded. More results of this kind are in appendix II.

Note also that if $\underline{M},\underline{N}\models$ZF and $j:\underline{M}\to_e\underline{N}$ and $j\neq id|M$ then there is an ordinal κ such that $j(\kappa)\neq\kappa$. For otherwise for all x $j(\text{rank}(x))=$ $=\text{rank}(x)$ so $\text{rank}(j(x))=\text{rank}(x)$ and an easy induction shows that for all $\alpha\in On_M$ $j|V^M_\alpha=id|V^M_\alpha$. The least such κ is called the critical point of j.

DESCRIPTIVE SET THEORY

The reals are identified with $P(\omega)$. A formula is arithmetic if all its quantifiers are restricted to range over ω: ϕ is Σ^1_1 if there is an arithmetic formula ψ such that $\phi(x) \Leftrightarrow (\exists a\in P(\omega))\psi(a,x)$. Generally, it is Σ^1_{n+1} if this holds with "Π^1_n" in place of "arithmetic"; and it is Π^1_{n+1} if $\neg\phi$ is Σ^1_{n+1}. $\Delta^1_n=\Sigma^1_n\cap\Pi^1_n$. Free variables may represent either natural numbers or reals: we may therefore also define Σ^1_n sets of reals and Σ^1_n reals. All notation is light-faced. Note especially that if ϕ is, say, Σ^1_1 and we write $\underline{A}\models\phi$ this means that there is a real in A satisfying the appropriate condition; in other words, ϕ is not treated as a second-order formula. Our coverage of descriptive set theory is very skimpy and the reader should consult Jech [25] for details.

NOTATION

Other than the above most notation can be found in the index of definitions. For the rest the following is a list of standard symbols that may not be immediately recognisable as such.

(i) $\setminus$ is complement, but - is negation.

(ii) $\cup$ and $\cap$ are used both for the binary and for the unary union and intersection.

(iii) $|$ denotes functional restriction, $f|X=\{\langle x,y\rangle \in f : x \in X\}$; $f''X$ denotes the range of $f|X$.

(iv) $\leftrightarrow$ indicates a bijection, $\cong$ an isomorphism.

(v) ∞ and On are used interchangeably.

(vi) $\emptyset$ is the null set; ϕ is phi.

(vii) $\square$ marks the end of a proof.

(viii) $\mathcal{P}$ is power set.

(ix) # is a sharp, † is a dagger and ¶ is a pistol.

(x) $H_\kappa=\{x:\overline{\overline{TC(x)}}<\kappa\}$.

The core model arises from the mixture of two techniques that had once seemed incompatible: fine-structure and iterated ultrapowers. Fine structure was designed for use in Gödel's L; but the major application of iterated ultrapowers was to measurable cardinals: and there are no measurable cardinals in L. First let us examine the two sources separately.

1: FINE STRUCTURE

Gödel's constructible universe L may briefly be defined as follows:

$L_0=\phi$;

$L_{\alpha+1}=\mathrm{Def}(L_\alpha)$;

$L_\lambda=\bigcup_{\alpha<\lambda} L_\alpha$ (λ a limit).

Then $L=\bigcup_{\alpha\in On} L_\alpha$. By Def(X) is meant the collection of sets first-order definable over X: that is, all sets $x\subseteq X$ such that for some formula ϕ of $\mathcal{L}$ with n+1 free variables and for $p_1 \ldots p_n \in X$

$x=\{t\in X:\langle X,\in\rangle\models\phi(t,p_1\ldots p_n)\}$.

Gödel's construction is to be thought of as going on within some model V of ZF: then $L\models$ZFC. So Consis(ZF) ⇒ Consis(ZFC).

L also has the important <u>condensation property</u>. Suppose κ is a cardinal and $X\prec L_\kappa$. Let $\pi:\underline{M}\cong\langle X,\in\rangle$ where M is transitive. Such an $\underline{M}$ exists by the Mostowski collapsing lemma. Then $M=L_\lambda$ for some λ. We could get by with conditions on κ and X much weaker than these.

Now consider any $a\subseteq\omega$, $a\in L$. For some cardinal κ $a\in L_\kappa$. Let $X\prec L_\kappa$ with $\bar{\bar{X}}=\aleph_0$, $a\in X$. $\omega\subseteq X$, of course. Let $\pi:\underline{M}\cong\langle X,\in\rangle$ with M transitive. Say $M=L_\lambda$. Then say $\bar{a}=\pi^{-1}(a)$. $\bar{a}\subseteq\omega$ and for all $n\in\omega$ $\pi(n)=n$ so $n\in\bar{a} \Leftrightarrow n\in a$. Hence $\bar{a}=a$: thus $a\in L_\lambda$. But it is easily seen that for infinite λ $\bar{\bar{L}}_\lambda=\bar{\bar{\lambda}}$. And $\bar{\bar{L}}_\lambda=\bar{\bar{X}}=\aleph_0$, so $\bar{\bar{\lambda}}=\aleph_0$: that is, $\lambda<\omega_1$. Hence $a\in L_{\omega_1}$. We have proved $P^L(\omega)\subseteq L_{\omega_1}$. If we carry out the argument in L we deduce that $2^{\aleph_0}\leq\bar{\bar{L}}_\omega=\aleph_1$. By Cantor's theorem $2^{\aleph_0}\geq\aleph_1$ so the continuum hypothesis holds in L. In fact an almost identical proof shows that for all infinite cardinals λ $P^L(\lambda)\subseteq L_{\lambda^+}$, so that GCH holds in L.

Gödel deduced that Consis(ZF) ⇒ Consis(ZF+GCH). In fact the condensation property can be used to deduce more powerful properties such as $\diamondsuit$ (see [6]).

A little close attention to the condensation property reveals many more details about the corollary that $P^L(\omega)\subseteq L_{\omega_1}$. We have already observed that we may replace ω_1 by ω_1^L: and this must be the best possible, since $L\models P(\omega)$ is uncountable. But that does not mean that every $\alpha<\omega_1$ must yield a new subset of ω - we may have $\omega<\alpha<\omega_1$ and $P(\omega)\cap L_\alpha=P(\omega)\cap L_{\alpha+1}$. This is called a gap. Indeed there can be gaps much longer than 1 - see [39].

The existence of a gap is equivalent to some form of comprehension: for it says that all subsets of ω definable over L_λ are already in L_λ. We could (and shall) define a measure of the failure of comprehension over L_λ as the least γ such that $L_\lambda\cap P(\gamma)\neq L_{\lambda+1}\cap P(\gamma)$. Such measures are clearly related to admissible set theory (KP).

In fact it pays to be even more precise. Maybe there are no new subsets of ω Σ_1-definable over L_λ, but there are such Σ_2-definable definable subsets? "We find such questions both interesting and important in their own right. Admittedly, however, the questions - and the methods used to solve them - are somewhat remote from the normal concerns of the set theorist. One might refer to "micro set theory" in contradistinction to the usual "macro set theory". Happily, micro set theory turns out to have non-trivial applications in macro set theory"(Jensen [26]). We shall return to the question of applications at the end of this section.

Jensen's paper just quoted contains a full fine-structural analysis of L. This involves putting "coarse" results into forms that involve "fine" definability distinctions. As an example consider the result of Gödel that if $a\subseteq\gamma$, $a\in L_{\lambda+1}\setminus L_\lambda$ then $\overline{\overline{\lambda}}=\overline{\overline{\gamma}}$. It turns out that if $a\in\Sigma_1(L_\lambda)\setminus L_\lambda$ then there is a function f Σ_1-definable over L_λ of a subset of γ onto λ. (We have to say "a subset of" because if we tried to add trivial values for $\beta\in\gamma\setminus\mathrm{dom}(f)$ we might end up with a Σ_2 function.)

It turns out to be convenient to replace the L hierarchy by a modified form, the J hierarchy. This leaves the total model the same but redistributes sets among the levels. One reason for doing this is that the pairing axiom fails in arbitrary L_λ, which makes for difficulties. As the text uses an odd definition it is worth giving the original here.

A function is called rudimentary if and only if it is finitely generated by the following schemata:

(a) $f(\vec{x})=x_i$;
(b) $f(\vec{x})=x_i\setminus x_j$;
(c) $f(\vec{x})=\{x_i,x_j\}$;
(d) $f(\vec{x})=h(\vec{g}(\vec{x}))$;

(e) $f(y,\vec{x}) = \bigcup_{z\in y} g(z,\vec{x})$.

Rudimentary functions were invented as part of a programme to generalise the notion of primitive recursive to arbitrary sets (see [17] and [32]).

Now the J hierarchy is defined as follows:

$J_0=\phi$;

$J_{\alpha+1}=R(J_\alpha\cup\{J_\alpha\})$;

$J_\lambda=\bigcup_{\alpha<\lambda} J_\alpha$ (λ a limit).

$R(X)$ denotes $\{f(\vec{x}): f \text{ rudimentary}, \vec{x}\in X\}$. Then•

(i) $L=\bigcup_{\alpha\in On} J_\alpha$;

(ii) $On\cap J_\alpha=\omega\alpha$ (whereas $On\cap L_\alpha=\alpha$);

(iii) $J_{\alpha+1}\cap P(J_\alpha)=Def(J_\alpha)$.

$J_\alpha=L_\alpha$ if and only if $\omega\alpha=\alpha$. (iii) shows that R is very like Def; (ii) shows that in general it is longer.

One of the most important technical devices of [26] is the "master code". One way of looking at this is to say that if there is some subset of γ in $\Sigma_1(J_\lambda)\setminus J_\lambda$ then there is a "universal" one. To put this more precisely, let us define the projectum. The Σ_n projectum of λ, ρ^n_λ, is the least γ such that $P(\omega\gamma)\cap\Sigma_n(J_\lambda)\not\subseteq J_\lambda$. We shall restrict ourselves to Σ_1 for a bit. As we have already said there is a Σ_1 function of a subset of $\omega\rho^1_\lambda$ onto J_λ. Indeed there is such a function uniformly, called the Σ_1 Skolem function. This is a partial Σ_1 function h with the property that given a fixed enumeration $\langle\phi_i: i<\omega\rangle$ of Σ_1 formulae with two free variables

$\exists y\phi_i(y,x) \rightarrow \phi_i(h(i,x),x)$.

It may help if we sketch a proof of the assertion that h is a surjection when restricted to $\omega\times\omega\rho^1_\lambda$ (in fact this is not quite exact because h may need a parameter argument as well, but we shall avoid this difficulty). Suppose $a\in P(\omega\rho^1_\lambda)\cap(\Sigma_1(J_\lambda)\setminus J_\lambda)$. Assume for simplicity that a has a parameter-free Σ_1 definition over J_λ:

$\beta\in a \leftrightarrow J_\lambda\models\phi(\beta)$.

Now let $X=h''(\omega\times\omega\rho^1_\lambda)$. It is easily seen that $X\prec_{\Sigma_1} J_\lambda$. Let $\pi:\underline{M}\cong\langle X,\in\rangle$ with M transitive. Now the condensation property holds in the J hierarchy for all levels - indeed there is a Π_2 sentence saying, in any transitive set, "I am a J_α" - so $M=J_\beta$ for some β. $\pi:J_\beta\to_{\Sigma_1} J_\lambda$. Now for $\gamma\in\omega\rho^1_\lambda$ $\pi(\gamma)=\gamma$, and $\underline{M}\models\phi(\gamma) \leftrightarrow J_\lambda\models\phi(\pi(\gamma)) \leftrightarrow J_\lambda\models\phi(\gamma)$, so $a\in\Sigma_1(J_\beta)$ If $\beta<\lambda$ we should have $a\in J_{\beta+1}\subseteq J_\lambda$, but $a\notin J_\lambda$, so $\beta=\lambda$. So $\pi:J_\lambda\to_{\Sigma_1} J_\lambda$. Furthermore given $x\in J_\lambda$ $\pi(x)\in X$ so $\pi(x)=h(i,\gamma)$ say, with $i\in\omega$ and $\gamma<\omega\rho^1_\lambda$. But then $x=h(i,\gamma)$ so $x\in X$; thus $X=J_\lambda$.

h is not total, of course. Now for our master code. Since there is a Σ_1 map of $\omega\rho^1_\lambda$ onto $J_{\rho^1_\lambda}$ it is possible to make it a subset of

$J_{\rho^1_\lambda}$ - we can then code it in $\omega\rho^1_\lambda$ if necessary. In general we only know that for some parameter p $h''(\omega\times(\omega\rho^1_\lambda\times\{p\}))=J_\lambda$: let p_λ be the least such p in the well-order of J_λ used to prove $L\models AC$. Then let $A=\{\langle i,x\rangle: i\in\omega \wedge x\in J_{\rho^1_\lambda} \wedge J_\lambda\models\phi_i(x,p_\lambda)\}$. For every $B\in\Sigma_1(J_\lambda)\cap P(J_{\rho^1_\lambda})$ there is a rudimentary function f such that $x\in B \Leftrightarrow f(x)\in A$. A is a Σ_1 master code.

Having obtained this result we find that the means are more interesting that the end. For A does not merely code subsets of $\omega\rho^1_\lambda$: it codes the whole of J_λ. For example, if $\pi:\underline{M}\to_{\Sigma_1}\langle J_{\rho^1_\lambda},A\rangle$ then there is a unique μ and a unique $\tilde{\pi}\supseteq\pi$ such that $\tilde{\pi}:J_\mu\to_{\Sigma_1}J_\lambda$ and, letting $\underline{M}=\langle M,B\rangle$, B is a Σ_1 master code for J_μ, $M=J_{\rho^1_\mu}$ and $\tilde{\pi}(p_\mu)=p_\lambda$. Also $\tilde{\pi}:J_\mu\to_{\Sigma_2}J_\lambda$. This is proved by reconstructing J_μ from B (see lemma 3.26). To generalise this coding to ρ^n_λ it is necessary to define a relativised projectum of a structure $\langle J_\rho,A\rangle$ and to show that $\rho^2_\lambda=\rho^1_{\rho^1_\lambda,A}$. Since relativised projecta will need detailed consideration in a more general context we shall say nothing more about them here.

In [26] the applications of these principles are all combinatorial. It is difficult to give a brief account of these that reveals the role played by fine structure, and we shall not try, for the applications that concern us are not combinatorial. Essentially the point is this: many nice theorems can be proved about Σ_1 maps, sets and formulae but do not generalise to Σ_n: ultraproducts are a source of such resuls, as we shall see. By fine structural analysis, Σ_{n+1} properties of J_λ are reduced to Σ_1 properties of $J_{\rho^n_\lambda}$, which are then handled nicely and returned to Σ_{n+1} form by a futher bit of fine structure. For example, for every $\Sigma_1(J_\lambda)$ relation R(x,y) there is a Σ_1 function r(x) such that

$$\exists yR(x,y) \Rightarrow R(x,r(x)).$$

(Σ_1 relations are said to be "Σ_1 uniformisable"). This is not so clear for Σ_{n+1} relations. We may prove it as follows, by induction on n. n=0 is given. Suppose R is Σ_{n+1}; recall that there is a Σ_n f mapping a subset of $J_{\rho^n_\lambda}$ onto J_λ. Say $\tilde{R}(x,y) \Leftrightarrow R(f(x),f(y))$: then $\tilde{R}\subseteq J_{\rho^n_\lambda}$ is $\Sigma_{n+1}(J_\lambda)$, hence $\Sigma_1(J_{\rho^n_\lambda},A^n)$ where A^n is the nth master code formed by the inductive process we have just described. Uniformise $\tilde{R}$ by $\tilde{r}$, so that

$$\exists y\tilde{R}(x,y) \Rightarrow \tilde{R}(x,\tilde{r}(x)).$$

Finally f^{-1} is a $\Sigma_n(J_\lambda)$ relation so by induction hypothesis has a uniformising function g; in other words g is an inverse function for f. Let $r(x)=f(\tilde{r}(g(x)))$. r is the required Σ_{n+1} function. For details of other combinatorial applications the reader should consult [26] or [6]. In section 5 of this introduction we shall see how the covering lemma uses fine structure.

In this book we are dealing all the time with relative constructibility. That is, given a set A we define a function as "rudimentary in A" provided that it is generated by (a)-(e) above and

(f) $f(\vec{x})=x_i \cap A$.

We define

$J^A_0=\phi$;

$J^A_{\alpha+1}=R_A(J^A_\alpha \cup \{J^A_\alpha\})$;

$J^A_\lambda=\bigcup_{\alpha<\lambda} J^A_\alpha$ (λ a limit)

where R_A is like R with "rudimentary in A" in place of "rudimentary". It is also possible to define an L^A_α hierarchy by using an operation Def_A replacing L by L_1 and $\langle X,\in\rangle$ by $\langle X,\in,X\cap A\rangle$. Or, indeed, for any number of $A_1 \ldots A_N$. $\bigcup_{\alpha\in On} J^A_\alpha$ is always called L[A]. The most striking fact about the J^A_α hierarchy is that it does not have the condensation property: if $X \prec J^A_\alpha$ and $\pi:\underline{M} \cong \langle X,\in,A\cap X\rangle$ with M transitive and $\underline{M}=\langle M,\in,\bar{A}\rangle$ then certainly for some β $M=J^A_\beta$. But $\bar{A}=A\cap J^A_\beta$ does not necessarily hold. But in the main application cited, the proof that there is a $\Sigma_1(J_\lambda)$ map of a subset of $J_{\rho^1_\lambda}$ onto J_λ it was essential that the transitive collapse should leave us in the same hierarchy. Of course if $\omega\rho^1_\lambda >$ $>$sup A this is possible so ordinary fine structure holds above sup A; this is not much consolation. In fact we speedily see that the Σ_1 map property fails. Suppose $a \subseteq \omega$ but $a \notin L$. Let $A=\{\omega_1+n : n\in a\}$. Then for $\alpha<\omega_1$ $A\cap J_\alpha=\phi$ so $J^A_\alpha=J_\alpha$: hence $a \notin J^A_{\omega_1}$. Also for all $\beta<\omega_1+n$ $A\cap\beta\in L$ as it is finite, so $J^A_{\omega_1+1}=J_{\omega_1+1}$ and $a\notin J^A_{\omega_1+1}$. But $a\in\Sigma_1(J^A_{\omega_1+1})$, for $a=\{n : \omega_1+n\in A\}$. Yet it is impossible that there should be any map of ω onto $J^A_{\omega_1+1}$, whether $\Sigma_1(\underline{J}^A_{\omega_1+1})$ or not.

Worse is to come. Suppose we had let $A=\{\langle\omega_1,n\rangle : n\in a\}$. Then we should have had $a\in J^A_{\omega_1+1}$: but since $A\cap J^A_{\omega_1}=\phi$, $\Sigma_\omega(\langle J^A_{\omega_1},A\rangle)=\Sigma_\omega(J^A_{\omega_1})\subseteq L$ so $a\notin\Sigma_\omega(J^A_{\omega_1})$. Thus there may be undefinable new subsets of ω in $J^A_{\omega_1+1}$: this means that the projectum is not a satisfactory index of formation of new subsets. And we no longer have $R_A(J^A_\lambda \cup \{J^A_\lambda\}) \cap P(J^A_\lambda)=$ $=Def_A(J^A_\lambda)$.

We may still define a master code as before; but, for example, the Σ_1 master code may no longer code all the Σ_1 subsets of $\omega\rho^1_{\lambda,A}$ over J_λ. For example in the first example above we may assume $\aleph^L_1=\aleph_1$; but no subset of ω could code $\aleph_1$ reals in the sort of coding we have used. We said, though, that the major interest of the master code was that it coded the whole of J_λ. What structure do these "inadequate" master codes code?

A fact about L that we did not state is that if A is the Σ_1 master code of J_λ then $\langle J_{\rho^1_\lambda},A\rangle$ is amenable, i.e. for all $x\in J_{\rho^1_\lambda}$ $x\cap A\in J_{\rho^1_\lambda}$. The coding would have made no sense otherwise, for important arguments in $\langle J_{\rho^1_\lambda},A\rangle$ would have failed. For example the Σ_1 master code of $\langle J_{\rho^1_\lambda},A\rangle$ would not be Σ_1 definable over $\langle J_{\rho^1_\lambda},A\rangle$. So we must know that for all $x\in J^A_{\rho^1_{\lambda,A}}$ $x\cap A\in J^A_{\rho^1_{\lambda,A}}$. For a start it is clear from the definition of $\rho^1_{\lambda,A}$ that $x\cap A\in J^A_\lambda$. But we cannot get any further than that. For example, take $a\notin L$ and $A=\{\omega_1+n:n\in a\}$ again. Take some $\alpha>\omega_1$ with $\rho^1_{\alpha,A}=\omega_1$. Let B be the Σ_1 master code of J^A_α. Then a is rudimentary in B so if $\langle J^A_{\rho^1_{\alpha,A}},B\rangle$ is amenable $a\in J^A_{\omega_1}=J_{\omega_1}\subseteq L$.

$J^A_{\rho^1_{\lambda,A}}$ is not a convenient structure to work with; it is too small. All the sets we were trying to capture in $J^A_{\rho^1_{\lambda,A}}$ were in J^A_λ and bounded subsets of $\omega\rho^1_{\lambda,A}$; and $\omega\rho^1_{\lambda,A}$ is a cardinal in J^A_λ so we could get by with $H=(H_{\omega\rho^1_{\lambda,A}})^{J^A_\lambda}$, the collection of sets in J^A_λ whose transitive closure in J^A_λ is of cardinality less than $\omega\rho^1_{\lambda,A}$ in J^A_λ; certainly $\langle H,B\rangle$ will be amenable, where B is the Σ_1 master code of J^A_λ.

On the other hand H is not a very tidy structure. It turns out, though, that H is $J^B_{\rho^1_{\lambda,A}}$, to which fine structural techniques can be applied. At least, they are equal subject to the acceptability constraint. For example, if H is to be included in $J^B_{\rho^1_{\lambda,A}}$ there cannot be more than $\omega\rho^1_{\lambda,A}$ bounded subsets of $\omega\rho^1_{\lambda,A}$ in J^A_λ. We should have no difficulties if GCH held in J^A_λ.

Here is another difficulty: to get $J^B_{\rho^1_{\lambda,A}}\subseteq H$ we need not only a cardinality restriction (which is easy) but also we need to know that $J^B_{\rho^1_{\lambda,A}}\subseteq J^A_\lambda$. This may fail. (This is exercise 2 of chapter 3. Here is a hint: suppose $\aleph_1=\aleph_1^L$, $2^{\aleph_0}=\aleph_1$ but $V\neq L$. Let $\langle x_\alpha:\alpha<\omega_1\rangle$ enumerate $P(\omega)$ and let $A=\{\langle\omega_1,\alpha,n\rangle:n\in x_\alpha\}$.)

Again GCH will prevent this. GCH can only be formulated if we have the power set axiom, which in general we certainly do not, so we formulate a weaker constraint, called acceptability:

$$P(\gamma)\cap J^A_{\lambda+1}\not\subseteq J^A_\lambda \Rightarrow \forall u\in PP(\gamma)\cap J^A_{\lambda+1}\ \ J^A_{\lambda+1}\models\bar{\bar{u}}\leq\bar{\bar{\gamma}}.$$

This implies a weak form of GCH: if $P(\gamma)$ is a set then $\overline{\overline{P(\gamma)}}=\bar{\bar{\gamma}}^+$; otherwise if $u\subseteq P(\gamma)$ then $\bar{\bar{u}}\leq\bar{\bar{\gamma}}$. Acceptability plainly fails in the case in the hint. (Actually as stated the acceptability property is not sufficiently uniform for the coding property claimed).

"Generalised fine structure" is fine structure that applies to all acceptable J^A_λ, i.e. to all J^A_λ in which the acceptability axiom holds. It is also generalised in that it applies to non well-founded structures. The reason for this latter extension should become apparent as we go along: in order to avoid excessive indexing we present the theory internally as an axiomatic development of sentences that say roughly "I am an acceptable J^A_α". It is always easy to convert back to the terminology of this introduction when dealing with transitive models. Since our structures may have several added predicates, $\underline{M}=\langle M,\in,A_1\ldots A_N\rangle$, it is more convenient to write ρ_M than to list all the predicates by writing $\rho_{\lambda,A_1\ldots A_N}$. Anyway, non transitive models of the axioms are not of the form $J^{A_1\ldots A_N}_\lambda$ so the other form would not work.

Generalised fine-structure preserves the coding property of master codes - that if A_N is the Σ_1 master code of $\underline{N}$ then $\underline{N}$ can be coded in $\langle H_{\rho^1_N},A_N\rangle$. But for $\underline{N}$ to be recovered exactly from the code another condition is necessary. Recall our counterexample in which $\rho^1_N=1$ but $\bar{\bar{N}}=\aleph_1$; it would not be reasonable to expect A_N to code all of N. Since A_N was defined to be $\{\langle i,x\rangle : i\in\omega \wedge x\in H_{\rho^1_N} \wedge \underline{N}\models\phi_i(x,p)\}$ for some p $\underline{N}$ will only be coded by A_N if every x in N is Σ_1 definable from parameters in $H_{\rho^1_N}\cup\{p\}$, that is, if $N=h_N''(\omega\times(\omega\rho^1_N\times\{p\}))$. Such $\underline{N}$ are called p-sound. All levels of the J_α hierarchy are p-sound for some p, but most of the structures that concern us are not.

Given an acceptable $\underline{N}$ we may form its master code even if it is not p-sound and then decode the master code to obtain a p-sound structure $\underline{N}'$. Apart from the fact that there is a Σ_1 map of $\underline{N}'$ into $\underline{N}$ there is little to be said about the relation between $\underline{N}$ and $\underline{N}'$ in general. Provisionally $\underline{N}'$ will be called the core of $\underline{N}$; later we shall give a rather different definition of this term. Although generalised fine structure has little to say about the relation of a structure to its core, the theory of iterated ultrapowers will reveal a very simple relation for the structures that we shall be examining.

Unfortunately we encounter considerable difficulties when we try to iterate this construction to the projectum ρ^n_N. The core in this case is defined differently. Another problem is that although we can form cores we cannot in general extend maps $\pi:\langle H_{\rho^1_N},A_N\rangle\to\underline{M}'$ to $\tilde{\pi}:\underline{N}\to\underline{M}$ with $\underline{M}'=\langle H_{\rho^1_M},A_M\rangle$; unless $\underline{N}$ is p-sound for some p the usual construction breaks down. This does not matter for the structures considered in this book; it is a problem when more general structures - with lots of measurable cardinals, for example - are taken into account. Generalised fine structure does provide an extension of embeddings lemma in this case but it involves a new hierarchy of formulae, called Σ^{*h}_n; this would take us too far afield.

The reader may ask whether it would not anyway be simpler to handle each case as it arises and not bother with a general theory. This raises the very general question of the necessity of fine structure. Recall Jensen's cited remark about the applicability of micro set theory: since those words were written most of the macro set theoretic results proved using fine structure have been proved without it. First there was the technique of "Silver machines", a slowed down construction of L that gave proofs of most of the combinatorial results proved using fine structure and of the covering lemma for L. Then Silver found an even simpler technique, needing nothing more than the notion of a Σ_n substructure. This is the method used in Silver's proof of the covering lemma (see [24]); it can certainly be generalised to the core model.

There are three reasons why we have nevertheless used fine structure in this book.

Firstly, like Jensen "we find such questions both interesting and important in their own right". The projectum is an excellent technical tool for describing various features of mice: and the simplification in not using fine structure is much less than in L. Solovay's crucial lemma - corollary 11.21, proved on the assumption $\omega\rho_N=\kappa$- is a fine-structural statement, but one with consequences that would have to be proved by some means or other.

Secondly, and this is the most important point, we have indicated that the Σ_n hierarchy is not susceptible to generalisation to models that are not p-sound. As soon as we have to look at such models - for example in part six when we consider mice with many measures - we find that it is the properties of the projectum that generalise and the properties of the Σ_n hierarchy that are lost. In fact this hierarchy must be replaced by a new Σ_n^{*h} hierarchy <u>whose definition requires the notion of projectum</u>. It is strange that a technique that was invented to facilitate the analysis of Σ_n formulae should turn out to be of more interest than the formulae themselves; but even a reading of the fine-structure of L soon reveals that the actual hierarchy of formulae is not very important provided we know that its union gives all formulae.

Thirdly, and this point would carry no weight if the second one fell, fine structure has proved a most productive research tool whereas the various simplifications have not yielded any new results, although they have given more elegant proofs of old ones. Admittedly there could be all sorts of subjective explanations of this; and admittedly it is difficult to make comparisons. We feel, though, that investigation of large core models can only really be tackled using the techniques of fine structure. If after the analysis is done it turns out that there is another way to produce the same results, so much the better. Had the aim of this book been simply to prove the covering lemma for K as briefly as possible we should have used Σ_n substructures. In summary, we feel that in a book intended as an aid to research, scenic detours are permissible, and indeed desirable if the direct route conceals important structure.

2: ITERATED ULTRAPOWERS

The question "what large cardinals are there?" is, although undecidable (unless there are none) surely a natural one. As soon as set theory is axiomatised by rules guaranteeing the existence of certain sets the question must arise whether all such rules have been included. In ZF set theory there are two ways of generating big sets: the power set axiom and replacement together with union. It is well known that from these two principles a regular strong limit cardinal that is uncountable cannot be obtained. We could consistently stop with these two operations - classical mathematics would be none the worse - but such a decision surely seems arbitrary. Not that these strong inaccessibles obviously exist; but if caution was to be exercised it should have been exercised a long way earlier. Anyone who is happy about unlimited application of the power set operation can feel few qualms about an inaccessible.

And anyone who admits one strong inaccessible can scarcely complain about two. But there is presumably some limit to this process; for example, is there to be a strong inaccessible κ that is the κth inaccessible? A more natural, although stronger, axiom is that the set of regular cardinals below κ is stationary: such a κ is called Mahlo. This is a new principle for generating large cardinals; and it would be possible, for example, to ask whether the set of Mahlo cardinals less than some κ is stationary in κ. This process could go on indefinitely.

It could be assumed consistently that it stopped, though. Not only is it impossible to prove the existence of 33 Mahlo cardinals from the existence of 32, but also it cannot be proved in ZFC+there are 32 Mahlo cardinals that (Consis ZFC+32 Mahlo cardinals) implies (Consis ZFC+33 Mahlo cardinals) (unless the former theory is inconsistent). Not only does the ad hominem argument have no technical force, it also does not take us very far in the direction we want to follow.

Perhaps the feeling that it is unnatural to stop generating cardinals by some process before the process has been exhausted can be made precise by looking at indescribability. This partly captures the sense that we must discontinue the large cardinal sequence at a point where more apparatus is needed for description of the property.

σ describes κ provided that $\langle V_\kappa, \in, U\rangle \models \sigma$ but for all $\alpha<\kappa$ $\langle V_\alpha, \in, U\cap V_\alpha\rangle \not\models \sigma$, where $U \subseteq V_\alpha$. κ is Δ-indescribable provided that no $\sigma \in \Delta$ describes κ. This classification has links with the old one; for example, κ is strongly inaccessible if and only if κ is Σ_ω-indescribable.

Perhaps that is indeed a reason for restricting generating principles to those of ZF. But if we are determined to go further then we can still use indescribability provided we admit formulae with higher order quantifiers. An m+1-th order quantifier over X ranges over $P^m(X)$. A formula with n alterations of m+1th order quantifiers beginning with a universal is said to be Π^m_n. On the present criterion if we are to admit any inaccessibles then we should not impose any restrictive criterion on cardinals weaker than Π^1_1-describability. Being Π^1_1 indescribable turns out to be the same as being weakly compact, a very natural large cardinal property with many equivalent definitions.

If we did not want to be restrictive at all we might investigate the existence of totally indescribable cardinals. For example, suppose j were an elementary embedding of V to itself, but not the identity. Then there is an ordinal κ such that $j|\kappa = id|\kappa$ but $j(\kappa)>\kappa$. Suppose σ described κ. Then $\langle V_\kappa, \in, U\rangle \models \sigma$ and $\forall\alpha<\kappa \langle V_\alpha, \in, U\cap V_\alpha\rangle \not\models \sigma$. Hence $\forall\alpha<j(\kappa)\langle V_\alpha, \in, j(U)\cap V_\alpha\rangle \not\models \sigma$. But since $j|\kappa = id|\kappa$ $j(U)\cap V_\kappa = U$. Contradiction! So κ is totally indescribable.

As far as the analysis using fine structure is concerned, embeddings are easier to deal with than properties of levels of the V hierarchy. It was noted in section 1 of this introduction that fine structure gives extension of embeddings properties for levels of the J^A hierarchy. But the beautifully simple large cardinal property mentioned - that there is a non-trivial elementary embedding of the universe into itself - is not consistent. There are weaker forms that are not known to be inconsistent, and one of them will concern us closely. A cardinal is said to be critical if it is the critical point of some elementary embedding of the universe into an inner model. This does not yield total indescribability, for if the inner model is called M the former contradiction turns into $\langle V^M_\kappa, \in, j(U)\cap V^M_\kappa\rangle \not\models \sigma$ but $\langle V_\kappa, \in, U\rangle \models \sigma$. Note that since $j|V_\kappa = id|V_\kappa$ $V_\kappa = V^M_\kappa$; indeed if $a \subseteq V_\kappa$ then $j(a)\cap V_\kappa = a$ so $V_{\kappa+1} = V^M_{\kappa+1}$. Thus σ cannot be Π^1_n for any n. Indeed if σ were Π^2_1 it would be downward absolute, so

κ is Π^2_1-indescribable. None of this discussion is intended to make the existence of these cardinals seem plausible; the cardinals that we are discussing will arise naturally in the study of inner models and covering lemmas. Unless they can be proved inconsistent they must be analysed however implausible they seem. But it does seem that critical cardinals are closely connected with other notions of large cardinality that arise naturally if not demonstrably.

For "κ critical" is equivalent to a property that had been invented long before indescribable cardinals came to be discussed. In measure theory a σ-additive measure on a set S is a function $\mu:P(S)\to[0,1]$ such that

(a) $\mu(\phi)=0$, $\mu(S)=1$;

(b) $X\subseteq Y \Rightarrow \mu(X)\leq\mu(Y)$;

(c) $\mu(\{a\})=0$ for all $a\in S$;

(d) if $X_n\subseteq S$ $(n\in\omega)$ and $n\neq m \Rightarrow X_n\cap X_m=\phi$ then $\mu(\bigcup_{n=0}^{\omega} X_n)=\sum_{n=0}^{\omega}\mu(X_n)$.

The question was asked: could there be a set S with a σ-additive measure μ such that $\mathrm{rng}(\mu)=\{0,1\}$? Certainly not for $S=\mathbb{R}$. In fact letting κ be the least cardinal with such μ it can be shown that κ is critical, hence inaccessible; indeed a great deal more than inaccessible. It is trivially true that if $\lambda>\kappa$ then λ also carries a σ-additive 2-valued measure. Suppose that (d) were replaced by

(d)' if $X_\alpha\subseteq S$ $(\alpha<\beta<\bar{\bar{S}})$ and $\alpha<\alpha'<\beta \Rightarrow X_\alpha\cap X_{\alpha'}=\phi$ then $\mu(\bigcup_{\alpha<\beta} X_\alpha)=1$ if and only if $\mu(X_\alpha)=1$ for some $\alpha<\beta$;

call κ measurable if κ carries a measure μ that satisfies (a)-(c) and (d)' and κ is uncountable; then κ is measurable if and only if κ is critical.

The discovery that measurable cardinals are critical was made using the model-theoretic ultrapower technique. To take an ultrapower of the universe, first let $V^\kappa=\{f:f:\kappa\to V\}$ where κ is measurable. Rather than working directly with μ it is customary to let $U=\{X\in P(\kappa):\mu(X)=1\}$ and call U a κ-additive measure on $P(\kappa)$. Let an equivalence relation $\sim$ be defined on V^κ by

$$f\sim g \Leftrightarrow \{\xi:f(\xi)=g(\xi)\}\in U$$

and let $M=V^\kappa/\sim$. Say $f\hat{E}g \Leftrightarrow \{\xi:f(\xi)\in g(\xi)\}\in U$ and let $E=\hat{E}/\sim$. σ-additivity implies that $\langle M,E\rangle$ is well-founded so we may take its Mostowski collapse $\pi:\langle\bar{M},\in\rangle\cong\langle M,E\rangle$. Let c_x denote $\{\langle\xi,x\rangle:\xi\in\kappa\}$. Then if $j(x)=\pi([c_x])$ $j:V\to_e M$ with κ critical. $j:V\to_e M$ is a special case of

$$\bar{M}\models\phi(\pi([f_1])\ldots\pi([f_n])) \Leftrightarrow \{\xi:\phi(f_1(\xi)\ldots f_n(\xi))\}\in U$$

which is Los' theorem; [f] denotes the $\sim$-class of f.

In fact we are not restricted to taking ultrapowers of the universe by U; we could take an ultrapower of any structure M. We usually insist at least that $\langle M,U\rangle$ be amenable. In this case we need only require that $\langle M,U\rangle\models$U is a σ-additive measure. But there is one difficulty: even if M is transitive we cannot show that its ultrapower is well-founded. For there may be a sequence $\langle f_i : i<\omega \rangle$ such that

(a) $\{\xi : f_{i+1}(\xi) \in f_i(\xi)\} \in U$ all $i \in \omega$;

(b) each $f_i \in M$;

but the sequence $\langle f_i : i \in \omega \rangle \notin M$ so that σ-additivity cannot be applied. If U really were a σ-additive measure this would follow, but in general we shall not be able to assume this. Hence our concern with non well-founded models in section 1. Also we can generally only prove $j : M \to_{\Sigma_1} M'$, if we are working with levels of some J^A hierarchy rather than models of ZFC. Our justification for fine structure was that Σ_1 has nice properties that Σ_{n+1} does not; this is the nice property that concerns us particularly.

There was a time when measurable cardinals seemed vast untameable beasts: all known results about them stated negative properties: they were bigger than such and such a cardinal. It was Silver's work on L[U] that began to make them respectable. If U is a normal measure on κ (normality is a slight generalisation of κ-additivity that it is helpful to assume) then $U \cap L[U]$ is a normal measure on κ in L[U]. But in L[U] GCH holds: the first really positive result about measurables that was known. (For a time it was hoped that large cardinals would settle GCH negatively; Silver's result showed that at least measurables do not). Silver transferred various comfortable results about L to L[U], and Solovay showed that L[U] has a fine structure, of which more later. Our concern now is with the work of Kunen and Gaifman.

Here is an obvious question: take the ultrapower of V by U, M say. Let $U' = j(U)$ where $j : V \to_e M$ is the ultrapower map. Could we repeat the process, taking the ultrapower of M by U'? It was with this in mind that we did not insist that U' be a normal measure in V, only in M'. We can repeat this any finite number of times. This gives models $\langle \langle M_n \rangle_{n<\omega}, \langle j_{nm} \rangle_{n \leq m<\omega} \rangle$ such that $M_0 = V$, j_{nn+1} is the ultrapower map of M_n to M_{n+1} and $\bar{j}_{mn} j_{km} = j_{kn}$ $(k \leq m \leq n)$. This is called a commutative system: it has a "direct limit" (a sort of union with points x,y such that $y = j_{nm}(x)$ identified). This direct limit is to be called M_ω, and having dealt with limit stages we can now go on forever. If we start with M rather than V there are two complications:

(a) If $U \notin M$ then the definition of U' fails. It can be repaired provided $\langle M, U \rangle$ is amenable;

(b) If some M_α is not well-founded then we must be able to take an ultrapower of a non-well-founded structure.

M is said to be iterable by U provided each M_α is well-founded.

V is always iterable by U.

Kunen's paper ([35]) proved using these methods that if $U \cap L[U]$ is a normal measure on κ in $L[U]$ then $L[U]$ is an iterate of the unique $L[V]$ with V a normal measure on the least ordinal that carries a normal measure in some inner model; also that $U \cap L[U]$ is the only normal measure in $L[U]$. Not that there may not be two different normal measures U,V on κ: but $L[U]=L[V]$ and $U \cap L[U]= V \cap L[U]$.

Iterated ultrapowers yield indiscernibles. Let $\langle\langle \underline{M}_\alpha\rangle_{\alpha\in On}, \langle \pi_{\alpha\beta}\rangle_{\alpha\leq\beta\in On}\rangle$ be the iterated ultrapower of $\underline{M}$. Let $\underline{M}_\alpha = L[U_\alpha]$ where U_α is a normal measure on κ_α in $L[U_\alpha]$. Then for any formula ϕ and $\alpha_1<\ldots<\alpha_n<\theta, \beta_1<\ldots<\beta_n<\theta$, $x\in M$

$$\underline{M}_\theta \models \phi(\kappa_{\alpha_1}\ldots\kappa_{\alpha_n}, \pi_{0\theta}(x)) \Leftrightarrow \phi(\kappa_{\beta_1}\ldots\kappa_{\beta_n}, \pi_{0\theta}(x)).$$

In the case where M is an inner model these indiscernibles are closely related to some results of Silver that we shall discuss in a moment. If $\underline{M}$ is only some level of the J^U hierarchy (a premouse) then the stated result holds only when ϕ is Σ_1. However a combinatorial argument shows that if all of $\alpha_1\ldots\alpha_n, \beta_1\ldots\beta_n, \theta$ are limit ordinals then ϕ may be Σ_2. If they are limits of limits then ϕ may be Σ_3. And so on. We shall see several applications of this idea.

Suppose M is as before and θ is regular. Then $\kappa_\theta=\theta$. Furthermore given $X\subseteq\theta$, $X\in M$ then $X=\pi_{\alpha\theta}(\overline{X})$ for some $\overline{X}$ and some $\alpha<\theta$. By the indiscernibility result $\kappa_\beta\in X \Leftrightarrow \kappa_{\beta'}\in X$ $(\alpha\leq\beta\leq\beta'<\theta)$. If $X\in U_\theta$ then $\kappa_\alpha\in X$ so $\{\kappa_\beta:\alpha\leq\beta<\theta\}\subseteq X$. It follows that $U_\theta=\{Y\in M: Y\subseteq\theta \wedge (\exists C)(C$ is closed unbounded in $\theta \wedge C\subseteq Y)\}$. This is true irrespective of the choice of U. This is the heart of the proof that if U and V are normal measures on κ in $L[U], L[V]$ then $U\cap L[V]=V\cap L[U]$. If X is not in M but is Σ_1 definable over M the proof still works. But if X is Σ_n definable then the limit point trick must be used.

3: MICE

If Silver tamed measurable cardinals he also showed that from the point of view of L they are wild. Suppose there is a measurable cardinal. Then there is an embedding of V to M, $j:V\to_e M$, other than the identitity. So, since $L^M=L$, $j:L\to_e L$ (j is a class, but, since a non-trivial $j:V\to_e V$ is impossible, cannot be a class of L; that is, there is no formula ϕ such that for some $p\in L$ $j(x)=y \Leftrightarrow L\models\phi(x,y,p)$). Hence $V\neq L$ (so there are no measurable cardinals in L. This is the most indirect proof of the fact imaginable).

Scott ([51]) had already proved this result directly. One might expect that the existence of a measurable cardinal would only affect the region of L above the critical point of j. So Rowbottom's result ([49]) - that if there is a measurable cardinal then $P(\omega)\cap L$ is countable - is unexpected. How can $j:L\to_e L$ with so vast a critical point affect $P(\omega)$?

Although κ may have been chosen least with $j:V\to_e M$ with critical point κ it may not be the least critical point of a $j:L\to_e L$. We shall show that if there is a $j:L\to_e L$ which is non-trivial then there is such a j with countable critical point. Since the critical point of j must be an L-cardinal and $j(\omega_1^L)=\omega_1^L$ it is clear that $\omega_1^L<\omega_1$ so that $P(\omega)\cap L$ is countable.

Suppose, then, that $j:L\to_e L$ with κ critical. Define $U=\{X\subseteq\kappa:X\in L \wedge \wedge\ \kappa\in j(X)\}$. Then U is a normal measure on κ in L (it is <u>not</u> claimed that $U\in L$). $\langle L_{\kappa^+},U\rangle$ can be shown to be amenable; take a countable $X\prec\langle L_{\kappa^+},U\rangle$. Let $\pi:\langle J_\gamma,\bar{U}\rangle\cong\langle X,U\cap X\rangle$. Let $\bar{\kappa}=\pi^{-1}(\kappa)$. Then take the iterated ultrapower of L by $\bar{U}$: it can be shown that L is iterable by $\bar{U}$ so that the iterated ultrapower is of the form $\langle\langle L\rangle_{\alpha\in On},\langle\pi_{\alpha\beta}\rangle_{\alpha\leq\beta\in On}\rangle$ Let $\kappa_\alpha=\pi_{0\alpha}(\bar{\kappa})$. Then the class $C=\{\kappa_\alpha:\alpha\in On\}$ will be indiscernibles for L. C will contain all uncountable cardinals and for every x in L there is a term t of L and there are $\gamma_1\ldots\gamma_n$ such that $L\models x=t(\kappa_{\gamma_1}\ldots\kappa_{\gamma_n})$. C are called the Silver indiscernibles for L.

Silver's construction was rather different: this proof was discovered by Kunen.

If we fix some enumeration $\{\phi_n:n\in\omega\}$ of m(n)-place formulae we may let $0^\#=\{n:L\models\phi_n(\aleph_1\ldots\aleph_{m(n)})\}$. $0^\#$ is a non-constructible real. In fact every constructible real is recursive in $0^\#$.

The statement that $0^\#$ exists - which is equivalent to the statement that there is a non-trivial $j:L\to_e L$ -says that the universe is about as unlike L as could be imagined. For example, suppose κ is a definable ordinal in L, say $\lambda=\kappa \Leftrightarrow L\models\phi(\lambda)$. If $\kappa\geq\aleph_1$ then $x=\aleph_1$ satisfies $(\forall\gamma)(\phi(\gamma) \Rightarrow \gamma\geq x)$, so by indiscernibility all uncountable cardinals will satisfy this (in L). But that is impossible. Thus κ must be countable; so $\aleph_\omega^L$, $(\aleph_{\aleph_1})^L$, the smallest Mahlo cardinal in L and so on are all countable.

This procedure can be generalised to any real a; $a^\#$ exists is equivalent to the statement that there is a non-trivial $j:L[a]\to_e L[a]$ If $a^\#$ exists then the universe is as unlike L[a] as can be imagined.

In particular, define a sequence 0^n by $0^0=0$, $0^{n+1}=(0^n)^\#$. Let $0^\omega=\{\langle n,m\rangle:n\in 0^m\}$: given a suitable coding of ordered pairs 0^ω can be regarded as a real. This can go on for some while, depending on how suitable our suitable codings must be: certainly up to 0^α for any $\alpha<(\omega_1)^L$. But plainly it cannot go right up to 0^{ω_1} which has instead to be coded as a subset of ω_1. This is not too complicated and the 0 sequence can be iterated right through the ordinals, giving a model L[A] where $A=\{\langle\beta,\alpha\rangle:\beta\in 0^\alpha\}$. If there is a non-trivial $j:L[A]\to_e L[A]$ then (surprisingly?) there is one with a countable critical point, so that the embedding can be coded by a real and we start all over again.

Informally the process can be described as follows: take some model M. There cannot be a $j:M \to_e M$ in M (other than the identity), but there may be one in the universe; if there is then add it to M and call the new model M'. Repeat indefinitely, or until a "rigid" model is obtained. This is all very well informally, but adding a proper class (and choosing which one to add) involves difficulties, and the coding we have discussed is needed to make the definition precise; even then the construction may break down long before we reach a rigid model.

We saw that the existence of $0^\#$ was linked to the existence of a certain normal measure on $P(\kappa) \cap L$. Call this U: U is not in L for L must not contain a non-trivial $j:L \to_e L$, and anyway L has no normal measures. Nor will U be a normal measure in L[U], for $X \in U \Rightarrow X \in L$ or $\kappa \setminus X \in L$, whereas $0^\# \in L[U]$ so $P(\kappa) \cap L[U] \neq P(\kappa) \cap L$. U will be a normal measure in an intermediate structure, as we shall see in a moment.

It turns out that there is a remarkable connection between the stages of this process of iteration by adding elementary embeddings and the fine structure of L[U]. This theory - invented by Solovay - must now be summarised.

Recall that in generalised fine structure the condensation lemma fails. Suppose κ is measurable in L[U], U a normal measure on κ in L[U] and L[U] is acceptable. Take $X \prec J_\beta^U$ and let $\pi:\underline{M} \cong J_\beta^U|X$; assume X is countable, M transitive. Suppose $\underline{M} = J_{\bar\beta}^{\bar U}$. $\bar U$ will be a normal measure on $\bar\kappa = \pi^{-1}(\kappa)$ in $\underline{M}$, but may not be a normal measure in $L[\bar U]$. On the other hand $\underline{M}$ is an iterable premouse. Working in L[U] take the κth iterate of $\underline{M}$; this will be of the form $J_{\hat\beta}^F$, where $F = \{X \in L[U] \cap P(\kappa) : \exists C \in L[U]$ C is closed unbounded in $\kappa \wedge C \subseteq X\}$. But then $F \subseteq U$ so $J_{\hat\beta}^F = J_{\hat\beta}^U$. If β were some simply definable ordinal - $\kappa+1$ say - then $\hat\beta$ would satisfy the same definition. Then $J_{\kappa+1}^U$ would be an iterate of $\underline{M}$. This adds a great deal to generalised structure. Recall our definition of the core of a structure; first code $\underline{N}$ in the projectum and then uncode the projectum to get $\underline{N}'$. We have shown that $J_{\kappa+1}^U$ is an iterate of its core. This result is of such importance that it is worth establishing a more general condition for it than $\beta = \kappa+1$.

This condition is $\omega\rho_N^1 \leq \kappa$. First of all, since the condensation lemma holds for $X \prec J_\beta^U$ with $\kappa \subseteq X$ we deduce as in ordinary fine-structure that for some p $N = J_\beta^U = h_{J_\beta^U}"(\omega \times (\kappa \times \{p\}))$ when $\omega\rho_N^1 \leq \kappa$. If $\omega\rho_N^1 = \kappa$ the result is proved, for then $J_\beta^U = h_N"(\omega \times (\omega\rho_N^1 \times \{p\}))$. Suppose then that $\omega\rho_N^1<\kappa$ and let p be such that there is $A\in(\Sigma_1(\underline{N})\cap P(\omega\rho_N^1))\setminus N$ which is Σ_1-definable in parameter p. Let $X=h_N"(\omega\times(\omega\rho_N^1\times\{p\}))$ and let $\pi:\underline{N}'\cong\underline{N}|X$. $\underline{N}'$ is an iterable premouse: let $\underline{M}'$ be its κth iterate. As above $\underline{M}'=\underline{J}^U_{\hat\beta}$ for some $\hat\beta$. Suppose $\hat\beta<\beta$. Then firstly A is Σ_1-definable over $\underline{N}'$ in parameter $\pi^{-1}(p)$, since $\omega\rho_N^1\subseteq X$. But secondly let $\sigma:\underline{N}'\to_{\Sigma_1}\underline{M}'$ be the iteration map. Then since $\omega\rho_N^1<\kappa$ $\omega\rho_N^1<\pi^{-1}(\kappa)$ which is the critical point of σ, so $\sigma|\omega\rho_N^1=\mathrm{id}|\omega\rho_N^1$. Hence A is Σ_1-definable over $\underline{J}^U_{\hat\beta}$ in $\sigma\pi^{-1}(p)$. So $A\in J^U_\beta$, which is a contradiction. A similar argument - using $P(\omega\rho_N^1)\cap N'=P(\omega\rho_N^1)\cap J^U_{\hat\beta}$ - prevents $\hat\beta>\beta$. So $\hat\beta=\beta$. Thus $\underline{N}$ is the κth iterate of $\underline{N}'$: and $\underline{N}'=h_{N'}"(\omega\times(\omega\rho_N^1\times\{\pi^{-1}(p)\}))$. This $\underline{N}'$ is called the core of $\underline{N}$. And the indiscernibles C_N given by the iteration of $\underline{N}'$ to $\underline{N}$ turn out to be Π_1-definable over $\underline{N}$. By the limit point trick they may be thinned to get Σ_n-indiscernibles C^n for all n: and each of these sets will be in $J^U_{\beta+1}$. Indeed each will be in $\Sigma_\omega(\underline{J}^U_\beta)$: yet any set X that is Σ_n-definable over J^U_β can be "measured" by C^n, because $X\in U \Leftrightarrow (\exists\gamma<\kappa)C^n\setminus\gamma\subseteq X$. It follows that $U\cap J^U_{\beta+1}=F\cap J^U_{\beta+1}$ and that if we just close J^U_β under functions rudimentary in $U\cap J^U_\beta$ the result is nevertheless closed under functions rudimentary in U, so that it must be $J^U_{\beta+1}$. Hence $P(\kappa)\cap J^U_{\beta+1}=P(\kappa)\cap\Sigma_\omega(J^U_\beta)$ - a result that is also true in the fine structure of L, although it is not true in general fine structure.

Obviously this is not going to work for all β, irrespective of whether or not $\omega\rho_{J^U_\beta}<\kappa$. We said, though, that the aim of fine-structure was to transfer the nice properties of Σ_1 to all Σ_n; yet there are difficulties here. Suppose $\omega\rho_N^n<\kappa$. Take the usual $\underline{N}'$ and the map $\pi:\underline{N}'\to_{\Sigma_{n+1}}\underline{N}$ given by generalised fine-structure with $\pi|\omega\rho_N^n=\mathrm{id}|\omega\rho_N^n$. Let $\underline{M}'$ be the κth iterate of $\underline{N}'$ with iteration map σ. Then $\underline{M}'$ will be of the form $J^U_{\hat\beta}$. And $\hat\beta\leq\beta$. But when we try to get a contradiction from $\hat\beta<\beta$ we encounter the fact that although $\sigma:\underline{N}'\to_{\Sigma_1}\underline{M}'$ we do not know that $\sigma:\underline{N}'\to_{\Sigma_{n+1}}\underline{M}'$. And it is not difficult to obtain an example where $\hat\beta$ is actually less than β.

In this case we replace the ordinary iteration map by a fine-structural one. Suppose n>1 is least with $\omega\rho_N^n\leq\kappa$. Then using the fact that fine structure holds above κ we show that $N=h"(\omega\times(\omega\rho_N^1\times\{p\}))$ for some p, and so on while $\omega\rho_N^m>\kappa$. Let $\underline{N}'=J^A_{\omega\rho_N^{n-1}}$ where A is the n-1th master code of $\underline{N}$. Then $\underline{N}'$ is a premouse, so we may take its ultrapower by U, $\eta:\underline{N}'\to_{\Sigma_1}\underline{M}'$ and then use the extension of embeddings lemma to get $\hat\eta\supseteq\eta$, $\hat\eta:\underline{N}\to_{\Sigma_{n+1}}\underline{M}$. This process may of course be iterated: $\underline{M}$ is called an n-iterate of $\underline{N}$. Now the gap in the theory may be filled by replacing "iterate" by "n-iterate" throughout. J^U_β is an n-iterate of its core, where n is least with $\omega\rho^{n+1}_{J^U_\beta}\leq\kappa$. 0-iterate means iterate. If for this n $\omega\rho^{n+1}_{J^U_\beta}<\kappa$ then $\Sigma_\omega(J^U_\beta)\cap P(\kappa)=J^U_{\beta+1}\cap P(\kappa)$.

Now consider the other possible cases. Of course if $\alpha<\kappa$ then $J^U_\alpha=J_\alpha$ and if $\alpha\geq\kappa^+$ then $P(\kappa)\cap L[U]\subseteq J^U_\alpha$ so nothing interesting happens. If $\kappa\leq\alpha<\kappa^+$ and $\omega\rho^n_{J^U_\alpha}>\kappa$, all n, then $U\cap J^U_{\alpha+1}\subseteq J^U_\alpha$ as there are no new subsets of κ and $\Sigma_\omega(J^U_\alpha)\cap P(\kappa)=J^U_{\alpha+1}\cap P(\kappa)$. The remaining possibility is that $\omega\rho^n_{J^U_\alpha}=\kappa$ for some n. This case must occur - indeed it occurs when $\alpha=\kappa$ - and gives us the "unpredictable" bits of the measure. In other words we do not expect $\Sigma_\omega(J^U_\alpha)\cap P(\kappa)=J^U_{\alpha+1}\cap P(\kappa)$. This again puts in doubt the suitability of the projectum as an index of formation of new subsets, but it can be shown in this case that for $\gamma<\kappa$ $P(\gamma)\cap J^U_\alpha=P(\gamma)\cap J^U_{\alpha+1}$. In particular $\omega\rho^{n+m}_{J^U_\alpha}=\kappa$ for all $m\geq 0$. The essential property of the projectum - "if there is a new subset of γ then $\omega\rho^n_\alpha\leq\gamma$ for some n" - is preserved. Even so there will be a lot of undefinable new subsets of κ.

Solovay used this theory to show that various combinatorial results from L hold in L[U] (L[U] is always acceptable, by a result of Silver). Later we shall use them to get a covering lemma. Our present interest is in the sharp series. $0^\#$ was equivalent to a certain normal measure U; we looked at the amenable structure $\langle J_{\kappa^+},U\rangle$. Now consider $\underline{J}^U_{\kappa+1}$, with U the measure given by $0^\#$. $U\cap L= =\{X:\kappa\in j(X)\}$ as usual. Since $J^U_{\kappa+1}$ is the closure of J_κ under functions rudimentary in U and $\langle J_{\kappa^+},U\rangle$ is amenable $J^{U^\kappa}_{\kappa+1}\subseteq L_{\kappa^+}$. Clearly equality does not hold. On the other hand some iterate $\underline{N}$ of $J^U_{\kappa+1}$ will contain L_{κ^+} and $P(\kappa)\cap J^U_{\kappa+1}=P(\kappa)\cap N$: so $P(\kappa)\cap L=P(\kappa)\cap J^U_{\kappa+1}$. $J^U_{\kappa+1}$ is an easier structure to handle than $\langle J_{\kappa^+},U\rangle$. $0^\#$ is Σ_1-definable over $J^U_{\kappa+1}$, the iteration indiscernibles of $J^U_{\kappa+1}$ are exactly the Silver indiscernibles for L, and $0^\#\notin L$ so $0^\#\notin J^U_{\kappa+1}$. Hence $\rho_{J^U_{\kappa+1}}=1$.

In other words the existence of $0^\#$ gives an iterable premouse $\underline{N}$ with $\rho_N=1$. But also any iterable premouse generates an embedding of L to L that is not the identity. (The requirement of iterability is not redundant. Although it is a theorem of ZFC that if U is a normal measure then iterated ultrapowers by U are well-founded, it is not necessarily the case that $J^U_{\kappa+1}$ is a model of ZFC; indeed it is not.)

$J^U_{\kappa+1}$ is a simple example. We may transplant the whole of Solovay's analysis of U that are normal measures in L[U] to the analysis of structures $N=J^U_\alpha$ that are acceptable premice with $\omega\rho^n_N\leq\kappa$ (some n, U a measure on κ in N) and all of whose n-iterates are well-founded (this is stronger that iterability). Such structures are called "mice". $J^U_{\kappa+1}$ is a mouse.

We had to say "acceptable" because although L[U] is acceptable we do not know that arbitrary premice are; indeed they are not. But we shall see that iterable premice are always acceptable; the proof puts together all of the fine-structural results that we have discussed.

4: THE CORE MODEL

In a moment we shall return to the sharp series. First we need a lemma extracted from the proof that if $J^U_{\beta+1}$ is an iterable premouse, U a normal measure on κ in $J^U_{\beta+1}$ and $\rho^n_{J^U_\beta}=\kappa$ then for all $\gamma<\kappa$ $P(\gamma)\cap J^U_\beta = P(\gamma)\cap J^U_{\beta+1}$: this says that in this case $\underline{M}_\kappa = \langle H^{J^U_\beta}_\kappa, A^n_{J^U_\beta}\rangle$ is a model of ZFC. This is not an inner model, but we can get such M_κ of arbitrary length by iterating $J^U_{\beta+1}$. Then $M=\cup M_\kappa$ is an inner model. It is called $K_{J^U_{\beta+1}}$. If $\beta=\kappa$ then it is just L. Generally $P(\kappa)\cap M = P(\kappa)\cap J^U_{\beta+1}$.

The interesting relation with the sharp sequence is this: if there is an inner model with a measurable cardinal and we enumerate all β with $\kappa\leq\beta<\kappa^+$ with $\omega\rho^n_\beta=\kappa$, some n, and consturct $K_{J^U_{\beta+1}}$ then we find that we have enumerated precisely the sharp sequence of models. The sharp sequence was a bit vague, but this is quite precise.

An equivalent formulation of $K_{J^U_{\beta+1}}$ exists: $K_{J^U_{\beta+1}}$ is the union of mice which are constructible from $J^U_{\beta+1}$ but from which $J^U_{\beta+1}$ is not constructible, together with L in case there are no mice. (This is more general, K_M being definable for a large class of mice $\underline{M}$). So $K_{J^U_{\beta+1}}$ is the union of mice in $K_{J^U_{\beta+1}}$ (with L). We can turn this into a completely general definition: the <u>core model</u> is the union of all mice together with L. In the text a different definition, which makes $K\models$ZFC immediate, is used. Also $K\models$GCH.

K_M was a definite model but K depends on the size of the universe. For example if $0^\#$ does not exist then K=L: if $0^\#$ exists but $0^{\#\#}$ does not then $K=L[0^\#]$. The possible values of K are the sharp sequence of models, up to a point. Suppose that there is an inner model with a measurable cardinal. Then K has a maximal form: that is, if N is an inner model with a measurable cardinal then $K^N=K$. So K is not K_M for any mouse M in this case. In fact K arises from the inner model with the measurable just as K_M arose from M: $K=\bigcup_{i<\infty} H^{L[U_i]}_{\kappa_i}$ where $L[U_i]$ is the ith iterate of L[U] and U_i is a normal measure on κ_i in $L[U_i]$(recall that if U is taken at a minimal point then <u>all</u> inner models L[V] with V a normal measure in L[V] are of the form $L[U_i]$). Alternatively $K=\bigcap_{i<\infty} L[U_i]$.

Suppose that $j:K \to_e K$ is non-trivial. What happens when we try to extend the sharp sequence by adding j to K? As usual j defines a normal measure U on $P(\kappa)\cap K$, where κ is the critical point of j, such that $X \in U \Leftrightarrow \kappa \in j(X)$. Form L[U]. If $\alpha \leq (\kappa^+)^K$ then $J_\alpha^U \subseteq K$, so U is a normal measure in $J^U_{(\kappa^+)}K$. But suppose $\alpha > \kappa^+$ and $\omega\rho^n_{J_\alpha}U=\kappa$: then J_α^U would be a mouse (this is a gross over-simplification; n-iterability requires a lot of work) but not in K, and this is impossible. Therefore $P(\kappa)\cap L[U] \subseteq K$, so U is a normal measure in L[U]. In other words, if there is a non-trivial $j:K \to_e K$ then there is an inner model with a measurable cardinal. It needs to be added that the unsimplified proof will give a normal measure not on κ but on some larger point.

Just as the core model contains $0^\#$ and its immediate generalisations, so other large cardinal axioms weaker than measurable hold in K provided that they hold in V. For example partition cardinals $\kappa \to (\alpha)^{<\omega}$ and Ramsey cardinals (these are defined in the text), as well as those (such as weakly compact) that relativise to L.

5: THE COVERING LEMMA

Sections 3 and 4 broke off the historical development in order to discuss the core model in its own right: but the core model was developed in order to prove the covering lemma over K; which in turn was motivated by consideration of the singular cardinal hypothesis.

At the risk of covering very familiar ground, we discuss briefly the history of GCH. Since finite sums and products of cardinals are trivially calculated it must have been infuriating for early set theorists that exponents of 2 could not be found - not even the simplest, $2^{\aleph_0}$. And since it was known that $2^{\aleph_\alpha} > \aleph_\alpha$ for all α $2^{\aleph_\alpha} = \aleph_{\alpha+1}$ was the simplest possible answer. Skolem seems to have been the first to advocate the view that there would be no answer on the basis of the ZFC axioms, and this was proved by Gödel([18]) and Cohen([4]). Gödel, as we have seen, proved GCH in L: Cohen's proof used - indeed, inaugurated - the method of forcing. Cohen gave a model in which $2^{\aleph_0}=\aleph_2$: Solovay([56]) showed

(a) that $2^{\aleph_0}$ can be "anything it ought", that is, any κ such that $cf(\kappa) \neq \aleph_0$;

(b) that a similar result can be proved for any regular κ: $cf(2^\kappa) > \kappa$ is essential by König's inequality.

Then Easton ([13]) showed that given a class function G satisfying

(i) $\alpha \leq \beta \Rightarrow G(\alpha) \leq G(\beta)$;

(ii) $cf(\aleph_{G(\alpha)}) > \aleph_\alpha$

which is absolute for ZFC models there is a model of ZFC in which, for regular $\aleph_\alpha$, $2^{\aleph_\alpha}=\aleph_{G(\alpha)}$: the values of $2^{\aleph_\alpha}$ for singular $\aleph_\alpha$ are determined by the values for regular $\aleph_\alpha$ in Easton's model.

Thus the matter stood for some time (Easton obtained his results in 1963-64) and the general view was that some forcing construction would be found that removed the regularity restriction from Easton's result. This was wrong, though, for Silver proved ([55]) that provided $cf(\aleph_\beta)$ is uncountable and $\aleph_\beta$ is singular, then $\forall\alpha<\beta(2^{\aleph_\alpha}=\aleph_{\alpha+1}) \Rightarrow 2^{\aleph_\beta}=\aleph_{\beta+1}$. Actually his result is much broader than this, but it is the fact that there is any restriction at all that is surprising. The proof of [55] rests heavily on the uncountable cofinality assumption: for example, the question whether $2^{\aleph_n}=\aleph_{n+1}$ for all finite n and $2^{\aleph_\omega}=\aleph_{\omega+2}$ are possible in the same model remained open.

Jensen's "Marginalia to a Theorem of Silver" extended these results. On the assumption that $0^\#$ does not exist Jensen firstly showed that for $\aleph_\beta$ singular of uncountable cofinality a very general rule for the determination of $2^{\aleph_\beta}$, the singular cardinal hypothesis, holds, and also that V is L-like in various other ways: for example if $\aleph_\beta$ is singular of uncountable cofinality then $\aleph_\beta$ is singular in L. Then (Marginalia II) he removed the restriction that $cf(\aleph_\beta)$ be uncountable: so, provided that $0^\#$ does not exist, $\forall n\in\omega(2^{\aleph_n}=\aleph_{n+1}) \Rightarrow$ $\Rightarrow 2^{\aleph_\omega}=\aleph_{\omega+1}$. Finally (Marginalia III) he adapted his earlier proofs to get the covering lemma: if $X\subseteq On$ is uncountable then there is $Y\in L$ with $X\subseteq Y$ and $\overline{\overline{Y}}=\overline{\overline{X}}$, still on condition that $0^\#$ does not exist (if $0^\#$ exists the result fails; it also may fail for countable X). In [8] these are all written up with many simplifications.

It will be helpful to summarise the proof, although it is given in full in the text. Perhaps first of all one extra remark about the subsequent literature is needed. There have been many further simplifications to the proof of [8]. At the end of section 1 we said that most of the results proved using fine structure have later been proved without it; this is a case in point. Silver machines gave a short proof of the covering lemma, and subsequently Silver found an elementary proof that used only the idea of Σ_n substructures. We shall not repeat our reasons for nevertheless using fine structure; we have incorporated Silver's other simplifications.

Now for the summary. We are to suppose that for all $\gamma<\tau$ and uncountable $X\subseteq\gamma$ there is $Y\in L$ with $X\subseteq Y$ and $\overline{\overline{Y}}=\overline{\overline{X}}$; we have to show that this holds for τ. If τ is not a cardinal in L then the result holds because it holds for $X\subseteq(\overline{\overline{\tau}})^L$. If $X\subseteq\tau$, $\overline{\overline{X}}=\overline{\overline{\tau}}$ then let $Y=\tau$. So take some X cofinal in τ with $\overline{\overline{X}}<\tau$. Let $Y\prec J_\tau$ with $\overline{\overline{Y}}=\overline{\overline{X}}$ and $X\subseteq Y$. We cannot quite prove $Y\in L$; some modification is needed.

Let $\pi:J_{\bar\tau}\cong J_\tau|Y$, then. The main lemma needed states, roughly, that provided $\bar{\tau}$ is a cardinal in J_β there will be a map $\tilde{\pi}:J_\beta \to_{\Sigma_1} J_{\beta'}$ with $\tilde{\pi}\supseteq\pi$ for some β'. Essentially this is done by coding J_β as a direct system: for every $\gamma<\beta$ and $\delta<\bar{\tau}$ let $X_{\gamma\delta}=h_{J_\gamma}{}''(\omega\times\delta)$. Let $\sigma_{\gamma\delta}:\underline{N}_{\gamma\delta}\cong J_\gamma|X_{\gamma\delta}$ where $N_{\gamma\delta}$ is transitive. Then by the condensation lemma $\underline{N}_{\gamma\delta}=J_{\alpha_{\gamma\delta}}$, say: and because $\bar{\tau}$ is a cardinal in J_β $\alpha_{\gamma\delta}<\bar{\tau}$. Then if we define $\langle\gamma,\delta\rangle\leq_I\langle\gamma',\delta'\rangle \leftrightarrow \gamma\leq\gamma' \wedge \delta\leq\delta'$, and for $\langle\gamma,\delta\rangle\leq_I\langle\gamma',\delta'\rangle$ let $\sigma_{\gamma\delta\gamma'\delta'}=\sigma^{-1}_{\gamma'\delta'}\sigma_{\gamma\delta}$ then $\sigma_{\gamma\delta\gamma'\delta'}:\underline{N}_{\gamma\delta}\to_{\Sigma_0}\underline{N}_{\gamma'\delta'}$ and $\sigma_{\gamma\delta\gamma'\delta'}\in J_\beta$. Let $I=\beta\times\bar{\tau}$; then $\langle J_\beta,\langle\sigma_i\rangle_{i\in I}\rangle$ is the direct limit of $\langle\langle\underline{N}_i\rangle_{i\in I},\langle\sigma_{ij}\rangle_{i\leq_I j}\rangle$. Let $\underline{N}^*_i=\pi(\underline{N}_i)$, $\sigma^*_{ij}=\pi(\sigma_{ij})$ and let $\langle M,\langle\sigma^*_i\rangle_{i\in I}\rangle$ be the direct limit of $\langle\langle\underline{N}^*_i\rangle_{i\in I},\langle\sigma^*_{ij}\rangle_{i\leq_I j}\rangle$. Is M well-founded? This is the most difficult technical point of the proof, and requires careful choice of Y; let us assume that Y has been chosen so that M is well-founded. Let $M=J_{\beta'}$, then, and let $\tilde{\pi}:J_\beta\to_{\Sigma_1}J_{\beta'}$ be defined so that $\tilde{\pi}\sigma_i=\sigma^*_i\pi$ for all $i\in I$. So $\tilde{\pi}\supseteq\pi$.

Now suppose that $\bar{\tau}$ is a cardinal in L. Then the above argument gives $\tilde{\pi}\supseteq\pi$, $\tilde{\pi}:L\to_{\Sigma_1}L$: that is, $0^\#$ exists. So we may suppose that $\bar{\tau}$ is not a cardinal in L. Let $\beta+1$ be least with some $f:\gamma\to\bar{\tau}$, $\gamma<\bar{\tau}$ in $J_{\beta+1}$, f a surjection. So $\bar{\tau}$ is a cardinal in J_β. For some n $\omega\rho^n_\beta<\bar{\tau}$, as there is a new well-order of some $\gamma<\bar{\tau}$ in $J_{\beta+1}$. Take n least such. Let $\rho=\rho^{n-1}_\beta$ (so $\omega\rho\geq\bar{\tau}$) and let $\tilde{\pi}:J_\rho\to_{\Sigma_1}J_{\rho'}$ extend π as usual. Let $\hat{\pi}:J_\beta\to_{\Sigma_n}J_{\beta'}$ be the extension of $\tilde{\pi}$ given by the extension of embeddings lemma.

Now we may prove the result. There is p such that $J_\rho = =h''(\omega\times(\omega\rho^n_\beta\times\{p\}))$, where h is the Σ_1 Skolem function of $\langle J_\rho,A^{n-1}_\beta\rangle$. Let h' be the Σ_1 Skolem function of $\langle J_{\rho'},A^{n-1}_{\beta'}\rangle$; let $\tilde{\rho}=\sup\pi''\omega\rho^n$. Then $\tilde{\rho}<\tau$ since $\omega\rho^n_\beta<\bar{\tau}$; let $Y'=h'''(\omega\times(\tilde{\rho}\times\{\tilde{\pi}(p)\}))$; so $Y'\in L$. Moreover $Y\subseteq Y'$, for $Y=\mathrm{rng}(\pi)$ and $Y'\supseteq\mathrm{rng}(\tilde{\pi})$. Finally, in L, $\bar{\bar{Y}}'<\tau$. So there is $f\in L$ such that $f:\gamma\leftrightarrow Y'$ with $\gamma<\tau$. Let $X'=f^{-1}{}''X$: since $X'\subseteq\gamma$ there is $Z\supseteq X'$ with $Z\subseteq\gamma$, $\bar{\bar{Z}}=\bar{\bar{X}}$ and $Z\in L$. Let $Z'=f''Z$. Z' is the required cover for X and the covering lemma for L is proved.

Now let us see what this implies about singular cardinals. A sample application shows that if β is a singular cardinal then β is singular in L. For suppose X is cofinal in β and $\bar{\bar{X}}<\beta$. Then there is $Y\in L$ with $X\subseteq Y\subseteq\beta$ and $\bar{\bar{Y}}=\bar{\bar{X}}+\aleph_1$: hence since $\beta>\bar{\bar{X}}$ and $\beta>\aleph_1$, $L\models\beta>\bar{\bar{Y}}$. So $L\models\beta$ singular. It follows also that if β is a singular cardinal then $(\beta^+)^L=\beta^+$: all provided $0^\#$ does not exist.

As for cardinal arithmetic, the following can be shown: if there is $\gamma<\beta$ such that for all cardinals δ with $\gamma<\delta<\beta$ $2^\gamma=2^\delta$ then $2^\beta=2^\gamma$ (this does not use the covering lemma: it is a theorem of ZFC due to Bukovsky and Hechler). Otherwise $2^\beta=(\sup_{\gamma<\beta}2^\gamma)^+$ (Here β is a singular cardinal and $0^\#$ does not exist). Thus 2^β for singular cardinals β is determined wholly by 2^α for $\alpha<\beta$, α regular. Indeed all exponential calculations are fixed by the values of 2^{α} for regular α. Given any infinite cardinals κ,λ

$\kappa^{\lambda}=2^{\lambda}$ if $2^{\lambda}\geq\kappa$;
$=\kappa$ if $2^{\lambda}<\kappa$ and $\lambda<cf(\kappa)$;
$=\kappa^{+}$ otherwise.

Jech([25]) gives a brief formula that implies all of this: the singular cardinal hypothesis (SCH) is the claim that for all singular cardinals β, if $2^{cf(\beta)}<\beta$ then $\beta^{cf(\beta)}=\beta^{+}$.

A simple lemma enables us to convert these results into impossibility results about forcing. If V is a generic extension of L then $0^{\#}$ does not exist. So we cannot by forcing over L get a model in which SCH fails. We cannot even add a generic subset of a singular cardinal β without adding a subset of some smaller γ. For suppose $X\subseteq\beta$, $X\notin L$ but for all $\gamma<\beta$ $P(\gamma)\subseteq L$. Let $f:\beta\to\bigcup_{\gamma<\beta}P(\gamma)$ be a constructible bijection. Let C be cofinal in β with $\overline{\overline{C}}<\beta$ and let $X'=\{f^{-1}(X\cap\gamma):\gamma\in C\}$. Then $\overline{\overline{X}}'<\beta$ so there is $Y\in L$ with $Y\supseteq X'$ and $\overline{\overline{Y}}=\overline{\overline{X}}'+\aleph_1$. Hence $L\models\overline{\overline{Y}}<\beta$. Let $g\in L$ map Y one-one onto $\gamma=(\overline{\overline{Y}})^{L}$. Then $g''X'\subseteq\gamma$, so $g''X'\in L$. Hence $X'\in L$, so $X=\bigcup_{\gamma\in X'}f(\gamma)\in L$.

It was already known that by forcing over a model with a measurable cardinal one could add a cofinal subset of the measurable of order-type ω without adding any new bounded subsets of the measurable.

The covering lemma can only be generalised by changing the model L: none of the conditions can be weakened. If $a^{\#}$ does not exist ($a\subseteq\omega$) then the covering lemma holds over L[a]. This helps us to get a little way along the sharp sequence. Since we have just seen that the lemma may fail over L[U] in a generic extension (U a normal measure in L[U]) there must be doubts as to how far we can go beyond this.

Hence the core model: the furthest point we can reach in the sharp sequence without getting an inner model with a measurable cardinal. The original non-fine-structural motivation for the core model was that although the covering lemma fails over L[U] in a generic extension, yet we could iterate the measure out of the universe: $K=\bigcap_{i<\infty}L[U_i]$. The term core model refers to this characterisation of K as the core of the $L[U_i]$ that is left when all the U_i are removed. This characterisation only works, though, when there is an inner model with a measurable cardinal. The general definition gets around the problem.

From the proof of the covering lemma for L it can be seen that failure of covering gives an embedding of the relevant model to itself non-trivially and hence advances one stage up the sharp sequence; but the core model already contains the whole sharp sequence; so the only possibility is that the coding of the sequence by the core model has reached its end, i.e. that there is an inner model with a measurable cardinal. So we hope to prove the following: if there is no inner model with a measurable cardinal and X is an uncountable set of ordinals then there is $Y \in K$ with $X \subseteq Y$ and $\overline{\overline{X}} = \overline{\overline{Y}}$.

There is one snag. At a crucial point the covering lemma for L used the fact that there is p such that $J_\beta = h_{J_\beta}$ "$(\omega \times (\omega\rho_\beta \times \{p\}))$. This fails in generalised fine structure. If we know that J^U_β is a mouse - call it $\underline{N}$ and assume for simplicity that U is normal on κ with $\omega\rho_{\underline{N}} \leq \kappa$ - then all we know is that there are indiscernibles C for $\underline{N}$ such that $N = h_N$"$(\omega \times (\omega\rho_N \times \{p\} \times C))$. We can always ensure, in the cases that concern us, that the order-type of C is less than or equal to ω. If it is less then C is finite, hence in N, and it may be added onto the parameter. But if C is infinite then a whole new construction is needed, giving directly an inner model with a measurable cardinal.

We can deduce that if there is no inner model with a measurable cardinal then SCH holds. Surprisingly we can go further: although one may generically add C of order-type ω cofinal in a measurable to L[U] (a "Prikry sequence") that is, in a sense, all the generic mischief that one can do. For it follows from the proof of the covering lemma over K that either the covering lemma holds with L[U] in place of L (where L[U] is the inner model with U a normal measure on the minimal point κ) or there is a Prikry sequence for U, C, such that the covering lemma holds with L[C] in place of L, or there is a non-trivial $j: L[U] \to_e L[U]$ with critical point above κ. This last possibility has an $0^\#$-type equivalent, called $0^\dagger$. $0^\dagger$ does not exist in any generic extension of L[U]. So despite initial appearances SCH holds provided that $0^\dagger$ does not exist.

6: BEYOND

We identified $0^\#$ with the mouse $J^V_{\kappa+1}$, where V is normal on κ in $J^V_{\kappa+1}$. Similarly we may identify $a^\#$ with an a-mouse $J^{aV}_{\kappa+1}$ with $a \subseteq \omega$: or indeed $a^\#$ for any $a \subseteq On$ may be identified with $J^{aV}_{\kappa+1}$ provided $a \subseteq \kappa$. So in particular $0^\dagger$ can be identified with $U^\#$. But there is more to it than that: $J^{UV}_{\kappa+1}$ is a double premouse and could be iterated by U as well as by V. The theory of these "double mice" is very similar to that of single mice and one gets a double core model K^+. For this we insist in all double mice that the projectum ρ^n drops to the point κ on which U is a normal measure, for some n. If there is no non-trivial $j: K^+ \to_e K^+$ then the covering lemma holds over K^+. If there is such j, with critical point κ say, then as usual we get a normal measure U on $P(\lambda)\cap K^+$ for some $\lambda>\kappa$ which is normal in L[U] (subject to the usual technicalities). That is not very thrilling: K^+ is full of such models. To get anywhere we must add U to K^+ by taking K^U as the union of all double mice whose bottom measure is U. There is some difficulty in deciding which U to add; we ignore this. See if the covering lemma holds over K^U, or some Prikry generic extension of K^U. If not then there is an embedding of K^U $j:K^U \to_e K^U$ whose critical point is above κ. This gives an inner model with two measures L[U,V]. Does the covering lemma hold over L[U,V] adding, if necessary, Prikry sequences to U or V or both? If not then there is $j:L[U,V]\to_e L[U,V]$ with critical point high up, and this gives $0^{\dagger\dagger}$, as we may call it. All of the models K^+, K^U, L[U,V] and the generic extensions satisfy GCH, so from the failure of SCH we may deduce that $0^{\dagger\dagger}$ exists. And so on for any finite number of †, the provisions about Prikry sequences and the like getting more and more complicated.

For a limit number of † we have to be careful. When we are looking at mice with ω measures it no longer suffices, if we are to generalise the proof from L, that each set of indiscernibles be finite; their union must be finite, that is, the mouse must be a finite iteration of its core. Although elaborate coding takes one a little further at the time of writing it does not go beyond the first measurable that is a limit of a sequence of measurables whose order-type is inaccessible. At what stage is there such a riot of Prikry sequences that SCH fails?

To get some idea about this, go to the other extreme of the universe, the very large cardinal axioms. A cardinal κ is β-supercompact provided there is $j:V\to_e M$ with κ critical and $M^\beta\subseteq M$, i.e. M is closed under β-sequences. κ is supercompact provided it is β-supercompact for all β. Clearly supercompact cardinals are measurable: in fact the first supercompact (if there is one) is much greater than the first measurable. Silver showed (see Menas [41]) that if there is a supercompact cardinal κ then there is a generic extension in which κ is measurable and $2^\kappa>\kappa^+$. By adding a Prikry sequence to this model we may make $cf(\kappa)=\omega$ but preserve all bounded subsets of κ and all cardinalities. Hence $2^\kappa>\kappa^+$. On the other hand since κ measurable $\Rightarrow$ κ a strong limit $2^\gamma<\kappa$ for all $\gamma<\kappa$. So SCH fails in this model: $2^\kappa>(\sup_{\gamma<\kappa} 2^\gamma)^+$. On the other hand GCH already fails dramatically below κ by a result of Scott([21]). Magidor obtained a model in which $2^{\aleph_n}<\aleph_\omega$, all $n<\omega$, but $2^{\aleph_\omega}>\aleph_{\omega+1}$: using an even larger cardinal he was able to replace $2^{\aleph_n}<\aleph_\omega$ by $2^{\aleph_n}=\aleph_{n+1}$ ([38]). By Silver's theorem $\aleph_\omega$ could not be replaced by

$\aleph_{\omega_1}$. Some scrappy details are in appendix I.

After Cohen's result on the independance of GCH appeared there were many who hoped that large cardinal axioms would nevertheless settle the problem. The idea was that large cardinal axioms should be taken as true whenever they could (as far as was known) consistently be assumed: the reason for this being a heuristic assumption that set theorists should aim for maximum "richness" of sets. (Similar arguments may be used about forcing extensions, to justify, for example, the conclusion that "really" V is not L. This gave no clue about GCH, though, for just as it may be destroyed in a generic extension so it may be restored by one). In fact large cardinals have little effect on GCH: Scott showed that GCH cannot fail for the first time at a measurable cardinal, but no other very significant restriction on regular powers of 2 ever emerged. It is anyway hard to believe that anyone would have been persuaded to change their views about the power of the continuum by anything as recherché as a supercompact cardinal.

The interesting point is that SCH, a weak form of GCH, is heavily influenced by large cardinal assumptions. For

(i) Consis(ZFC+there is a supercompact) ⇒ Consis(ZFC+-SCH);

(ii) Consis(ZFC+-SCH) ⇒ Consis(ZFC+there is a measurable).

There is a wide gap between measurables and supercompacts; in part six of this book we try to set up some markers in it. What we should like to prove is that -SCH is equiconsistent with some large cardinal axiom: that is, to obtain axiom X such that

Consis(ZFC+there is an X cardinal) ⇔ Consis(ZFC+-SCH).

(We cannot hope to get there is an X cardinal ⇔ -SCH). There is no concealing the fact that we are very far from getting such an X.

PART ONE: FINE STRUCTURE

The analysis of part one is a general theory of relative constructibility subject to the "acceptability" constraint in definition 3.1. Although it is called fine structure, really chapters 1 and 2 are coarse structure; fine structure begins when we define (definition 3.5) the projectum.

In chapter one we consider a theory equivalent to "I am rud closed". The main technical result of the chapter is a finite axiomatisation of R. Our treatment is purely internal, all theorems being proved in R, with liberal use of classes. This would be a paradoxical way of developing the subject were it not that straightforward accounts already exist; our approach avoids excessive indexing and gives a few neat proofs later on.

Chapter 2 strengthens R to R^+, whose transitive models are $J_\alpha^{\vec{A}}$. R^+ enables us in addition to do various inductive arguments that failed in R; division by ω for example. The axiom of choice holds in R^+ as well. Throughout the chapter the reader should remember that the intended models are the transitive ones.

Chapter three contains the heart of fine structure theory. Again we strengthen the theory, this time to RA, by adding a form of GCH. In part three of the book we shall verify that the theory RA holds in the structures that make up the core model. In RA the central constructions are the master code A^p and the projectum ρ.

Chapter four moves into ZFC so as to use the Mostowski collapsing lemma. It also tidies up terminology for the rest of the book and contains the "extension of embeddings" lemmas.

The attraction of this general theory is that when we come in part six to generalised core models we do not have to change the fine structure, although the upward extension of embeddings lemma needs substantial generalisation. This is a modular fine structure theory that can be plugged in to any acceptable context.

On the other hand this makes the going rather hard for a reader with no knowledge of fine structure. It would be better for a first approach to read [26] up to section 4 and then to return to this part. The reader who is familiar with fine structure may prefer to use part one for reference and to go straight to part two.

1: A WEAK SET THEORY

Let L_N be the language L augmented by N unary predicate symbols $A_1 \ldots A_N$.

Definition 1.1. *The theory R^N is the deductive closure of the following axioms:*

(i) $\forall x \forall y(x=y \Leftrightarrow \forall z(z \in x \Leftrightarrow z \in y))$ *(extensionality)*

(ii) $\forall x(x \neq \phi \Rightarrow \exists y(y \in x \wedge x \cap y=\phi))$ *(foundation)*

(iii) $\forall x \forall y \exists z \forall t(t \in z \Leftrightarrow t=x \vee t=y)$ *(pairing)*

(iv) $\forall x \exists y \forall t(t \in y \Leftrightarrow \exists z(z \in x \wedge t \in z))$ *(union)*

(v)$_\phi$ $\forall u \forall \vec{x} \forall w \exists y \forall z(z \in y \Leftrightarrow \exists t \in w\ z=\{s \in u: \phi(s,t,\vec{x})\})$

where the formula ϕ is Σ_0. *(Σ_0 closure)*

The theory is usually referred to simply as R.

The development of chapters 1-3 takes place in the theory R or in strenghthened forms of R.

Lemma 1.2. *If ϕ is Σ_0 then*

$$\forall u \forall \vec{x} \exists y \forall t(t \in y \Leftrightarrow t \in u \wedge \phi(t,\vec{x})) \quad (1.1)$$

Proof: By (v)$_\phi$ there is a set y' such that

$$\forall z(z \in y' \Leftrightarrow \exists t \in u\ z=\{s \in u: \phi(s,\vec{x})\}). \quad (1.2)$$

Let $y=\bigcup y'$; then y satisfies

$$\forall t(t \in y \Leftrightarrow t \in u \wedge \phi(t,\vec{x})) \quad (1.3)$$

so (1.1) holds. □

The main result of chapter 1 is that the theory R is finitely axiomatisable.

Definition 1.3 *The functions $F_1 \ldots F_{17+N}$ are defined by*

$F_1(x,y)=\{x,y\}$

$F_2(x,y)=\bigcup x$

$F_3(x,y)=x \setminus y$

$F_4(x,y)= x \times y$

$F_5(x,y)=\mathrm{dom}(x)$

$F_6(x,y)=\{x''\{z\}: z \in y\}$

$F_7(x,y)=\{\langle u,v,w\rangle: u \in x \wedge \langle v,w\rangle \in y\}$

$F_8(x,y)=\{\langle u,w,v\rangle: \langle u,v\rangle \in x \wedge w \in y\}$

$F_9(x,y)=\{\langle u,v\rangle \in x \times y: u=v\}$

$F_{10}(x,y)=\{\langle u,v\rangle \in x\times y:\ u\in v\}$
$F_{11}(x,y)=\langle x,y\rangle$
$F_{12}(x,y)=\langle x,v,w\rangle$ if $y=\langle v,w\rangle$; 0 otherwise
$F_{13}(x,y)=\langle u,y,v\rangle$ if $x=\langle u,v\rangle$; 0 otherwise
$F_{14}(x,y)=\{\langle x,v\rangle,w\}$ if $y=\langle v,w\rangle$; 0 otherwise
$F_{15}(x,y)=\{\langle u,y\rangle,v\}$ if $x=\langle u,v\rangle$; 0 otherwise
$F_{16}(x,y)=\{\langle x,y\rangle\}$
$F_{17}(x,y)=\{t:\langle y,t\rangle\in x\}$
$F_{17+i}(x,y)=A_i\cap x \quad (1\le i\le N)$.

F_i are called basis functions.

<u>Lemma 1.4</u>. *F_i is a total function whenever $0<i\le 17+N$.*
Proof: (1) By axiom (iii).
(2) By axiom (iv).
(3) $x\smallsetminus y=\{s\in x: s\notin y\}$; use lemma 1.2.
(4) Let $X=\{\{u,v\}: u\in x\cup y\wedge v\in x\cup y\}$. X exists by (iv) and (v). Then $x\times y=\{\{s\in X: s\in\langle u,v\rangle\}: u\in x\wedge v\in y\}$ which is a set by (v).
(5) $\mathrm{dom}(x)=\{u\in\bigcup\bigcup x: (\exists v\in\bigcup\bigcup x)\langle u,v\rangle\in x\}$.
The rest are left to the reader. □

The definition of each F_i can be written as a Σ_0 formula of the language L_N; that is, there is a Σ_0 formula ϕ_i such that

$$F_i(x,y)=z \Leftrightarrow \phi_i(x,y,z). \tag{1.4}$$

Let R' (or, to be precise, R'^N) be the deductive closure of (i) and (ii) together with the axioms $(vi)_i$:

$$(vi)_i \quad \forall x\forall y\exists z\phi_i(x,y,z) \tag{1.5}$$

for $0<i\le 17+N$. To show that R is finitely axiomatisable it suffices to show R=R'. By lemma 1.4 $R'\subseteq R$. For the reverse it would suffice to show $(v)_\phi\in R'$ for all Σ_0 ϕ.

To this end we prove:

<u>Lemma 1.5</u>. *Suppose $\phi(x_1\ldots x_n)$ is a Σ_0 formula of L_N. There is a function F_ϕ which is a composition of $F_1\ldots F_{17+N}$ such that*

$$R'\vdash F_\phi(a_1\ldots a_n)=\{\langle x_1\ldots x_n\rangle\in a_1\times\ldots\times a_n:\phi(x_1\ldots x_n)\} \tag{1.6}$$

Proof: For convenience we call a formula which is Σ_0 *orderly* provided that whenever the variable v_i occurs free in the scope of a quantifier $\forall v_j$ or $\exists v_j$ then $i<j$. There is no loss of generality in proving (1.6) for orderly ϕ since every formula is equivalent (in predicate calculus) to an orderly one.

Our first aim is to dispose of dummy variables: recall our convention that although all the free variables of $\phi(x_1\ldots x_n)$ must lie among $x_1\ldots x_n$, not all of $x_1\ldots x_n$ need be free in $\phi(x_1\ldots x_n)$.

(1) If $\phi(x_1\ldots x_n)$ is $\psi(x_1\ldots x_{n-1})$ and F_ψ satisfies (1.6) then there is F_ϕ satisfying (1.6).

Proof: $F_\phi(a_1\ldots a_n)=F_\psi(a_1\ldots a_{n-1})\times a_n$
$=F_4(F_\psi(a_1\ldots a_{n-1}),a_n)$. □(1)

(2) If $\phi(x_1\ldots x_n)$ is $\psi(x_1\ldots x_{n+1})$ and F_ψ satisfies (1.6) then there is F_ϕ satisfying (1.6).

Proof: $F_\phi(a_1\ldots a_n)=\mathrm{dom}(F_\psi(a_1\ldots a_n,\{0\}))$
$= F_5(F_\psi(a_1\ldots a_n,F_1(F_3(a_1,a_1),F_3(a_1,a_1))),a_1)$

Hence by induction:

(3) If $\phi(x_1\ldots x_n)$ is $\psi(x_1\ldots x_m)$ and F_ψ satisfies (1.6) then there is F_ϕ satisfying (1.6).

Next the propositional connectives:

(4) If F_ϕ satisfies (1.6) then there is $F_{-\phi}$ satisfying (1.6).

Proof: $F_{-\phi}(a_1\ldots a_n)= (a_1\times\ldots\times a_n)\setminus F_\phi(a_1\ldots a_n)$ □(4)

(5) If $\phi(x_1\ldots x_n)$ and $\psi(x_1\ldots x_n)$ each have F_ϕ,F_ψ satisfying (1.6) then, letting $\chi(x_1\ldots x_n)\Leftrightarrow\phi(x_1\ldots x_n)\wedge\psi(x_1\ldots x_n)$ there is F_χ satisfying (1.6).

Proof: $F_\chi(a_1\ldots a_n)=F_\phi(a_1\ldots a_n)\cap F_\psi(a_1\ldots a_n)$
$=F_\phi(a_1\ldots a_n)\setminus(F_\phi(a_1\ldots a_n)\setminus F_\psi(a_1\ldots a_n))$ □(5)

The next two claims use F_7 and F_8 to manipulate ordered sequences.

(6) If $\psi(x_1\ldots x_n)$ has F_ψ satisfying (1.6) and $\phi(x_1\ldots x_{n+1})$ results from ψ by replacing each free occurrence of x_n by x_{n+1} then there is F_ϕ satisfying (1.6).

Proof: If n=1 then $\phi(x_1,x_2)\Leftrightarrow\psi(x_2)$ so $F_\phi(a_1,a_2)=a_1\times F_\psi(a_2)$. Otherwise
$F_\phi(a_1\ldots a_{n+1})=F_8(F_\psi(a_1\ldots a_{n-1},a_{n+1}),a_n)$ □(6)

(7) If $\psi(x_1,x_2)$ has F_ψ satisfying (1.6) and $\phi(x_1\ldots x_n)$ arises from ψ by replacing free x_1 by x_{n-1} and free x_2 by x_n $(n\geq 2)$ then there is F_ϕ satisfying (1.6).

Proof: We may assume n>2. Then
$F_\phi(a_1\ldots a_n)=F_7(a_1\times\ldots\times a_{n-2},F_\psi(a_{n-1},a_n))$. □(7)

We turn to the atomic formulae.

(8) There is $F_{x_1=x_2}$ satisfying (1.6).

Note: It no longer matters whether $x_1=x_2$ is $\phi(x_1,x_2)$ or $\phi(x_1\ldots x_m)$, m>2, by (3).

Proof: $F_{x_1=x_2}$ is F_9. □(8)

(9) There is $F_{x_n=x_{n+1}}$ satisfying (1.6).

Proof: By (8) and (7) □(9)

Now an induction starting from (9) and using (6) yields:

(10) For all m,n there is $F_{x_m=x_n}$ satisfying (1.6).

Proof: If m<n this follows by induction on n-m. If m>n use $F_{x_n=x_m}$.
If n=m then $F_{x_n=x_m}(a_1\ldots a_n)=a_1\times\ldots\times a_n$. □(10)

Using F_{10} we can establish by similar arguments

(11) For m<n there is $F_{x_m \in x_n}$ satisfying (1.6).

Now

(12) For all m,n there is $F_{x_m \in x_n}$ satisfying (1.6).

Proof: Suppose $\phi(x_1 \ldots x_m)$ is $x_m \in x_n$, for by (11) we may assume $m \geq n$. Say

$$\psi(x_1 \ldots x_{m+1}) \Leftrightarrow x_n = x_{m+1} \wedge x_m \in x_{m+1}.$$

By (10), (11) and (5), together with (3), there is F_ψ satisfying (1.6). Let

$$F_{x_m \in x_n}(a_1 \ldots a_m) = \mathrm{dom}(F_\psi(a_1 \ldots a_m, a_n)) \qquad \square(12)$$

(13) For i>0, $i \leq N$ there is F_{A_i} satisfying (1.6).

Proof: Suppose $\phi(x_1 \ldots x_n) \Leftrightarrow A_i(x_j)$. Then

$$F_{A_i}(a_1 \ldots a_n) = a_1 \times \ldots \times a_{j-1} \times F_{17+i}(a_j, a_j) \times a_{j+1} \times \ldots \times a_n. \qquad \square(13)$$

Hence

(14) If $\phi(x_1 \ldots x_n)$ is atomic then there is F_ϕ satisfying (1.6)

It remains only to consider bounded quantifiers. We may of course restrict ourselves to the existential quantifier.

(15) If $\psi(x_1 \ldots x_k)$ has F_ψ satisfying (1.6) and $\phi(x_1 \ldots x_n)$ is $\exists x_k \in x_j \psi(x_1 \ldots x_k)$, where k>n, then there is F_ϕ satisfying (1.6).

Note: The assumption that k>n is justified by the fact that ϕ is orderly.

Proof: Let $\theta(x_1 \ldots x_k)$ be $x_k \in x_j$. So there is $F_{\theta \wedge \psi}$ satisfying (1.6); and

$$F_{\theta \wedge \psi}(a_1 \ldots a_{k-1}, \cup a_j) = \{\langle x_1 \ldots x_k \rangle : x_k \in x_j, x_i \in a_i \ (1 \leq i \leq k) \text{ and } \psi(x_1 \ldots x_k)\}.$$

Then let $F_\phi(a_1 \ldots a_n) = \mathrm{dom}^{k-n}(F_{\theta \wedge \psi}(a_1 \ldots a_n, a_1 \ldots a_1, \cup a_j))$ $\square$(15)

Lemma 1.5 is proved. $\square$

<u>Lemma 1.6</u>. *If $\phi(x_1 \ldots x_n)$ is a Σ_0 formula then there is F, a composition of F_i $(0 < i \leq 17+N)$ such that*

$$F(a, x_1 \ldots x_{i-1}, x_{i+1} \ldots x_n) = \{x_i \in a : \phi(x_1 \ldots x_n)\} \qquad (1.7)$$

Proof: Let F_ϕ be as in lemma 1.5. Then

$$\mathrm{dom}^{n-i}(F_\phi(\{x_1\} \ldots \{x_{i-1}\}, a, \{x_{i+1}\} \ldots \{x_n\}) = \\ = \{\langle x_1 \ldots x_i \rangle \in \{x_1\} \times \ldots \times \{x_{i-1}\} \times a : \phi(x_1 \ldots x_n)\}. \qquad (1.8)$$

Applying rng ($\mathrm{rng}(x) = F_2(F_6(x, \mathrm{dom}(x)), x)$) gives the required set. $\square$

<u>Lemma 1.7</u>. *If $1 \leq i \leq 17+N$ then there is a function F_i^* which is a composition of basis functions such that $F_i^*(u,v) = F_i{}''u \times v$.*

Proof: Suppose there were a function G_i such that

$$G_i(u,v) = \{\langle\langle x,y \rangle, t\rangle : x \in u \wedge y \in v \wedge t \in F_i(x,y)\}.$$

Then $F_i^*(u,v) = F_6(G_i(u,v), u \times v)$. So it suffices to show that G_i is a composition of basis functions.

(1) $\{\langle\langle x,y\rangle,t\rangle : x\in u\wedge y\in v\wedge t\in\{x,y\}\} = ((u\times v)\times u)\cup((u\times v)\times v)$

(2) $\{\langle\langle x,y\rangle,t\rangle : x\in u\wedge y\in v\wedge t\in\bigcup x\} = \{\langle\langle x,y\rangle,t\rangle\in(u\times v)\times\bigcup\bigcup u : t\in\bigcup x\}$

Since $t\in\bigcup x$ is Σ_0 the result follows from lemma 1.6. We shall not repeat this remark with the other clauses.

(3) $\{\langle\langle x,y\rangle,t\rangle : x\in u\wedge y\in v\wedge t\in x\setminus y\} = \{\langle\langle x,y\rangle,t\rangle\in(u\times v)\times\bigcup u : t\in x\setminus y\}$

(4) $\{\langle\langle x,y\rangle,t\rangle : x\in u\wedge y\in v\wedge t\in x\times y\} = \{\langle\langle x,y\rangle,t\rangle\in(u\times v)\times(\bigcup u\times\bigcup v) : t\in x\times y\}$

(5) $\{\langle\langle x,y\rangle,t\rangle : x\in u\wedge y\in v\wedge t\in \mathrm{dom}(x)\} = \{\langle\langle x,y\rangle,t\rangle\in(u\times v)\times\bigcup\bigcup\bigcup u : t\in\mathrm{dom}(x)\}$

(6) $\{\langle\langle x,y\rangle,t\rangle : x\in u\wedge y\in v\wedge t\in x''\{z\}\} = \{\langle\langle x,y\rangle,t\rangle\in(u\times v)\times\bigcup\bigcup\bigcup x : t\in x''\{z\}\}$

(7) $\{\langle\langle x,y\rangle,t\rangle : x\in u\wedge y\in v\wedge\exists\langle s,z\rangle\in y\,\exists w\in x\ t=\langle w,s,z\rangle\} =$

$= \{\langle\langle x,y\rangle,t\rangle\in(u\times v)\times(\bigcup u\times\bigcup\bigcup\bigcup v\times\bigcup\bigcup\bigcup v) : \exists\langle s,z\rangle\in y\,\exists w\in x\, t=\langle w,s,z\rangle\}$

The remainder are easily deduced. □

It follows by induction that:

Lemma 1.8. *If F is a composition of basis functions and* $F^*(u_1\ldots u_n) = F''u_1\times\ldots\times u_n$ *then* F^* *is a composition of basis functions.*

Lemma 1.9. $R'=R$.

Proof: We saw that it is sufficient to prove $(v)_\phi\in R'$ for all Σ_0 ϕ. Given ϕ let F be as in lemma 1.6 with

$$F(u,t,x_1\ldots x_m)=\{s\in u:\phi(s,t,x_1\ldots x_m)\} \tag{1.9}$$

Then $F^*(\{u\},w,\{x_1\}\ldots\{x_m\}) = \{F(u,t,x_1\ldots x_m) : t\in w\}$

$= \{\{s\in u:\phi(s,t,x_1\ldots x_m)\} : t\in w\}$ □

It follows that R may be axiomatised by a single Π_2 axiom.

Definition 1.10. *A function F is rudimentary provided F is a composition of basis functions.*

We have asserted that for each i greater than 0 with $i\leq 17+N$ there is a Σ_0 formula ϕ_i such that

$$z=F_i(x,y)\leftrightarrow\phi_i(x,y,z) \tag{1.10}$$

If N=0 we can prove (1.10) for all rudimentary functions. So assume N=0 until further notice. We actually prove a stronger assertion.

Definition 1.11. *A function F is substitutable provided that whenever* $\phi(x_1\ldots x_n,y)$ *is* Σ_0 *there is a* Σ_0 *formula* ψ *such that*

$$\phi(x_1\ldots x_n,F(z_1\ldots z_m))\leftrightarrow\psi(x_1\ldots x_n,z_1\ldots z_m) \tag{1.11}$$

Lemma 1.12. *All rudimentary functions are substitutable.*

Proof: It will suffice to consider only Σ_0 formulae of the form

$\forall t \in y \chi(t, x_1 \ldots x_n)$ where χ is Σ_0.

For from this the result follows by induction on Σ_0 structure.

Also substitutable functions are closed under composition. So we need only consider the basis functions.

Note that

$x_i \in F(z_1 \ldots z_m) \Leftrightarrow -(\forall t \in F(z_1 \ldots z_m))-(x_i = t)$.

We shall only consider a few cases.

(1) $\forall t \in \{z_1, z_2\} \chi(t, x_1 \ldots x_n) \Leftrightarrow \chi(z_1, x_1 \ldots x_n) \wedge \chi(z_2, x_1 \ldots x_n)$

(2) $\forall t \in \cup z_1 \chi(t, x_1 \ldots x_n) \Leftrightarrow (\forall s \in z_1)(\forall t \in s) \chi(t, x_1 \ldots x_n)$

(3) $\forall t \in z_1 \setminus z_2 \chi(t, x_1 \ldots x_n) \Leftrightarrow (\forall t \in z_1)(t \in z_2 \vee \chi(t, x_1 \ldots x_n))$

(4) $\forall t \in z_1 \times z_2 \chi(t, x_1 \ldots x_n) \Leftrightarrow \forall t_1 \in z_1 \forall t_2 \in z_2 \chi(\langle t_1, t_2 \rangle, x_1 \ldots x_n)$

using

(5) $\forall t \in \langle z_1, z_2 \rangle \chi(t, x_1 \ldots x_n) \Leftrightarrow \chi(\{z_1\}, x_1 \ldots x_n) \wedge \chi(\{z_1, z_2\}, x_1 \ldots x_n)$

which in turn uses (1). The reader may check the others. □

<u>Corollary 1.13</u>. *If F is rudimentary then there is a Σ_0 formula ϕ such that*

$z = F(x_1 \ldots x_n) \Leftrightarrow \phi(x_1 \ldots x_n, z)$

If $N \neq 0$ then corollary 1.13 may fail. For example, $\{x, y\} \in A$ is not necessarily Σ_0 in L_1. For this reason we need to consider a broader class of relations.

<u>Definition 1.14</u> *A relation R is rudimentary provided that for some rudimentary function F*

$$R(x_1 \ldots x_n) \Leftrightarrow F(x_1 \ldots x_n) \neq \phi \tag{1.12}$$

<u>Lemma 1.15</u> *Every Σ_0 relation is rudimentary.*

Proof: Let ϕ be a Σ_0 formula. Then by lemma 1.6 there is a rudimentary function F' such that

$$F'(a, x_1 \ldots x_n) = \{x_{n+1} \in a : \phi(x_1 \ldots x_n) \wedge x_{n+1} = x_{n+1}\}. \tag{1.13}$$

Let $F(x_1 \ldots x_n) = F'(\{0\}, x_1 \ldots x_n)$ □

The function in (1.12) can, informally speaking, be obtained "effectively" from ϕ. To formalise this idea we introduce a coding of the rudimentary functions.

Let $Q = \{q$: q is a finite function $\wedge\ q \neq \phi\ \wedge\ \mathrm{dom}(q) \subseteq 2^{<\omega}\ \wedge$
$\wedge\ (s = t|\mathrm{dom}(s), t \in \mathrm{dom}(q) \Rightarrow s \in \mathrm{dom}(q))\ \wedge\ \mathrm{rng}(q) \subseteq 17+N+1\ \wedge$
$\wedge\ (q(s) \neq 0 \Rightarrow s^\frown 0, s^\frown 1 \in \mathrm{dom}(q))\ \wedge\ (q(s) = 0 \Rightarrow s^\frown 0, s^\frown 1 \notin \mathrm{dom}(q))\}$

Let $\alpha_q = \{f$: f is a function $\wedge\ \mathrm{dom}(f) = \{s \in \mathrm{dom}(q) : q(s) = \phi\}\}$.

Let F_q be a function defined by

$F_q(f) = z \Leftrightarrow f \in \alpha_q\ \wedge\ (\exists g)(\mathrm{dom}(g) = \mathrm{dom}(q)\ \wedge$
$\wedge (\forall s \in \mathrm{dom}(g))(q(s) = \phi \Rightarrow g(s) = f(s)\ \wedge$
$\wedge\ \ q(s) \neq \phi \Rightarrow g(s) = F_{q(s)}(g(s^\frown 0), g(s^\frown 1)))\ \wedge$
$\wedge\ g(0) = z)$

The clause with $F_{q(s)}$ must, of course, be written out as a finite disjunction.

For all q in Q F_q is rudimentary; and $F_q(f)=z$ is a Δ_1 formula. Now let τ be some function from $\{s\in dom(q):q(s)=0\}$ to n. Let

$F^n_{q,\tau}(x_1\ldots x_n)=F_q(f)$ where $f(s)=x_{\tau(s)}$, all s in $dom(\tau)$.

Then every rudimentary function is of the form $F^n_{q,\tau}$. And $F^n_{q,\tau}(x_1\ldots x_n)=z$ is a Δ_1 property of $x_1\ldots x_n,z,q,\tau$. For any ϕ we can therefore find q_ϕ,τ_ϕ such that

$$\phi(x_1\ldots x_n)\Leftrightarrow F^n_{q_\phi,\tau_\phi}(x_1\ldots x_n)\neq\phi\ . \tag{1.14}$$

Fix such q_ϕ,τ_ϕ for all Σ_0 ϕ (informally, do so "effectively").

<u>Definition 1.16</u>. *The Σ_0 satisfaction relation, $\models^n_{\Sigma_0}\phi(x_1\ldots x_n)$ is the relation $R(\langle q_\phi,\tau_\phi\rangle,\langle x_1\ldots x_n\rangle)\Leftrightarrow F^n_{q_\phi,\tau_\phi}(x_1\ldots x_n)\neq\phi$.*

This is a Δ_1 relation.

<u>Definition 1.17</u> *The Σ_m satisfaction relation, $\models^n_{\Sigma_m}\phi(x_1\ldots x_n)$ is the relation $\exists y_1\forall y_2\ldots Qy_m R(\langle q_\phi,\tau_\phi\rangle,y_1,\ldots y_m,\langle x_1\ldots x_n\rangle)$*

We identify Σ_0 formulae ϕ with $\langle 0,\langle q_\phi,\tau_\phi\rangle\rangle$. If ϕ is $\exists y_1\forall y_2\ldots Qy_m\psi(y_1\ldots y_m,x_1\ldots x_n)$ then we identify ϕ with $\langle m,\langle q_\psi,\tau_\psi\rangle\rangle$. The Σ_m satisfaction relation is itself Σ_m. We shall never have to refer to this representation explicitly: and later on we shall code the formulae by natural numbers. The superscript n may usually be omitted. Note that we have shown that every rudimentary relation is Δ_1.

This is perhaps the moment to mention the role of classes in this theory. As we do not have a Σ_1 comprehension axiom there may be Σ_1 definable bounded subsets of the ordinals - which, by the way, are defined in the usual way in R - that are not sets. They may even be like ordinals themselves: they may be initial segments of On. Such segments cannot have a largest element, of course.

<u>Definition 1.18</u>. *A Σ_n limit segment (η) is a Σ_n class of ordinals (i.e. $\{\alpha:\phi(\alpha,x_1\ldots x_n)\}$, for some Σ_n ϕ and some $x_1\ldots x_n$) such that*

(i) $\alpha<\beta\in\eta\Rightarrow\alpha\in\eta$

(ii) $\alpha\in\eta\Rightarrow\alpha+1\in\eta$.

Every Σ_0 limit segment is either On or an ordinal.

It is not even precluded that ω might have a Σ_n limit segment properly included in it. In our applications this will be disallowed, however.

Exercises

1. X is rudimentarily closed provided

$x,y\in X \Rightarrow F_i(x,y)\in X$

for all i with $0<i\leq 17+N$. Is it the case that X is rudimentarily closed if and only if $\langle X,\in\cap X^2,\dot{A}_1\cap X\ldots A_N\cap X\rangle \models R$?

2. For which α does $L_\alpha \models R$?

3. Show that F_9 is definable from $F_1\ldots F_8$ and F_{10}.

4. If X is a set, can it be proved in R that there is a smallest set $Y\supseteq X$ such that Y is rudimentarily closed?

5. $\langle M,A\rangle$ is amenable provided that for all $x\in M$ $A\cap x\in M$. Show that $\langle M,A\rangle\models R$ if and only if $M \models R$ and $\langle M,A\rangle$ is amenable.

6. Show that if u is transitive and $0<i\leq 17+N$ then $\bigcup\bigcup\bigcup F_i``u^2\subseteq u$.

The theory R is not very strong. Although it is possible to define ordinal numbers in it there is practically no ordinal arithmetic. One way to strengthen R would be to add axioms making R closer to ZFC; for example, KP set theory. We want a theory that works in all levels of the constructible hierarchy, however, and not merely at admissible levels. We shall add to R axioms that say $V=L[A_1 \ldots A_N]$.

Definition 2.1. $S(u)=u\cup\bigcup_{i=1}^{17+N} F_i{}''u^2$

By lemma 1.7 S is rudimentary, so $R\vdash \forall u\exists z\, z=S(u)$

Lemma 2.2. *If u is transitive then so is S(u).*

Proof: This is where the apparently redundant F_{11} to F_{17} are used. We check each F_i.

(1) $x,y\in u \Rightarrow \{x,y\}\subseteq u\subseteq S(u)$.

(2) $x\in u \Rightarrow \bigcup x\subseteq u\subseteq S(u)$ (as u is transitive).

(3) $x,y\in u \Rightarrow x\smallsetminus y\subseteq u\subseteq S(u)$

(4) $x,y\in u, s\in x, t\in y \Rightarrow \langle s,t\rangle=F_{11}(s,t)\in S(u)$.

(5) $x,y\in u \Rightarrow \mathrm{dom}(x)\subseteq u\subseteq S(u)$.

(6) $x,y\in u,\ z\in y,\ t=x''\{z\} \Rightarrow t=F_{17}(x,z)$.

(7) $x,y\in u, u'\in x, \langle v',w'\rangle\in y \Rightarrow \langle u',v',w'\rangle=F_{12}(u',\langle v'w'\rangle)$

(8) $x,y\in u,\ \langle u',v'\rangle\in x,\ w'\in y \Rightarrow \langle u',w',v'\rangle=F_{13}(\langle u',v'\rangle,w')$

(9) $F_9(x,y)\subseteq F_4(x,y)$; (10) $F_{10}(x,y)\subseteq F_4(x,y)$.

(11) $x,y\in u, s\in\langle x,y\rangle \Rightarrow s=\{x\}=F_1(x,x)$ or $s=\{x,y\}=F_1(x,y)$.

(12) $x,y\in u,\ y=\langle v',w'\rangle,\ s\in\langle x,v',w'\rangle \Rightarrow$
$\Rightarrow s=\{\langle x,v'\rangle\}=F_{16}(x,v')$ or $s=\{\langle x,v'\rangle,w'\}=F_{14}(x,y)$.

(13) $x,y\in u, x=\langle u',v'\rangle,\ s\in\langle u',y,v'\rangle \Rightarrow$
$\Rightarrow s=\{\langle u',y\rangle\}=F_{16}(u',y)$ or $s=\{\langle u',y\rangle,v'\}=F_{15}(x,y)$.

(14) $x,y\in u,\ y=\langle v',w'\rangle,\ s\in\{\langle x,v'\rangle,w'\} \Rightarrow s=w'$ or $s=\langle x,v'\rangle=F_{11}(x,v')$.

(15) $x,y\in u,\ x=\langle u',v'\rangle,\ s\in\{\langle u',y\rangle,v'\} \Rightarrow s=v'$ or $s=\langle u',y\rangle=F_{11}(u',y)$.

(16) $x,y\in u,\ s\in\{\langle x,y\rangle\} \Rightarrow s=F_{11}(x,y)$.

(17) $x,y\in u,\ s\in\{t:\langle y,t\rangle\in x\}\Rightarrow\langle y,s\rangle\in x\Rightarrow s\in u$; (17+i) $x,y\in u \Rightarrow A_i\cap x\subseteq u$. □

Definition 2.3. *(i)* $f=S|x \Leftrightarrow x\in On \wedge f$ *is a function* $\wedge\ \mathrm{dom}(f)=x\ \wedge$
$\wedge\ (0\in x \Rightarrow f(0)=\phi)\ \wedge\ (\forall\nu\in x)(\nu=\mu+1 \Rightarrow f(\nu)=S(f(\mu)\cup\{f(\mu)\})\ \wedge$

$\wedge\ (\forall\lambda\in x)(lim(\lambda) \Rightarrow f(\lambda)=\bigcup_{\xi<\lambda} f(\xi))$

$(ii)\ y=S_\nu \Leftrightarrow (\exists f)(f=S|dom(f)\ \wedge\ y=f(\nu))$

Definition 2.4. *R^+ is R together with the axioms*

$(i)\ \forall\nu\exists f\ f=S|\nu$

$(ii)\ \forall x\exists\nu\exists y\ (y=S_\nu \wedge x\in y)$

For later use we also define

Definition 2.5. *$rud(u)=\bigcup_{n\in\omega} S_n(u)$, where $S_n(u)$ is defined inductively by*

$S_0(u)=u\cup\{u\}$

$S_{n+1}(u)=S(S_n(u)\cup\{S_n(u)\})$

Clearly rud(u) is rudimentarily closed. But it need not be a set even if u is; and it may not be the smallest rudimentarily closed $X\supseteq u\cup\{u\}$ (unless we restrict X to be a set).

From now on we work in R^+.

Lemma 2.6. *Each S_ν is transitive.*

Proof: (This is a model for various subsequent inductions)

Suppose not. Then letting $E=\{\nu : S_\nu$ is not transitive$\}$, $E\neq\phi$.

Claim E has a least member.

Proof: Take $\gamma\in E$. If $E\cap\gamma=\phi$ γ is the least member. Otherwise let $f=S|\gamma$. Then

$$E\cap\gamma=\{\nu<\gamma:\ f(\nu) \text{ is not transitive}\}. \tag{2.1}$$

This is a set and therefore has a least member. □(Claim)

Let γ be least, then. $\gamma\neq 0$, of course, nor is γ a successor by lemma 2.2. but if γ is a limit then $S_\gamma=\bigcup_{\nu<\gamma} S_\nu$ and each S_ν is transitive so S_γ is transitive. Contradiction! □

Similarly

Lemma 2.7. $\gamma\leq\gamma' \Rightarrow S_\gamma\subseteq S_{\gamma'}$.

Lemma 2.8. *If λ is a limit ordinal then S_λ is rudimentarily closed.*

Proof: If $x,y\in S_\lambda$ then $x,y\in S_\gamma$, some $\gamma<\lambda$; for $0<i\leq 17+N$ $F_i(x,y)\in S_{\gamma+1}\subseteq S_\lambda$ □

Let $\underline{S}_\lambda=\langle S_\lambda, \in\cap S_\lambda^2, A_1\cap S_\lambda,\ldots,A_N\cap S_\lambda\rangle$. Clearly $\underline{S}_\lambda\models R$ (λ a limit).

Lemma 2.9. *$lim(\lambda) \Rightarrow \lambda=On\cap S_\lambda$.*

Proof: If not then we could use the technique of lemma 2.6 to obtain a least λ for which the lemma failed.

Case 1 λ is a limit of limit ordinals

Then $On\cap S_\lambda = On\cap \bigcup\{S_{\lambda'}: \lambda'<\lambda \text{ and } \lambda' \text{ a limit}\}$

$= \bigcup\{On\cap S_{\lambda'}: \lambda'<\lambda \text{ and } \lambda' \text{ a limit}\}$

$= \bigcup\{\lambda': \lambda'<\lambda\}$

$= \lambda.$

Case 2 There is a maximal $\lambda'<\lambda$ which is a limit ordinal.

Suppose $\lambda\not\subseteq S_\lambda$. Then there is a least $\nu<\lambda$ such that $\nu\notin S_\lambda$. Since $\lambda'=On\cap S_{\lambda'}\subseteq S_\lambda, \nu\geq\lambda'$. But $\lambda'=On\cap S_{\lambda'}$ and $S_{\lambda'}\in S_\lambda$ so $\lambda'\in S_\lambda$. Thus $\nu>\lambda'$; so ν is not a limit ordinal; say $\nu=\mu+1$. Then $\mu\in S_\lambda$; but then $\mu\cup\{\mu\}\in S_\lambda$ as $\underline{S}_\lambda\models R$. Contradiction!

Suppose, then, that $On\cap S_\lambda\not\subseteq\lambda$. Then there is $\gamma\in(On\cap S_\lambda)\setminus\lambda$; since $On\cap S_\lambda$ is transitive, $\lambda\in S_\lambda\cap On$. By the usual argument there is a least $\nu<\lambda$ with $\lambda\subseteq S_\nu\cap On$. And since $S_{\lambda'}\cap On=\lambda', \nu>\lambda'$. Say $\nu=\mu+1$. So $\lambda\not\subseteq S_\mu$. Now by exercise 6 in chapter 1 $\bigcup\bigcup\bigcup S_\nu\subseteq S_\mu$. But since λ is a limit ordinal $\lambda\subseteq\bigcup\bigcup\bigcup S_\nu$; contradiction! □

Lemma 2.10. *$lim(\lambda) \Rightarrow \underline{S}_\lambda\models R^+$*

Proof: Again we may assume that λ is the least where this fails and derive a contradiction.

Case 1 λ is a limit of limit ordinals.

R^+ is axiomatised by a single Π_2 axiom, so this is impossible.

Case 2 There is a maximal $\lambda'<\lambda$ which is a limit ordinal.

Suppose there is $\nu<\lambda$ such that $S|\nu\notin S_\lambda$. We may take ν least such. $\nu\geq\lambda'$, since otherwise $S|\nu\in S_{\lambda'}\subseteq S_\lambda$. And $S|\lambda'=\bigcup\{f\in S_{\lambda'}: \underline{S}_{\lambda'}\models f=S|dom(f)\}$ so $S|\lambda'\in S_\lambda$. So ν is a successor. Say $\nu=\mu+1$. Then $S|\mu\in S_\lambda$. But $S|\nu=S|\mu\cup\{\langle\mu,S_\mu\rangle\}$ and $S_\mu\in S_\nu\subseteq S_\lambda$. So $S|\nu\in S_\lambda$. Contradiction!

But given $x\in S_\lambda$ there is $\nu<\lambda$ with $x\in S_\nu$ so $\underline{S}_\lambda\models\exists\nu\exists y(y=S_\nu\wedge x\in y)$. Thus $\underline{S}_\lambda\models R^+$. □

In ordinary set theory it would now be straightforward to call $S_{\omega\lambda}$ J_λ; however, in R^+ we must first develop the necessary ordinal arithmetic.

Definition 2.11. *$\tau=\nu+n \Leftrightarrow \tau,\nu\in On \wedge n\in\omega \wedge$*

$\wedge\ \exists f(dom(f)\subseteq\omega \wedge f(0)=\nu \wedge (\forall i+1\in dom(f))(f(i+1)=f(i)+1) \wedge \tau=f(n))$

This is a Σ_1 formula. Generally we cannot prove that $\nu+n$ exists. Suppose for some ν it did not; then $\{n:\nu+n \text{ exists}\}$ is a Σ_1 limit segment of ω.

Definition 2.12. *The theory R^+_ω is R^+ together with the axiom schema stating that every definable limit segment of ω is either 0 or ω.*

From now on we work in R^+_ω. Note that the theory is no longer Π_2 axiomatisable. Now we get:

Lemma 2.13. *For all ν and all $n\in\omega$ $\nu+n$ exists.*
Proof: $\{n:\nu+n$ exists$\}$ is ω. □

R^+_ω strengthens R^+ by a strong form of induction on ω.

Lemma 2.14. *Suppose $(\forall n\in\omega)(\nu+n<\tau)$; then $\exists\lambda(\lambda=\nu+\omega)$ (by which we mean $\lambda=\sup_{n\in\omega}\nu+n$)*
Proof: Let $E=\{\xi<\tau$: for all $n\in\omega$ $\nu+n<\xi\}$.
Claim $\eta=\nu+n \Leftrightarrow \underline{S}_\tau\models\eta=\nu+n$
Proof: ($\Rightarrow$) Suppose $\eta=\nu+n$. Let f be as in lemma 2.11. Then $f\in S_\tau$ by an easy induction on n.

($\Leftarrow$) Obvious. □(Claim)

Thus $E=\{\xi<\tau$: for all $n\in\omega$ $\underline{S}_\tau\models\nu+n<\xi\}$ is a set and has a least member which is the required λ. □

Corollary 2.15. *If there is a largest limit ordinal λ then $On=\{\lambda+n:n\in\omega\}\cup\lambda$.*
Proof: Otherwise there is $\tau\in On\smallsetminus\{\lambda+n:n\in\omega\}\cup\lambda$. So $\lambda+\omega$ exists, but $\lambda+\omega>\lambda$, $\lambda+\omega$ is a limit ordinal. □

Definition 2.16. *(i) $f=D|\nu \Leftrightarrow f$ is a function $\wedge$ $\nu\in On$ $\wedge$ $dom(f)=\nu$ $\wedge$*

$$\wedge(\forall\xi+1<\nu)(f(\xi+1)=f(\xi)) \wedge (\forall\lambda<\nu)((\lambda=0 \vee lim(\lambda)) \Rightarrow f(\lambda)=\bigcup_{\xi<\lambda}(f(\xi)+1))$$

(ii) $\tau=[\nu/\omega]\Leftrightarrow\exists f(f=D|dom(f) \wedge \tau=f(\nu))$.

This is a Σ_1 formula. $[\nu/\omega]$ is the unique τ such that $\nu=\omega\tau+k$, some $k\in\omega$.

Lemma 2.17. *$\forall\nu[\nu/\omega]$ exists.*
Proof: It suffices to show $\forall\nu D|\nu$ exists.
Claim If $\lambda=0$ or $lim(\lambda)$ then $(\forall\nu<\lambda)D|\nu\in S_\lambda$.
Proof: Suppose not. As usual we may get a least λ where the claim fails. $\lambda\neq0$, then; and if λ is a limit of limit ordinals then the claim holds. So there is a maximal $\lambda'<\lambda$, λ' a limit ordinal. Thus by lemma 2.14 $\lambda=\{\lambda'+n:n\in\omega\}\cup\lambda'$. Now $D|\nu\in S_{\lambda'}$ for all $\nu<\lambda'$, and

$$D|\lambda'=\{\langle\nu,x\rangle:\underline{S}_{\lambda'}\models x=[\nu/\omega]\} \tag{2.2}$$

so $D|\lambda'\in S_{\lambda}$. Then if n is least such that $D|\lambda'+n\notin S_{\lambda'}$, $n=m+1$, say. But then $D|\lambda'+m\in S_{\lambda'}$ and $D|\lambda'+n=(D|\lambda'+m)\cup\{\langle m,[\lambda'/\omega]\rangle\}$ so $D|\lambda'+n\in S_\lambda$; contradiction! □(Claim)
Now if On is a limit of limit ordinals the lemma follows by the claim. Otherwise there is a largest limit ordinal λ and $On=$ $=\{\lambda+n:n\in\omega\}\cup\lambda$ by lemma 2.15, and the result is proved just as in the claim. □

Definition 2.18. *(i)* $\tau=\omega\nu \leftrightarrow (\tau=0\wedge\nu=0)\vee(\tau$ *is a limit and* $\nu=[\tau/\omega])$

(ii) $\hat{On}$ *denotes* $\{\gamma:\omega\gamma$ *exists*$\}$

(iii) $J_\nu=S_{\omega\nu}$ *(provided* $\omega\nu$ *exists)*

It is immediate from lemma 2.10 that $\underline{J}_\nu\models R^+$. Clearly every limit ordinal τ is of the form $\omega\nu$.

In R^+_ω the structure of limit segments is simplified by

Lemma 2.19. *If* $\eta\subseteq On$ *is a bounded* Σ_n *limit segment which is not an ordinal then* η *has no largest limit ordinal.*

Proof: Suppose it did and call it λ.

Claim $\lambda+n\in\eta$ for all $n\in\omega$.

Proof: Let $k=\{n:\lambda+n\in\eta\}$. k is a Σ_n limit segment of ω so $k=\omega$ □(Claim)

Hence $\lambda+\omega\subseteq\eta$. So $\lambda+\omega$ exists and is a limit ordinal; thus $\lambda+\omega=\eta$.

So η is an ordinal. □

Lemma 2.20. *If there is a largest limit ordinal* $\omega\nu$ *then* $\hat{On}=\nu+1$. *Otherwise* $\hat{On}$ *is a limit segment.*

Proof: Clear □

If η is a limit segment we let $\omega\eta$ denote $\{\gamma:\exists\nu\in\eta\ \gamma<\omega\nu\}$. So $\omega\eta$ is a limit segment. By lemma 2.19 every Σ_n limit segment that is bounded but not an ordinal is of the form $\omega\eta$.

Lemma 2.21. $\underline{S}_{\omega\eta}\models R^+$ $(\omega\eta\subseteq On)$

Proof: If $\omega\eta$ is an ordinal then this is lemma 2.10. If $\omega\eta$ is On $V=S_{\omega\eta}$. Otherwise $S_{\omega\eta}$ is a union of models of R^+ by lemma 2.19 and lemma 2.10. □

We can extend definition 2.18 to limit segments by allowing $J_\eta=S_{\omega\eta}$ provided $\eta\subseteq\hat{On}$. So $V=J_{\hat{On}}$. Also $\underline{J}_\eta=\underline{S}_{\omega\eta}$.

THE AXIOM OF CHOICE

Just as in ZFC $V=L[A]\rightarrow AC$, so $R^+\vdash AC$. The essential technical lemma is

Lemma 2.22. *Suppose u is a set with a well-order r. Then S(u) can be well-ordered. Indeed there is a rudimentary function W such that W(u,r) well-orders S(u).*

Proof: $W(u,r)=\{\langle x,y\rangle\in(S(u))^2: (x,y\in u\wedge xry)\vee(x\in u\wedge y\notin u)\vee$ (2.3)

$\vee(x\notin u\wedge y\notin u\wedge(\exists s\in u)(\exists t\in u)(\exists i\le 17+N)(x=F_i(s,t)\ \wedge$

$\wedge(\forall s'\in u)(\forall t'\in u)(s'rs\rightarrow(\forall j\le 17+N)\ y\ne F_j(s',t')\wedge$

$\wedge(s'=s\wedge t'rt\rightarrow(\forall j\le 17+N)\ y\ne F_j(s',t')))\ \wedge$

$\wedge(\forall j\leq i)(y\neq F_j(s,t)))))\}$.

Since the defining clause is Σ_0 W is rudimentary by 1.6. □

Note that W(u,r) is an end-extension of r. We let $f=W|\nu$ be the formula

$$f \text{ is a function} \wedge \mathrm{dom}(f)=\nu \wedge \nu\in On \wedge \tag{2.4}$$
$$\wedge(\forall\xi+1<\nu)(f(\xi+1)=W(S_\xi\cup\{S_\xi\},f(\xi)\cup\{\langle t,S_\xi\rangle:t\in S_\xi\}))\wedge$$
$$\wedge(\forall\xi<\nu)(\lim(\xi) \Rightarrow f(\xi)=\bigcup_{\xi'<\xi} f(\xi')) \wedge (\nu\neq 0 \Rightarrow f(0)=\phi)$$

And $r=W_\nu$ means $\exists f(\nu\in\mathrm{dom}(f) \wedge f=W|\mathrm{dom}(f) \wedge r=f(\nu))$

Lemma 2.23. *(i) For all ν W_ν exists*

(ii) For all ν W_ν is a well-order of S_ν.

Proof: Very like lemma 2.17. First of all establish that if λ is a limit ordinal then for $\nu<\lambda$ $W_\nu\in S_\lambda$. If this failed there would be a least limit λ where it failed. λ cannot be 0 or a limit of limits; hence there is a largest limit $\lambda'<\lambda$. $W_\nu\in S_{\lambda'}$ for $\nu<\lambda'$ and $W_{\lambda'}$ is $\underline{S}_{\lambda'}$-definable so $W_{\lambda'}\in S_\lambda$. If $\lambda'+n+1$ is least with $W_{\lambda'+n+1}\notin S_\lambda$ we derive a contradiction by rudimentary closure of S_λ. Then the result is proved unless $On=\lambda\cup\{\lambda+n:n\in\omega\}$, some limit λ, in which case it is proved by rudimentary closure.

(ii) is then a trivial induction. □

Corollary 2.24. *Every set can be well-ordered. Indeed there is a well-ordering of the universe that is a Σ_1 class.*

The well-ordering we have constructed is called the canonical well-ordering.

Lemma 2.25. *Suppose R is a Σ_1 relation. Then there is a Σ_1 function r such that*

$$\exists yR(x_1\ldots x_n,y) \Rightarrow R(x_1\ldots x_n,r(x_1\ldots x_n)) \tag{2.5}$$

for all $x_1\ldots x_n$.

Proof: Suppose $R(x_1\ldots x_n,y) \Leftrightarrow \exists zR'(x_1\ldots x_n,y,z)$. Let $r'(x_1\ldots x_n)\simeq$ $\simeq$ the least $\langle y,z\rangle$ such that $R'(x_1\ldots x_n,y,z)$ in the canonical well-ordering. Then, provided R' is Σ_0,

$$r'(x_1\ldots x_n)=\langle y,z\rangle \Leftrightarrow R'(x_1\ldots x_n,y,z) \wedge (\exists\gamma)(\langle y,z\rangle\in S_\gamma \wedge \tag{2.6}$$
$$\wedge(\forall y',z'\in S_\gamma)(\langle\langle y',z'\rangle,\langle y,z\rangle\rangle\in W_\gamma \Rightarrow -R'(x_1\ldots x_n,y,z))$$

So r' is Σ_1. Say

$$r(x_1\ldots x_n)=y \Leftrightarrow \exists zr'(x_1\ldots x_n)=\langle y,z\rangle . \tag{2.7}$$

(2.7) is a Σ_1 definition and r satisfies (2.5) □

Corollary 2.26. *If f is a Σ_1 function then there is a Σ_1 function g such that $f\circ g=id|rng(f)$*

Proof: $R(y,x) \Leftrightarrow f(x)=y$ is a Σ_1 relation. So there is g' such that

$$\exists x R(y,x) \Rightarrow R(y,g(y))$$
$$\Rightarrow f(g(y))=y.$$

Let $g=g'|rng(f)$. □

SKOLEM FUNCTIONS

The Kuratowski pair that we have used up to now is usually satisfactory, but sometimes we require a pairing function under which the ordinals are closed. The Gödel pairing function is such; but generally if we are working in R^+_ω we cannot assert that every pair of ordinals has a Gödel pair. Our official pairing function will nevertheless be an adaptation of the Gödel pairing function.

Definition 2.27. *(i)* $<_G$ *is the well-ordering of* On^2 *determined by*

$$\langle\alpha,\beta\rangle <_G \langle\alpha',\beta'\rangle \Leftrightarrow \max(\alpha,\beta)<\max(\alpha',\beta') \vee \qquad (2.8)$$
$$\vee\ (\max(\alpha,\beta)=\max(\alpha',\beta') \wedge \alpha<\alpha') \vee$$
$$\vee\ (\max(\alpha,\beta)=\max(\alpha',\beta') \wedge \alpha=\alpha' \wedge \beta<\beta')$$

(ii) $f=p|\gamma \Leftrightarrow dom(f)=\gamma \wedge f$ *is a function* $\wedge\ \forall\alpha,\beta<\gamma(\alpha<\beta \Rightarrow f(\alpha)<_G f(\beta)) \wedge$
$\wedge\ \forall\alpha<\gamma\forall\beta\in\alpha^2(\beta<_G f(\alpha) \Rightarrow \beta\in rng(f))$

(iii) $\langle\alpha,\beta\rangle=p(\gamma) \Leftrightarrow (\exists f)(f=p|dom(f) \wedge f(\gamma)=\langle\alpha,\beta\rangle)$

p is a Σ_1 function that enumerates the Gödel pairs.

Lemma 2.28. $\forall\alpha\exists f\ f=p|\alpha$.

Proof: Observe that $\langle\alpha,\beta\rangle=p(\gamma) \Rightarrow \max(\alpha,\beta)\leq p(\gamma)$. From this the technique of lemma 2.17 enables us to deduce that if λ is a limit ordinal then $\forall\nu\in\lambda\exists f\in S_\lambda\ f=p|\nu$, and hence to obtain the result by the usual adaptation. □

$p|\alpha:\alpha\to\alpha^2$, but is not, generally, surjective.

Definition 2.29. *(i)* $Q=\{\alpha:p(\alpha)=\langle 0,\alpha\rangle\}$.

(ii) α *is G-closed if and only if* α *is a limit point of* Q.

Lemma 2.30. *There is a* Σ_1 *map of On onto* On^2.

Proof: Although the strategy is quite standard we give the proof in detail.

Claim If λ is a limit ordinal then there is a $\Sigma_1(\underline{S}_\lambda)$ map of λ onto λ^2.

Proof: Suppose not. Let $E=\{\lambda:\ \lambda$ is a limit ordinal and there is no $\Sigma_1(\underline{S}_\lambda)$ map of λ onto $\lambda^2\}$. So $E\neq\phi$. Take some $\lambda\in E$; if $\lambda\cap E=\phi$ then λ was minimal. Otherwise since $\Sigma_1(\underline{S}_{\lambda'})\subseteq S_\lambda$ when λ' is an ordinal less than λ, we may use definition 1.17 to set

$E\cap\lambda=\{\lambda'<\lambda$: λ' is a limit and there is no $f\in S_\lambda$ such that $f\in\Sigma_1(\underline{S}_{\lambda'})$, $f:\lambda'\to\lambda'^2$ onto$\}$

so that $E\cap\lambda$ is a set. So there is a least λ where the claim fails. Clearly $\lambda\notin Q$.

Case 1 λ is a limit of limit ordinals.

Suppose $p(\lambda)=\langle\nu,\tau\rangle$. Then $\nu,\tau<\lambda$. And $p|\lambda$ is $\Sigma_1(\underline{S}_\lambda)$ and maps λ one-one onto $\{z:z<_G\langle\nu,\tau\rangle\}$. Let $y=\{z:z<_G\langle\nu,\tau\rangle\}$; suppose λ' is a limit and $\nu,\tau<\lambda'$. Then $y\in\Sigma_1(\underline{S}_{\lambda'})\subseteq S_\lambda$ and $p|\lambda$ maps λ one-one into λ'^2.

By inductive hypothesis there is $g\in\Sigma_1(\underline{S}_{\lambda'})$ mapping λ' onto λ'^2; so $g\in S_\lambda$. So by corollary 2.26 there is $g'\in\Sigma_1(\underline{S}_{\lambda'})$ mapping λ'^2 one-one into λ'. So $g'\in S_\lambda$.

Let $f(\langle\mu,\mu'\rangle)=g'(\langle g'p(\mu),g'p(\mu')\rangle)$, for $\mu,\mu'<\lambda$. So f is $\Sigma_1(\underline{S}_\lambda)$ and maps λ^2 one-one into λ'.Let $v=\text{rng}(f)$; then $v=g'"(g'"y)^2$, so $v\in S_\lambda$

Let

$$h(\gamma)=f^{-1}(\gamma) \quad \text{if } \gamma\in v \qquad (2.9)$$
$$=\langle 0,0\rangle \quad \text{otherwise}$$

Then $h:\lambda\to\lambda^2$ maps λ onto λ^2. Contradiction!

Case 2 There is a maximal limit $\lambda'<\lambda$.

Note that this is the only other case because $\omega\in Q$.

There is a $\Sigma_1(\underline{S}_{\lambda'})$ map $g:\lambda'\to(\lambda')^2$ onto, and so by corollary 2.26 a $\Sigma_1(\underline{S}_{\lambda'})$ map $g':(\lambda')^2\to\lambda'$ one-one. So $g'\in S_\lambda$. Since $\lambda=\{\lambda'+n:n\in\omega\}\cup\lambda'$ there is a bijection $j:\lambda\leftrightarrow\lambda'$ $\Sigma_1(\underline{S}_\lambda)$. Let $f(\mu,\mu')=g'(\langle j(\mu),j(\mu')\rangle)$. So $f:\lambda^2\to\lambda'$ is one-one. Let $v=\text{rng}(f)$; then $v=\text{rng}(g')$, so $v\in S_\lambda$. h is then defined as in (2.9). Contradiction! □(Claim)

As usual the lemma follows from the claim by applying the relevant case to On. □

It is important to notice that the proof of lemma 2.30 was not uniform. In the definition (2.9), furthermore, we had to use a parameter v,and there is no way in general of making the map parameter free and keeping it Σ_1. At times when we are being sensitive about parameters we shall use the Gödel pairing function.

Definition 2.31. *η is a Σ_n-cardinal provided*

(i) η is a limit segment of On;

(ii) there is no Σ_n map F with $\nu\in\eta$ such that $\eta\subseteq\{F(i):i<\nu$ and $i\in\text{dom}(F)\}$.

Lemma 2.32. *If η is a Σ_1 cardinal then η is closed under p (so if η is an ordinal, $\eta\in Q$).*

Proof: Note that η must be a limit of limit ordinals. But if the lemma failed there would be $\lambda'\in\eta$ with λ' a limit and $p"\eta\subseteq(\lambda')^2$. But then $p^{-1}|(\lambda')^2$ is a Σ_1 function with $\eta\subseteq\text{rng}(p^{-1}|(\lambda')^2)$. And there is a map of $\lambda'\leftrightarrow\lambda'^2$. Contradiction! □

Let s be a map of ω one-one onto $\{\langle q,\tau,n\rangle : q\in\Omega \wedge$ $\wedge\ \tau:\{\hat{s}\in dom(q):q(\hat{s})=0\}\to n, n<\omega\}$. Then the relation $R^m(i,\langle x_1\ldots x_n\rangle) \Leftrightarrow$ $\Leftrightarrow s(i)=\langle q_\phi,\tau_\phi,n\rangle \wedge \models^n_{\Sigma_m} \phi(x_1\ldots x_n)$ is Σ_m provided s is picked Σ_1; such an s is fixed form now on. ϕ_i is such that $s(i)=\langle q_{\phi_i},\tau_{\phi_i},n\rangle$, some n; ϕ_i is sometimes called s(i); R denotes R^1.

Definition 2.33. *$h(i,\langle\vec{x}\rangle)=y$ is the Σ_1 parameter free function that uniformises $R'(i,\langle\vec{x}\rangle,y) \Leftrightarrow R(i,\langle\vec{x},y\rangle)$ given by the proof of lemma 2.25. h is called the Σ_1 Skolem function.*

Lemma 2.34. *$h''(\omega\times X^{<\omega}) \prec_{\Sigma_1} V$.*

Proof: (V denotes $\langle V,A_1\ldots A_N\rangle$ of course)

Let $Y=h''(\omega\times X^{<\omega})$. Let $y_1\ldots y_n\in Y$. Say $y_i=h(j_i,\langle x_{i1}\ldots x_{ik_i}\rangle)$ where $j_i\in\omega, 0<i\leq n,\ 0<j\leq k_i \Rightarrow x_{ij}\in X$.

Suppose $\exists z\phi(z,y_1\ldots y_n)$ where ϕ is Σ_0. Let

$$\phi'(x_{11}\ldots x_{n1}\ldots x_{nk_n},z) \Leftrightarrow \phi(z,h(j_1,\langle x_{11}\ldots x_{1k_1}\rangle)\ldots \ldots h(j_n,\langle x_{n1}\ldots x_{nk_n}\rangle)) \tag{2.10}$$

and suppose $\phi'=s(i)$. Then

$$\phi'(x_{11}\ldots x_{n1}\ldots x_{nk_n},z) \Leftrightarrow R(i,\langle x_{11}\ldots x_{nk_n},z\rangle) \Leftrightarrow R'(i,\langle x_{11}\ldots x_{nk_n}\rangle,z) \tag{2.11}$$

Thus $\exists zR'(i,\langle x_{11}\ldots x_{nk_n}\rangle,z)$, so by definition

$$R'(i,\langle x_{11}\ldots x_{nk_n}\rangle,h(i,\langle x_{11}\ldots x_{nk_n}\rangle)). \tag{2.12}$$

Let $z'=h(i,\langle x_{11}\ldots x_{nk_n}\rangle)$; so $z'\in Y$ and by (2.12) $R'(i,\langle x_{11}\ldots x_{nk_n}\rangle,z')$ □

Definition 2.35. *h(X) denotes $h''(\omega\times X^{<\omega})$.*

Strictly speaking this is ambiguous, but in practice we always recognise functional applications of h by the natural number argument.

Lemma 2.36. *V=h(On).*

Proof: The usual strategy.

Claim If λ is a limit ordinal and $x\in S_\lambda$ then there are $i\in\omega$ and $\alpha_1\ldots\alpha_n<\lambda$ such that $\underline{S}_\lambda\models x=h(i,\langle\alpha_1\ldots\alpha_n\rangle)$.

Proof: Suppose the claim failed. There will be a least λ where it fails.

Case 1 λ is a limit of limit ordinals.

Suppose $x\in S_{\lambda'}, \lambda'<\lambda$, λ' a limit. Say $i\in\omega, \alpha_1<\ldots<\alpha_n<\lambda$ and $\underline{S}_{\lambda'}\models x=h(i,\langle\alpha_1\ldots\alpha_n\rangle)$. Then $\underline{S}_\lambda\models x=h(i,\langle\alpha_1\ldots\alpha_n\rangle)$.

Case 2 There is a largest limit $\lambda'<\lambda$.

Then $\lambda=\lambda'\cup\{\lambda'+n:n\in\omega\}$. Suppose n is least such that there is $x\in S_{\lambda'+n+1}$ and x violates the claim. Say $x=F_i(y,z)$ with $0<i\leq 15+N$, $y,z\in S_{\lambda'+n}\cup\{S_{\lambda'+n}\}$. It suffices to show that x is Σ_1 definable from ordinals in $\underline{S}_\lambda$. But $x=F_i(y,z)$ is Σ_0, so it suffices to show y,z Σ_1 definable. If $y,z\in S_{\lambda'+n}$ this follows by the minimality assumption. But

$$\underline{S}_\lambda\models y=S_{\lambda'+n} \Leftrightarrow \exists f(f=S|\mathrm{dom}(f) \wedge f(\lambda'+n)=y) \quad (2.13)$$

is a Σ_1 definition of $S_{\lambda'+n}$ from ordinals. This establishes the claim. □(Claim)

The lemma follows from the claim as usual. □

So there is a parameter free map of a subset of $\omega\times On^{<\omega}$ onto V.

Lemma 2.37. *There is a Σ_1 (with parameters) map of On onto $\omega\times On^{<\omega}$.*

Proof: By lemma 2.30 there is a Σ_1 map F of On onto On^2. Define by induction on $n<\omega$:

$$F_1(\alpha)=\alpha$$

$$F_{n+1}(\alpha)=\langle\beta_1\ldots\beta_{n-1},\gamma,\gamma'\rangle \text{ where } F_n(\alpha)=\langle\beta_1\ldots\beta_n\rangle \text{ and } \langle\gamma,\gamma'\rangle=F(\beta_n)$$

Plainly for all $n\in\omega$ $F_n:On\to On^n$ onto. Say $G(\alpha)=F_n(\beta)$ if $F(\alpha)=\langle n,\beta\rangle$ with $n\in\omega$; 0 otherwise. Then $G:On\to On^{<\omega}$ is onto. And

$$G(\alpha)=s \Leftrightarrow \exists n\in\omega\exists\beta(F(\alpha)=\langle n,\beta\rangle\wedge \exists f(\mathrm{dom}(f)=n \wedge f(0)=\beta \wedge (\forall m<n)(\exists\beta_1\ldots\beta_m,\gamma,\gamma')(f(m)=\langle\beta_1\ldots\beta_m\rangle \wedge f(m+1)=\langle\beta_1\ldots\beta_{m-1},\gamma,\gamma'\rangle\wedge \wedge F(\beta_m)=\langle\gamma,\gamma'\rangle)) \wedge f(n-1)=s) \quad (2.15)$$

Hence G is Σ_1. Finally $G':On\to\omega\times On$ may be defined by $G'(\alpha)=F(\alpha)$ if $F(\alpha)\in\omega\times On$; $=\langle 0,0\rangle$ otherwise. Letting $G''(\alpha)=\langle n,G(\beta)\rangle$ where $G'(\alpha)=\langle n,\beta\rangle$ G" is our map. □

If F was the Gödel pairing function then G is parameter free. In this case G is by abuse of notation called the Gödel pairing function. Often we write $G(\prec\alpha_1\ldots\alpha_n\succ)=\langle\alpha_1\ldots\alpha_n\rangle$.

Corollary 2.38. *There is a Σ_1 map of a subset of On onto V.*

Lemma 2.39. *There is a Σ_1 map of On onto V.*

Proof: Let $h^*(\tau,\langle i,\vec{x}\rangle)=h(i,\vec{x})$ if $\exists z\in S_\tau H(i,\langle\vec{x}\rangle,y,z)$
=0 otherwise,

where $h(i,\langle\vec{x}\rangle)=y \Leftrightarrow \exists zH(i,\langle\vec{x}\rangle,y,z)$ is some Σ_1 parameter free definition of h with H Σ_0. h^* is Σ_1 and total.

Furthermore $h^*:On\times(\omega\times On^{<\omega})\to V$ onto. But using F and the map of lemma 2.37 there is a map of On onto $On\times(\omega\times On^{<\omega})$. □

MISCELLANEOUS RESULTS IN R_ω^+

Lemma 2.40. *TC(x) exists for all x.*

Proof: Let $f=TC|y$ be the formula

$$f \text{ is a function} \wedge \mathrm{dom}(f)=y \wedge y \text{ is transitive} \wedge \wedge (\forall z\in y) f(z)=z\cup\bigcup_{t\in z} f(t) \tag{2.16}$$

Then $y=TC(x) \Leftrightarrow \exists f(f=TC|\mathrm{dom}(f) \wedge f(x)=y)$. So it suffices to show $\forall\nu\exists f\ f=TC|S_\nu$.

Claim $\forall\nu<\lambda\exists f\in S_\lambda\ f=TC|S_\nu$ when λ is a limit.

Proof: If not there is a least λ where the claim fails and $\lambda=\lambda'\cup\{\lambda'+n:n\in\omega\}$ for some limit $\lambda'<\lambda$. Since (2.16) is a Σ_1 definition and $TC|S_\nu\in S_\lambda$, for $\nu<\lambda'$, $TC|S_{\lambda'}\in S_\lambda$. Let n be least such that $TC|S_{\lambda'+n+1} \notin S_\lambda$.

Let $f_0=TC|S_{\lambda'+n}$. Let

$$f_{i+1}=\{\langle x,y\rangle:x\in S_{\lambda'+n+1} \wedge x\subseteq \mathrm{dom}(f_i) \wedge y=x\cup\bigcup_{z\in x} f_i(z)\} \tag{2.17}$$

Then for $x\in S_{\lambda'+n+1}$ we have $x\in\mathrm{dom}(f_{k+1})$ and $f_{k+1}(x)=TC(x)$, where k is such that $\bigcup^k x\subseteq S_{\lambda'+n}$. But by exercise 6 in chapter 1 if $x\in S_{\lambda'+n+1}$ then $\bigcup\bigcup\bigcup\bigcup x\subseteq S_{\lambda'+n}$. So $TC|S_{\lambda'+n+1}=f_5$, which is in S_λ. □(Claim)

As usual the lemma follows from the claim □

Lemma 2.41. *Suppose λ is the maximal limit ordinal and that $A_i\subseteq S_\lambda$ for $0<i\leq N$. Then every $x\subseteq S_\lambda$ is definable over S_λ.*

Proof: Suppose $x=F(y_1\ldots y_n)$ for $y_1\ldots y_n\in S_\lambda\cup\{S_\lambda\}$ with f rudimentary: this is possible because each $S_{\lambda+n}$ is rudimentary in $S_\lambda\cup\{S_\lambda\}$.
Now there is a function F' which is a composition of $F_1\ldots F_{17}$ such that

$$x=F(y_1\ldots y_n) \Leftrightarrow x=F'(y_1\ldots y_n,A_1\ldots A_N).$$

By lemma 1.12 there is a Σ_0 formula ϕ such that

$$t\in F(y_1\ldots y_n) \Leftrightarrow \phi(t,y_1\ldots y_n,A_1\ldots A_N)$$

and we may assume

$$t\in F(y_1\ldots y_n) \Leftrightarrow \phi(t,y_1\ldots y_{n-1},S_\lambda,A_1\ldots A_N)\ (y_1\ldots y_{n-1}\in S_\lambda)$$

By induction on Σ_0 structure of ϕ we may produce ψ such that

$$\phi(t,y_1\ldots y_{n-1},S_\lambda,A_1\ldots A_N) \Leftrightarrow \underline{S}_\lambda\models\psi(t,y_1\ldots y_{n-1}).$$

For

$$\phi(t,y_1\ldots y_{n-1},S_\lambda,A_1\ldots A_N) \Leftrightarrow \langle S_\lambda\cup\{S_\lambda\}\cup\{A_1\ldots A_N\},\in,A_1\ldots A_N\rangle\models \models\phi(t,y_1\ldots y_{n-1},S_\lambda,A_1\ldots A_N)$$

since $S_\lambda\cup\{S_\lambda\}\cup\{A_1\ldots A_N\}$ is transitive: the construction of ψ by induction is left to the reader: the only non-trivial clause replaces $(\exists y\in S_\lambda)$ by $\exists y$. □

Lemma 2.42. *If λ is a limit ordinal then for all $k\in\omega$ there is f mapping λ onto $S_{\lambda+k}$.*

Proof: By induction on k. k=0 is corollary 2.38. Otherwise let

$f:\lambda\to S_{\lambda+k}$ be onto. Define

$$g(\alpha,\gamma,\gamma')=f(\gamma) \text{ if } \alpha=0 \qquad (2.18)$$
$$=F_\alpha(f(\gamma),f(\gamma')) \text{ if } 0<\alpha\leq 17+N$$
$$=0 \text{ otherwise} \qquad (\alpha,\gamma,\gamma'<\lambda)$$

So $g:\lambda^3\to S_{\lambda+k+1}$. Then using F_3 of (2.14) defined over $\underline{S}_\lambda$, $gF_3:\lambda\to S_{\lambda+k+1}$ onto. □

Lemma 2.43. *Suppose λ is the maximal limit ordinal and $x\subseteq P(S_\lambda)$, and for $0<i\leq N$ $A_i\subseteq S_\lambda$. There is p such that $x\subseteq\Sigma_p(\underline{S}_\lambda)$.*

Proof: This is a uniform form of lemma 2.41. Suppose $x\in S_{\lambda+k}$, $k\in\omega$. So $x\subseteq S_{\lambda+k}$ as $S_{\lambda+k}$ is transitive. Say $f:\lambda\to S_{\lambda+k}$ onto by lemma 2.42. We may assume $x\neq\phi$. So letting

$$X=\{\langle\gamma,t\rangle:t\in f(\gamma) \wedge f(\gamma)\in x\} \qquad (2.19)$$

$X\subseteq S_\lambda$ and so by lemma 2.41 $X\in\Sigma_p(\underline{S}_\lambda)$ say. Then for $y\in x$ there is $\gamma<\lambda$ such that $y=f(\gamma)$; then $y=\{t:\langle\gamma,t\rangle\in X\}$ which is Σ_p. □

Exercises

1. Show that rank(x) exists for all x.

2. Show that the condition $A_i\subseteq S_\lambda$ in lemma 2.42 is necessary.

3. Is there $n\in\omega$ such that there is a parameter free Σ_n map of On onto V provably in R^+_ω?

4. A limit segment η is Σ_n regular provided there is no Σ_n map F such that for some $\nu\in\eta$ $\forall\gamma\in\eta\exists\gamma'<\nu(\gamma<F(\gamma')\wedge F(\gamma')\in\eta)$.
Does η Σ_1 regular imply that η is a Σ_1 cardinal? Vice-versa? If η=On?

3: ACCEPTABILITY AND THE PROJECTUM

Not all models of R_ω^+ have nice fine structure. The possibility of fine structure depends on our being able to code all the bounded subsets of γ which are Σ_1 by a single one, provided that γ is the least ordinal for which the Σ_1 comprehension axiom fails. This is assured by the following definition.

Definition 3.1. *RA is the theory R_ω^+ together with the axiom of acceptability which states:*

Suppose λ is a limit ordinal and $\delta<\lambda$ such that for some $a\subseteq\delta$, $a\in S_{\lambda+\omega}\setminus S_\lambda$; suppose $u\in S_{\lambda+\omega}$. Then there is $f=\langle f_\xi:\delta\leq\xi<\lambda\rangle\in S_{\lambda+\omega}$ such that

$$f_\xi:\xi\to\{\xi\}\cup(P(\xi)\cap u) \text{ onto} \tag{3.1}$$

In this chapter we work in RA.

Lemma 3.2. *If λ is a limit then $\underline{S}_\lambda\models RA$.*

Proof: This is immediate. □

The axiom of acceptability can be thought of as the best form of GCH we can get without assuming the power set axiom.

Lemma 3.3. *Suppose ω is a set. Let ν^+ denote the least cardinal $>\nu$, or On if ν is the largest cardinal. Then there is $\nu'\leq\nu^+$ and a Σ_1 class $\langle a_{ij}^\nu:\omega\leq\nu\wedge i<\nu\wedge\omega j<\nu'\rangle$ such that*

(a) $\{a_{ij}^\nu:\ i<\nu\wedge\omega j<\nu^+\}=P(\nu)$

(b) $\langle a_{ij}^\nu:\ i<\nu\wedge\omega j<\tau\rangle$ is a set for all $\tau<\nu'$.

Proof: Let $\Gamma_\nu=\{\xi\in On:\ \nu\leq\omega\xi\wedge P(\nu)\cap S_{\omega\xi+\omega}\not\subseteq S_{\omega\xi}\}$. $\xi\in\Gamma_\nu$ is a Σ_1 property. Let $\langle\xi_i^\nu:i<\nu^*\rangle$ enumerate Γ_ν. Then

$$\tau=\xi_i^\nu \Leftrightarrow \exists f(\text{ f is a function } \wedge \text{dom}(f)=i \wedge \text{rng}(f)=\Gamma_\nu\cap S_\tau \wedge \wedge \text{ f is strictly increasing }) \wedge \tau\in\Gamma_\nu \tag{3.2}$$

This is Σ_1 as $\Gamma_\nu\cap S_\tau$ is uniformly Σ_1 in τ. The technique of chapter 2 establishes $\langle\xi_j^\nu:j<i\rangle\in S_{\omega\xi_i^\nu+\omega}$ for all $i<\nu^*$.

Let $f_{i,n}^\nu$ be the least $f:\nu\to P(\nu)\cap S_{\omega\xi_i^\nu+n}$ in the canonical well-order, where $\omega\leq\nu, i<\nu^*$ and $n\in\omega$; such an f exists by the axiom of acceptability. There is a map g mapping ν one-one onto $\omega\times\nu$; if ν is a limit this follows from lemma 2.30 in S_ν; but if $\nu=\nu''+n$ there is a map of ν'' one-one onto $\omega\times\nu''$ and the extra ω elements can easily be dealt with since $\nu \geq \omega$. Let g be the least such map. Let $a^\nu_{g(n,i),j} = f^\nu_{j,n}(i)$. So $P(\nu) = \{a^\nu_{ij}: i<\nu \wedge j<\nu^*\}$; and $\{a^\nu_{ij}: \omega \leq \nu \wedge \wedge\, i<\nu \wedge j<\nu^*\}$ is a Σ_1 class. $\langle f^\nu_{j',n}: j'<j, n\in\omega\rangle$ is a set, for it is Σ_1 definable over $\underline{S}_{\omega\xi^\nu_j+\omega}$, so $\langle a^\nu_{ij}: i<\nu \wedge j<\tau\rangle$ is a set for $\tau<\nu^*$. Set $\nu' = \omega\nu^*$.

Suppose $\omega\nu^* > \nu^+$. Then $\nu^+ \in On$, so is a cardinal. Let $\beta = [\nu^+/\omega]$; then there is a map of β onto ν^+; so $\beta = \nu^+$. That is, $\nu^+ = \omega\nu^+$. Hence $\nu^* > \nu^+$. For $i<\nu^+$ let b_i be the least $b \in P(\nu) \cap S_{\omega\xi^\nu_i+\omega} \smallsetminus S_{\omega\xi^\nu_i}$ in the canonical well-order. So $b: \nu^+ \to P(\nu)$ is one-one.

But $\{b_i: i<\nu^+\} \in S_{\omega\xi^\nu_{(\nu^+)}+\omega}$ so by acceptability there is f mapping ν onto $\{b_i: i<\nu^+\}$; but then $b^{-1}f: \nu \to \nu^+$ onto. Contradiction! □

The following corollary makes it clear why acceptability is a weak form of GCH.

<u>Corollary 3.4</u>. *Suppose* $\nu \geq \omega$. *Then either* $P(\nu)$ *exists and* $\overline{\overline{P(\nu)}} = \nu^+$ *or for all* $u \subseteq P(\nu)$ $\overline{\overline{u}} \leq \nu$.

Proof: If $P(\nu)$ is not a set then given $u \subseteq P(\nu)$ there is a limit λ with $u \in S_{\lambda+\omega}$ and $P(\nu) \cap S_{\lambda+\omega} \smallsetminus S_\lambda \neq \phi$. By the axiom of acceptability $\overline{\overline{u}} \leq \nu$. If $P(\nu)$ is a set then $\overline{\overline{P(\nu)}} \leq \nu^+$ by lemma 3.3. But then $\overline{\overline{P(\nu)}} = \nu^+$ by Cantor's theorem. □

<u>Definition 3.5</u>. *The projectum* ρ *is defined by*

$$\omega\rho = \{\nu: \text{ for all } \Sigma_1\ A \subseteq On\ A \cap \nu \text{ is a set}\} \tag{3.3}$$

To justify this definition we need

<u>Lemma 3.6</u>. $\{\nu$: *for all* Σ_1 $A \subseteq On$ $A \cap \nu$ *is a set*$\}$ *is a limit segment*

Proof: If $A \cap \nu$ is a set then $A \cap (\nu+1)$ is a set. □

ρ is either an ordinal or a limit segment of $\mathcal{O}n$ or $\mathcal{O}n$ itself. As far as we are concerned the use of $\omega\rho$ rather than ρ in (3.3) is a nuisance, but its use keeps our results in line with the literature. The smallest possible value of ρ is 1.

<u>Definition 3.7</u>. RA_ρ *is the theory RA together with the axiom stating that there is a* Σ_1 $A \subseteq On$ *such that for all x* $A \cap \omega\rho \neq x \cap \omega\rho$.

If ρ is an ordinal then RA and RA_ρ coincide. From now on we work in RA_ρ.

<u>Lemma 3.8</u>. $\omega\rho$ *is a* Σ_1 *cardinal*.

Proof: Suppose F is Σ_1, $\gamma\in\omega\rho$, dom(F)$\subseteq\gamma$ and $\omega\rho\subseteq$rng(f). Suppose $\beta\in\gamma^+\setminus\omega\rho$. So there is a function from γ whose range includes $\omega\rho$, g. Take A Σ_1 such that $A\cap\omega\rho\neq x\cap\omega\rho$, all x. Say $\delta\in A' \Leftrightarrow g(\delta)\in A$. Then A' is Σ_1, $A'\subseteq\gamma\in\omega\rho$ so A' is a set. But $A\cap\omega\rho=(g''A')\cap\omega\rho$ and g"A' is a set. Contradiction!

So $\gamma^+\setminus\omega\rho=\phi$; that is, $\gamma^+=\omega\rho$ or $\gamma^+\in\omega\rho$. Hence $\gamma^+\subseteq$rng(F) and by lemma 3.3 there is a Σ_1 map G of γ onto $P(\gamma)$. Let $A=\{\xi:\xi\notin G(\xi)\}$; A is a Σ_1 subset of γ, hence A is a set. Say A=G(ξ). Then $\xi\in G(\xi) \Leftrightarrow \xi\notin G(\xi)$; contradiction! □

<u>Corollary 3.9</u>. *If $\rho\neq On$ then $\omega\rho=\rho$ or $\rho=1$.*

<u>Corollary 3.10</u>. *$\omega\rho$ is closed under the Gödel pairing function.*
Proof: By lemma 3.8 and lemma 2.32. □

Hence we could have defined $\omega\rho$ as the set of ν for which given $A\subseteq(On)^{<\omega}$ which is Σ_1, $A\cap(\nu)^{<\omega}$ is a set.

In ordinary fine structure, in L for example, we iterate the definition of projectum by looking at the structure $\langle J_\rho,A\rangle$ where A codes up the Σ_1 subsets of $\omega\rho$;A is called the Σ_1 master code. In our present circumstances we could not be assured that $\langle J_\rho,A\rangle$ was amenable. We have to replace J_ρ by a structure that at least contains all bounded subsets of $\omega\rho$.

<u>Definition 3.11</u>. *H_δ denotes $\{x:TC(x)\in\delta\}$ (This is well-defined by lemma 2.40)*

<u>Definition 3.12</u>. *P denotes $\{p\in[On]^{<\omega}$: for some $A\subseteq On$ Σ_1 in parameter p there is no x such that $A\cap\omega\rho=A\cap x\}$.*

<u>Lemma 3.13</u>. *$P\neq\phi$*
Proof: There is some p' with the property ascribed to members of P, since we are working in RA_ρ. But there is a parameter free map G say of $(On)^{<\omega}$ onto V. If $p'=G(\langle\alpha_1...\alpha_n\rangle)$ then let $p=\{\alpha_1...\alpha_n\}$. □

On is not necessarily closed under the Gödel pairing function so p may not be reducible to a single ordinal. On the other hand $\omega\rho$ is closed by corollary 3.10 so $\omega\rho$ may be replaced by $(\omega\rho)^{<\omega}$ in Definition 3.12.

<u>Definition 3.14</u>. *$T^p=\{\langle i,x\rangle: i\in\omega \wedge x\in H_{\omega\rho} \wedge \models^2_{\Sigma_1} s(i)(x,p)\}$;*
$A^p=T^p\cap(\omega\times(\omega\rho)^{<\omega})$.

So A^p,T^p are Σ_1.

<u>Definition 3.15</u>. *Let* $B_1 \ldots B_k$ *be given. Define* $F_{17+N+j}(x,y)=B_j \cap x$ *where* $0<j\leq k$. *Then* $S_{B_1 \ldots B_k}(u)$ *denotes* $u \cup \bigcup\{F_i{}''u^2 : 0<i\leq 17+N+k\}$.

$S_\nu^{B_1 \ldots B_k}$ *is defined from* $S_{B_1 \ldots B_k}$ *as* S_ν *was defined in definition 2.3.*

$\underline{S}_\nu^{B_1 \ldots B_k} = \langle S, \in \cap S^2, A_1 \cap S, \ldots, A_N \cap S, B_1 \cap S, \ldots, B_k \cap S \rangle$ *where* $S = S_\nu^{B_1 \ldots B_k}$

$J_\nu^{B_1 \ldots B_k}$ *denotes* $S_\nu^{B_1 \ldots B_k}$ *and* $\underline{J}_\nu^{B_1 \ldots B_k}$ *denotes* $\underline{S}_\nu^{B_1 \ldots B_k}$.

We make no existence claims in general. However:

<u>Lemma 3.16</u>. *For all* $\nu \in \omega\rho$ $S_\nu^{A^P}$ *exists.*

Proof: $\delta=\lambda+\gamma$ means $\exists f($ f is a function $\wedge$ $\mathrm{dom}(f) \subseteq On$ $\wedge$ $f(0)=\lambda$ $\wedge$

$$\wedge\ (\forall \nu+1 \in \mathrm{dom}(f))(f(\nu+1)=f(\nu)+1)\ \wedge\ (\forall \gamma \in \mathrm{dom}(f))(\lim(\gamma) \Rightarrow$$
$$\Rightarrow f(\gamma)=\sup_{\nu<\gamma} f(\nu)))$$

<u>Claim 1</u> There is a limit segment η such that $On=\{\lambda+\gamma : \gamma \in \eta\} \cap \lambda$.

Proof: Let $E=\{\lambda' : \lambda'$ a limit ordinal $\wedge$ $\lambda' \geq \lambda$ $\wedge$ $(\exists \gamma'<\lambda')(\forall \gamma<\lambda')$

$$\underline{S}_{\lambda'} \not\models \gamma'=\lambda+\gamma \ \vee\ \gamma'<\lambda\}.$$

If $E \neq \phi$ then it has a minimal element λ'. Clearly $\lambda'>\lambda$. It is easily seen that λ' is not a limit of limit ordinals. Say $\lambda'=\lambda'' \cup \{\lambda''+n : n \in \omega\}$ where λ'' is a limit. If $\gamma'<\lambda''$ then $(\exists \gamma<\lambda'')\underline{S}_{\lambda''} \models \gamma'=\lambda+\gamma \vee \gamma'<\lambda$.
But then $y=\{\gamma : (\exists \gamma'<\lambda'')(\gamma'=\lambda+\gamma)\}$ is a set in $S_{\lambda'}$, so there is a least $\gamma \notin y$; and $\underline{S}_{\lambda'} \models \lambda''=\lambda+\gamma$.

Then there is a least $\lambda''+n+1$ such that $(\forall \gamma<\lambda')\underline{S}_{\lambda'} \not\models \lambda''+n+1=\lambda+\eta$. Suppose $\underline{S}_{\lambda'} \models \lambda''+n+1=\lambda+\overline{\gamma}+1$ and $\overline{\gamma}+1<\lambda'$ (we are using the associativity of addition, which may be seen to hold in R^+). Contradiction! So $E=\phi$. The claim follows by the usual adaptation. □(Claim 1)

Now suppose $a \in S_{\lambda+\omega} \setminus S_\lambda$ and $a \subseteq \gamma \in \omega\rho$. Suppose $On=\{\lambda+\gamma' : \gamma' \in \eta\} \cup \lambda$.

<u>Claim 2</u> $\gamma \in \eta$.

Proof: Suppose $\eta \subseteq \gamma$. We are going to deduce that there is a Σ_1 map of γ onto $P(\gamma)$. This is done by establishing that if λ' is a limit and $\lambda'>\lambda$ then there is a $\Sigma_1(\underline{S}_{\lambda'})$ map of γ onto $P(\gamma) \cap S_{\lambda'}$. Otherwise take a least λ' where this fails. If there were a largest limit $\lambda''<\lambda'$ then $P(\gamma) \cap S_{\lambda'} \neq P(\gamma) \cap S_{\lambda''}$; if $\lambda''=\lambda$ this is given, and if $\lambda''>\lambda$ there is a $\Sigma_1(\underline{S}_{\lambda''})$ map of γ onto $P(\gamma) \cap S_{\lambda''}$. But by acceptability we deduce the existence of a $\Sigma_1(\underline{S}_{\lambda'})$ map of γ onto $P(\gamma) \cap S_{\lambda'}$.

So suppose λ' is a limit of limits. If $\lambda<\mu<\lambda'$ let f_μ denote the least map of γ onto $P(\gamma) \cap S_\mu$ in the canonical well-order (μ a limit). So $\langle f_\mu : \lambda<\mu<\lambda' \rangle \in \Sigma_1(\underline{S}_{\lambda'})$. But there is $g \in S_{\lambda'}$ such that $g : \gamma \leftrightarrow \gamma^2$. Let

$$F(\nu)=f_{\lambda+\nu_1}(\nu_2) \text{ if } \lambda+\nu_1<\lambda' \text{ and } g(\nu)=\langle \nu_1,\nu_2 \rangle$$
$$=0 \text{ otherwise.}$$

Then F is a Σ_1 map, and since $\eta \subseteq \gamma, \lambda' \subseteq \{\lambda+\gamma' : \gamma'<\gamma\}$ $F : \gamma \to P(\gamma) \cap S_{\lambda'}$ onto. The usual argument now gives a Σ_1 map F of γ onto $P(\gamma)$. Let $A=\{\xi<\gamma : \xi \notin F(\xi)\}$. A is a set as $\gamma \in \omega\rho$. But $\xi \in A \Leftrightarrow \xi \notin A$. Contradiction!

□(Claim 2)

Claim 3 $S^{\bar a}_\gamma$ exists for all $\bar a \in S_{\lambda+\omega}$.

Proof: Suppose λ' is a limit ordinal, $\lambda'>\lambda$ and $\lambda+\gamma'<\lambda'$. Then $\underline{S}_{\lambda'} \models \exists f(f=S^{\bar a}|\gamma')$.

To see this suppose it failed at some least λ'. λ' cannot be a limit of limits; so $\lambda'=\lambda''\cup\{\lambda''+n:n\in\omega\}$ where λ'' is a limit. If $\lambda+\gamma'<\lambda''$ then $S^{\bar a}|\gamma'\in S_{\lambda'}$. Since $\bar a\in S_{\lambda'}$ it follows that $S^{\bar a}|\gamma'\in S_{\lambda'}$ where $\lambda+\gamma'=\lambda''$. So there is least $\gamma'+n+1$, where $\lambda+\gamma'=\lambda''$, such that $S^{\bar a}|\gamma'+n+1\notin S_{\lambda'}$. But then $S^{\bar a}_{\gamma'}\in S_{\lambda'}$ and an easy induction gives $S^{\bar a}_{\gamma'+n}\in S_{\lambda'}$, so $S^{\bar a}|\gamma'+n+1 = S^{\bar a}|\gamma'+n\cup\{\langle\gamma'+n,S^{\bar a}_{\gamma'+n}\rangle\}\in S_{\lambda'}$; contradiction!

Now it follows as usual that if $\lambda+\gamma$ exists then $S^{\bar a}|\gamma$ and hence $S^{\bar a}_\gamma$ exists. But $\gamma\in\eta$ by claim 2. □(Claim 3)

Now suppose γ is a limit ordinal and let $\bar a=A^p\cap\gamma$. Let λ be the least limit ordinal with $\bar a\in S_{\lambda+\omega}$. If $\gamma=\lambda$ then $S_{\lambda+\omega}\cap P(\gamma)\neq S_\lambda\cap P(\gamma)$. Otherwise let $f\in S_{\lambda}$, $f:\gamma\leftrightarrow(\gamma)^{<\omega}$. Then let $a=f^{-1}``\bar a$, so $a\in P(\gamma)\cap S_{\lambda+\omega}\setminus S_\lambda$. Hence by claim 3 S^{a}_γ exists. So $S^{A^p}_\gamma$ exists for all $\gamma\in\omega\rho$. □

Claim 2 of the preceding lemma is often of independant interest: it implies, for example, that if λ is the largest limit ordinal and the projectum of $\underline{S}_\lambda$ is 1 then the projectum ρ must also be 1.

Lemma 3.17. *Suppose that for all $a\subseteq\gamma<\tau$ S^a_γ exists. If τ is a $\Sigma_0$cardinal then $H_\delta=\cup\{S^a_\gamma : a\subseteq\gamma<\tau\}$.*

Proof: ($\supseteq$) As S^a_γ is transitive it suffices to show $S^a_\gamma\in\tau$ for limit γ. But there is a $\Sigma_1(\underline{S}^a_\gamma)$ map of γ onto S^a_γ so this is clear.

($\subseteq$) Suppose $x\in H_\delta$. Say $\nu\in\tau$ and $f:\nu\leftrightarrow TC(x)$, ν a cardinal. Let $e=\{p^{-1}(\langle\xi,\tau\rangle):f(\xi)\in f(\tau)\}$. $\nu\in Q$ so $e\subseteq\nu$.

We claim $x\in S^e_\tau$. For $\text{rank}|TC(x):TC(x)\to\text{rank}(x)$ onto so $\text{rank}(x)\in\tau$. (Here we are using the result of exercise 1 in chapter 2.) Define

$$g=\text{coll}|\gamma \Leftrightarrow \gamma \text{ is an ordinal} \wedge g \text{ is a function} \wedge \text{dom}(g)=\gamma \wedge$$
$$\wedge\ g(0)=\phi \wedge (\forall\lambda\in\gamma)(\lambda \text{ a limit} \Rightarrow g(\lambda)=\bigcup_{\xi<\lambda} g(\xi)) \wedge$$
$$\wedge\ (\forall\nu+1\in\gamma)(g(\nu+1)=\{\langle\zeta,g(\nu)``\{\delta:p^{-1}(\langle\delta,\zeta\rangle)\in e\}\rangle:\ \zeta<\nu \wedge$$
$$\wedge\ \{\delta:p^{-1}(\langle\delta,\zeta\rangle)\in e\}\subseteq\text{dom } g(\nu)\})$$

Each $g(\nu)$ is a function and $g(\nu)(\delta)=f(\delta)$ provided $\delta\in\text{dom}(g(\nu))$. Hence $g(\text{rank}(x))$ is f. By the usual argument if λ' is a limit, $\lambda'\in\tau$, $\nu<\lambda'$ then $\exists g\in S^e_{\lambda'}$, $g=\text{coll}|\nu$. Hence $f\in S^e_{\lambda'}$, λ' any limit above ν. So $x\in S^e_\tau$. □

Lemma 3.18. $J^{A^p}_\rho=\cup\{S^a_\gamma : a\subseteq\gamma\in\omega\rho\}$.

Proof: ($\subseteq$) For $\gamma\in\rho$ $S^{A^p}_\gamma=S^{A^p\cap(\omega\times(\gamma)^{<\omega})}_\gamma$ and $A^p\cap(\omega\times(\gamma)^{<\omega})$ is a set.

($\supseteq$) Firstly observe that if $a\in J^{A^p}_\rho$ then for all $\gamma\in\omega\rho$ $S^a_\gamma\in J^{A^p}_\rho$. For if $a\in S^{A^p}_\delta$ then $\delta+\gamma$ exists and $\delta+\gamma\in\omega\rho$ because $\omega\rho$ is a Σ_1 cardinal. By claim 3 in lemma 3.16 $S^a_\gamma\in J^{A^p}_\rho$. So it suffices to prove that

$a\subseteq\gamma\in\omega\rho \Rightarrow a\in J_\rho^{A^p}$. Well, $\gamma^+\subseteq\omega\rho$ so by lemma 3.3 there are $i\in\omega$, $\xi\in\omega\rho$ such that $a=h(i,\xi)$. Say

$$\phi(z,y,p) \leftrightarrow y\in h(i,z) \tag{3.4}$$

and suppose ϕ is ϕ_j (as a Σ_1 formula). Then

$$a=\{\zeta\in\gamma:\langle j,\langle\zeta,\xi\rangle\rangle\in A^p\}. \tag{3.5}$$

So $a\in J_\rho^{A^p}$. □

<u>Corollary 3.19</u>. *$H_{\omega\rho}=J_\rho^{A^p}$; $H_{\omega\rho}\subseteq h(\omega\rho\cup\{p\})$.*

<u>Definition 3.20</u>. *RA^+ is R_ω^+ together with the axiom of strong acceptability:*

For all limit λ and $\delta<\lambda$ such that $P(\delta)\cap S_{\lambda+\omega}\neq P(\delta)\cap S_\lambda$, $S_{\lambda+\omega}\models\overline{\overline{\lambda}}\leq\delta$.

It is easily seen, using lemma 2.42, that $RA^+\supseteq RA$.

<u>Lemma 3.21</u>. *$\underline{J}_\rho^{A^p}\models RA^+$*

Proof: Clearly $\underline{J}_\rho^{A^p}\models R_\omega^+$. Suppose $a\in S_{\lambda+\omega}^{A^p}\setminus S_\lambda^{A^p}$, $a\subseteq\delta,\lambda$ a limit. Say $a=h(i,\xi)$ where $\xi\in\delta^+$. Let ϕ be the formula

$$\phi(x,y,p) \leftrightarrow y\in h(i,x) \tag{3.6}$$

Suppose ϕ is ϕ_j. So $a=\{\zeta<\delta:\langle j,\langle\xi,\delta\rangle\rangle\in A^p\}$. It follows that $a\in S_{\delta^+}^{A^p}$. So $\lambda<\delta^+$. Let $f:\delta\to\lambda$ onto. It suffices to show that some such $f\in S_{\lambda+\omega}^{A^p}$.

Assume that f is minimal in the canonical well-order. Let ϕ be the formula

$$\phi(\delta,\lambda,\xi,\zeta,p) \leftrightarrow f \text{ is the minimal map of } \delta \text{ onto } \lambda \wedge f(\xi)=\zeta \tag{3.7}$$

So if ϕ is ϕ_j

$$f(\xi)=\zeta \leftrightarrow \langle j,\langle\delta,\lambda,\xi,\zeta\rangle\rangle\in A^p.$$

But then $f\in S_{\lambda+\omega}^{A^p}$. □

<u>Definition 3.22</u>. *$V^p=\underline{J}_\rho^{A^p}$.*

$V^p\models RA$, so within V^p we can define a projectum. But we cannot assume that $V^p\models RA_\rho$.

<u>Definition 3.23</u>. *$\rho_0=On$; $A^0=T^0=PA_0=\emptyset$; $V^0=V$.*

Suppose $V^{np},T^{np},A^{np},\rho^n$ and PA_n are defined for $p\in PA_n$. Then

$PA_{n+1}=\{\langle p_1\ldots p_{n+1}\rangle:\langle p_1\ldots p_n\rangle\in PA_n \wedge p_{n+1}\in(\omega\rho_n)^{<\omega}\}$

$\rho^{n+1}=\bigcap\{\text{projecta of } V^{np},\ p\in PA_n\}$;

$T^{n+1p}=\{\langle i,x\rangle: i\in\omega \wedge x\in H_{\omega\rho^{n+1}} \wedge V^{np'}\models^2_{\Sigma_1}\phi_i(x,p'')\}$ where $p\in PA_{n+1}$, $p=\langle p',p''\rangle$;

$A^{n+1p}=T^{n+1p}\cap(\omega\times(\omega\rho^{n+1})^{<\omega})$;

$V^{n+1p}=\underline{J}_\rho^{A^{n+1p}}$ (which is contained in $(V^{np'})^{p''}$).

CODING V IN V^p

Lemma 3.24. *(i) $\langle V^p, T^p\rangle$ is amenable;*

(ii) $\Sigma_1(V^p,T^p)\subseteq\Sigma_1(V^p)$;

(iii) $x=y\cap T^p$ is $\Sigma_1(V^p)$.

Proof: $x=y\cap T^p \leftrightarrow (\exists f)(\ \mathrm{dom}(f)\in On \wedge (\exists i)(\exists \tau_1\ldots\tau_k\in On)($"$h(i,\langle\tau_1\ldots\tau_k,p\rangle)$ is a function"$\in A^p \wedge$ "$\mathrm{dom}(f)=\mathrm{dom}(h(i,\langle\tau_1\ldots\tau_k,p\rangle)$"$\in A^p \wedge$ $\wedge (\forall\xi\in\mathrm{dom}(f))(f(\xi)=\{f(\zeta)\in TC(y):$"$h(i,\langle\tau_1\ldots\tau_k,p\rangle)(\zeta)\in$ $\in h(i,\langle\tau_1\ldots\tau_k,p\rangle)(\xi)$"$\in A^p) \wedge (\exists\gamma)(y=f(\gamma) \wedge x=\{f(\zeta)\in y:$ "$f(\zeta)=\langle i,z\rangle \wedge i\in\omega \wedge \models_{\Sigma_1}\phi_i(z,p)$"$\in A^p\})))$ in $\langle V^p,T^p\rangle$, where "$\phi_i(\vec{\gamma},p)$" denotes $\langle i,\langle\vec{\gamma}\rangle\rangle$. □

Before the next definition we need a little temporary notation. If ϕ is a Σ_1 formula with $\mathrm{rng}(\tau_\phi)\subseteq m-1$ and $s(i)$ is the Σ_1 formula of one variable obtained by replacing each occurrence of v_k by $h(((v_0)_k)_0,\langle((v_0)_k)_1,p\rangle)$ then "$\phi(\langle i_0,x_0\rangle\ldots\langle i_{m-1},x_{m-1}\rangle)$" denotes $\langle i,\langle\langle i_0,x_0\rangle\ldots\langle i_{m-1},x_{m-1}\rangle\rangle\rangle\in T$. Suppose $s(i)$ is the formula $(\exists y)s(j)(\langle x,p\rangle,y)$; then $D(j,x)$ denotes $\langle i,x\rangle \quad \in T$.

Definition 3.25. *The theory MC is formulated in L_{N+2} as follows (calling the predicate symbols $B_1\ldots B_N,A,T$):*

(i) "$\phi(\langle i_0,x_0\rangle\ldots\langle i_{m-1},x_{m-1}\rangle)$" $\Rightarrow D(i_0,x_0)\ldots D(i_{m-1},x_{m-1})$;

(ii) if $\mathrm{rng}(\tau_\phi)\subseteq m-1$ *and* $n\geq m$ *then* "$\phi(\langle i_0,x_0\rangle\ldots\langle i_{m-1},x_{m-1}\rangle)$" $\leftrightarrow$ $\leftrightarrow$ "$\phi(\langle i_0,x_0\rangle\ldots\langle i_{n-1},x_{n-1}\rangle)$";

(iii) $D(i,x) \Rightarrow$ "$\langle i,x\rangle=\langle i,x\rangle$";

(iv) "$\langle i_0,x_0\rangle=\langle i_1,x_1\rangle$" $\wedge$ "$\langle i_1,x_1\rangle=\langle i_2,x_2\rangle$" $\Rightarrow$ "$\langle i_0,x_0\rangle=\langle i_2,x_2\rangle$" $\wedge$ "$\langle i_1,x_1\rangle=$ $=\langle i_0,x_0\rangle$";

(v) "$\langle i_0,x_0\rangle=\langle i_1,x_1\rangle$" $\wedge$ "$\langle j_0,x_0\rangle=\langle j_1,x_1\rangle$" $\wedge$ "$\langle i_0,x_0\rangle\in\langle j_0,x_0\rangle$" $\Rightarrow$ $\Rightarrow$"$\langle i_1,x_1\rangle\in\langle j_1,x_1\rangle$";

(vi)$_k$ "$B_k(\langle i_0,x_0\rangle)$" $\wedge$ "$\langle i_0,x_0\rangle=\langle i_1,x_1\rangle$" $\Rightarrow$ "$B_k(\langle i_1,x_1\rangle)$" $(0<k\leq N)$;

(vii) $\mathrm{rng}(\tau_\phi)\subseteq m-1 \wedge D(i_0,x_0)\ldots D(i_{m-1},x_{m-1}) \wedge \phi\ \Sigma_0 \Rightarrow$ $\Rightarrow ($"$\phi(\langle i_0,x_0\rangle\ldots\langle i_{m-1},x_{m-1}\rangle)$" $\leftrightarrow$ $\neg$"$\neg\phi(\langle i_0,x_0\rangle\ldots\langle i_{m-1},x_{m-1}\rangle)$"$)$;

(viii) if ϕ is $\psi\wedge\chi$ then "$\phi(\langle i_0,x_0\rangle\ldots\langle i_{m-1},x_{m-1}\rangle)$" $\leftrightarrow$ "$\psi(\langle i_0,x_0\rangle\ldots$ $\ldots\langle i_{m-1},x_{m-1}\rangle)$"*and*"$\chi(\langle i_0,x_0\rangle\ldots\langle i_{m-1},x_{m-1}\rangle)$";

(ix) "$\phi(\langle i_0,x_0\rangle\ldots\langle i_m,x_m\rangle)$" $\Rightarrow$ "$\exists y\phi(\langle i_0,x_0\rangle\ldots\langle i_{m-1},x_{m-1}\rangle,y)$";

(x) "$\exists y\phi(\langle i_0,x_0\rangle\ldots\langle i_{m-1},x_{m-1}\rangle,y)$"$\Rightarrow$"$\phi(\langle i_0,x_0\rangle\ldots\langle i_{m-1},x_{m-1}\rangle,\langle j,\langle x_0\ldots x_{m-1}\rangle\rangle)$"

where $\phi_j(v_0\ldots v_{m-1},p,y) \leftrightarrow \phi(h(i_0,\langle v_0,p\rangle)\ldots h(i_{m-1},\langle v_{m-1},p\rangle),y)$;

Let $s(k_0)$ denote the formula $y=v_0$.

(xi) let $\forall y\phi(y)$ be a single Π_2 axiom for R^+ together with the acceptability axiom; then $D(i,x) \Rightarrow$ "$\phi(\langle i,x\rangle)$";

(xii) $D(k_0,x_0)$;

(xiii) "$\langle k_0,x_0\rangle=\langle k_0,x_1\rangle$" $\leftrightarrow x_0=x_1$;

(xiv) "$\langle k_0,x_0\rangle\in\langle k_0,x_1\rangle$" $\leftrightarrow x_0\in x_1$;

(xv) "$B_k(\langle k_0,x\rangle)$" $\leftrightarrow B_kx$;

(xvi) "$\langle i,x\rangle\in\langle k_0,y\rangle$" $\Rightarrow$ $(\exists z\in y)$("$\langle k_0,z\rangle=\langle i,x\rangle$");

Let $s(k_1)$ be the formula $y=v_1$;

(xvii) $D(\langle k_1,0\rangle)$;

(xviii) $D(\langle i,x\rangle)$ $\Rightarrow$ "$\langle i,x\rangle=h(i,\langle\langle k_0,x\rangle,\langle k_1,0\rangle\rangle)$";

(xix) $\langle i,x\rangle\in T \Leftrightarrow$ "$\phi_i(\langle k_0,x\rangle,\langle k_1,0\rangle)$";

(xx) $A=T\cap(\omega\times(On)^{<\omega})$.

The reader should check that $\langle V^p,B_1...B_N,A^p,T^p\rangle\models MC$. We shall prove the following converse.

Lemma 3.26. *Suppose MC holds. Then there is a class $\underline{M}\models RA_\rho$ and $\hat{p}\in M$ such that there is an isomorphism $\sigma:\langle (V^{\hat{p}})^M,(T^{\hat{p}})^M\rangle\cong V$ and $\underline{M}\models V=h(\omega\rho\cup\{\hat{p}\})$; furthermore if $\sigma':(V^{p'})^N\cong\langle V,B_1...B_k,A\rangle$ and $\underline{N}\models V=h(\omega\rho\cup\{\hat{p}'\})$ then there is $\pi:\underline{M}\cong\underline{N}$ with $\pi(\hat{p})=(\hat{p}')$ and $\pi\sigma^{-1}=\sigma'^{-1}$.*

Proof: Let $M'=\{\langle j,x\rangle:D(j,x)\}$. Define a relation I on M' by

$\langle i_0,x_0\rangle I\langle i_1,x_1\rangle \Leftrightarrow$ "$\langle i_0,x_0\rangle=\langle i_1,x_1\rangle$".

By (iii) and (iv) I is an equivalence relation. Let $M=M'/I$ and let $[j,x]$ denote some member of the equivalence class of $\langle j,x\rangle$. Say

$[i_0,x_0]E[i_1,x_1] \Leftrightarrow$ "$\langle i_0,x_0\rangle\in\langle i_1,x_1\rangle$";

$B'_k([i,x]) \Leftrightarrow$ "$B_k(\langle i,x\rangle)$".

Then E, B'_k are well-defined by (v), $(vi)_k$. Let $\underline{M}=\langle M,E,B'_1,...B'_k\rangle$.

Claim 1 $\underline{M}\models\phi([i_0,x_0]...[i_{m-1},x_{m-1}]) \Leftrightarrow$ "$\phi(\langle i_0,x_0\rangle...\langle i_{m-1},x_{m-1}\rangle)$" where ϕ is Σ_1.

Proof: By induction on structure. Atomic formulae are immediate from the definitions and (ii). $-$ follows by (vii) and $\wedge$ by (viii). If $\underline{M}\models\exists y\phi([i_0,x_0]...[i_{m-1},x_{m-1}],y)$ say $\underline{M}\models\phi([i_0,x_0]...[i_m,x_m])$. Then by (ix) "$\exists y\phi(\langle i_0,x_0\rangle...\langle i_{m-1},x_{m-1}\rangle,y)$". Conversely if "$\exists y\phi(\langle i_0,x_0\rangle...$ $...\langle i_{m-1},x_{m-1}\rangle)$" then suppose "$\phi(\langle i_0,x_0\rangle...\langle i_{m-1},x_{m-1}\rangle,\langle j,\langle x_0...$ $...x_{m-1}\rangle\rangle)$"; so $\underline{M}\models\phi([i_0,x_0]...[i_{m-1},x_{m-1}],[j,\langle x_0...x_{m-1}\rangle])$, so $\underline{M}\models\exists y\phi([i_0,x_0]...[i_{m-1},x_{m-1}],y)$. □(Claim 1)

By (ix) $\underline{M}\models R^+_\omega$. Define a map $\sigma:V\to M$ by $\sigma(x)=[k_0,x]$. By (xii) σ is well-defined and by (xiii)-(xv) $\sigma:V\cong\underline{M}|rng(\sigma)$. By (xvi) $rng(\sigma)$ is an E-initial segment of $\underline{M}$. Hence $\underline{M}$ is a model of R^+_ω and by (ix) $\underline{M}\models RA$. Let $\hat{\rho}=(\rho)^M$, $\hat{p}=[k_1,0]$.

Claim 2 σ"$On=\omega\hat{\rho}$.

Proof: Suppose $\omega\hat{\rho}\subseteq\sigma(\gamma)$. Let $A\in\Sigma_1(\underline{M})$ be such that $A\cap\sigma(\gamma)\notin M$. Suppose $\underline{M}\models\xi\in A \Leftrightarrow \phi(\xi,[i,x])$. Let $a=\{\zeta<\gamma:$"$\phi(\langle k_0,\zeta\rangle,\langle i,x\rangle)$"$\}$. For $\zeta<\gamma$

$\sigma(\zeta)\in\sigma(a) \Leftrightarrow \zeta\in a$

$\Leftrightarrow$ "$\phi(\langle k_0,\zeta\rangle,\langle i,x\rangle)$"

$\Leftrightarrow \underline{M}\models\phi([k_0,\zeta],[i,x])$ (claim 1)

$\Leftrightarrow \underline{M}\models\phi(\sigma(\zeta),[i,x])$

$\Leftrightarrow \sigma(\zeta)\in A$.

So $A\cap\sigma(\gamma)=\sigma(a)$. Contradiction!

Suppose then that $\sigma''\mathrm{On}\subseteq\omega\gamma\in\omega\hat{\rho}$. By (xviii) $\underline{M}\models V=h(\mathrm{rng}(\sigma)\cup\{[k_1,0]\})$. Observe that

$$\begin{aligned}
\langle i,\langle\vec{\delta}\rangle\rangle\in A &\leftrightarrow \langle i,\langle\vec{\delta}\rangle\rangle\in T && \text{(by (xx))}\\
&\leftrightarrow \text{"}\phi_i(\langle k_0,\langle\vec{\delta}\rangle\rangle,\langle k_1,0\rangle)\text{"} && \text{(by (xix))}\\
&\leftrightarrow \underline{M}\models\phi_i(\sigma(\vec{\delta}),\hat{\rho}) && \text{(claim 1)}\\
&\leftrightarrow \langle i,\sigma(\vec{\delta})\rangle\in T^{\hat{\rho}}\\
&\leftrightarrow \sigma(\langle i,\vec{\delta}\rangle)\in A^{\hat{\rho}}.
\end{aligned}$$

So $\underline{M}\models\mathrm{rng}(\sigma)\subseteq J_\gamma^{A^{\hat{\rho}}}$ ($J_\gamma^{A^{\hat{\rho}}}$ is a set in $\underline{M}$ as $\gamma\in\hat{\rho}$). But there is $f\in\Sigma_1(J_\gamma^{A^{\hat{\rho}}})\subseteq M$ mapping $\omega\gamma$ onto $J_\gamma^{A^{\hat{\rho}}}$ so $\underline{M}\models V=h(\omega\gamma\cup\{\hat{\rho}\})$. Hence there is a $\Sigma_1(\underline{M})$ map, g say, of a subset of $\omega\gamma$ onto M. Let $A=\{\xi<\omega\gamma:\xi\notin g(\xi)\}$; then $A\in\Sigma_1(\underline{M})$ so $A\in M$. Say $A=g(\xi)$; then $\xi\in A\leftrightarrow\xi\in g(\xi)\leftrightarrow\xi\notin A$. Contradiction!

(Claim 2) □

Now we have seen that $\langle i,x\rangle\in T\rightarrow\sigma(\langle i,x\rangle)\in(T^{\hat{\rho}})^M$ so $\sigma:V\cong\langle\underline{M}^{\hat{\rho}},(T^{\hat{\rho}})^M\rangle$. Since $\underline{M}\models V=h(\omega\rho\cup\{\hat{\rho}\})$, $\underline{M}\models RA_\rho$ (as in claim 2). The uniqueness claim is left to the reader. □

Note that although MC is tedious to state it has a single Π_1 axiom.

Definition 3.27. *MC_1 is the theory MC;*

Given MC_n, $n>0$, define MC_{n+1} as follows: Given a structure $\langle\underline{M},B_1\ldots B_k,A\rangle$ define a set T from A by the definition of lemma 3.24. Let $\forall y\psi(y)$ be the Π_2 sentence asserting that $\langle M,T\rangle$ is amenable (where ψ is Σ_1). Then MC_{n+1} is $MC+$"MC_n"$+\ \forall i,x(D(i,x)\rightarrow$ "$\psi(\langle i,x\rangle)$").

Exercises

1. Suppose that in definition 3.15 $F_{18}\ldots F_{17+N}$ had been omitted. Show that corollary 3.19 would still hold.

2. Show that lemma 3.16 is not a theorem of R^+_ω.

3. Suppose $\pi:\underline{M}\rightarrow_{\Sigma_0}\underline{N}$ cofinally and $\langle\underline{M},T\rangle$ is amenable. Let $T'=\bigcup_{x\in M}\pi(x\cap T)$. Show that $\pi:\langle\underline{M},T\rangle\rightarrow_{\Sigma_0}\langle\underline{N},T'\rangle$ and $\langle\underline{N},T'\rangle$ is amenable.

4. Repeat exercise 3 for $\pi:\underline{M}\rightarrow_{\Sigma_2}\underline{N}$.

5. Use MC_n to formulate an iterated version of lemma 3.26.

6. Suppose $P(\delta)\cap J_\lambda\neq P(\delta)\cap J_{\lambda+1}$, $\delta\in\omega\rho$. Show that $\omega\lambda+\delta$ exists. Show also that either $\sup\{\omega\lambda+\delta:\delta\in\omega\rho\}$ is bounded in On or $\phi\in P$.

4: MODELS OF R

It is more convenient if we use ZFC as our basic theory in this chapter.

Definition 4.1 *A model $\underline{M}=\langle M,E,A_1\ldots A_N\rangle$ of R is standard provided*

(i) if M' is an initial segment of M (that is, $x\in M'$ and $yEx \Rightarrow y\in M'$) and $E\cap(M')^2$ is well-founded then M' is transitive and $E\cap(M')^2=\in\cap(M')^2$;

(ii) $\omega+1$ is an initial segment of M (or $On_M=\omega$).

Lemma 4.2. *If $\underline{M}\models R^+$ is standard then $\underline{M}\models R^+_\omega$.*

Proof: Suppose $\underline{M}\models$ η is a non-empty limit segment of ω. Then η is a non-empty limit segment of ω. So $\eta=\omega$. So $\underline{M}\models\eta=\omega$. □

By considering only standard models we shall not be embarrassed by the logically complex axioms of definition 2.12. The next lemma is a form of the Mostowski collapsing lemma:

Lemma 4.3. *Suppose $\underline{M}\models R$ and let $z=\{n:\underline{M}\models n\in\omega\}$. If $E\cap z^2$ is well-founded then $\underline{M}$ is isomorphic to a standard model of R.*

Proof: Set $T(x)=\{z:zEy_n$, some n$\}$ where $\underline{M}\models y_n=\cup^n x$. Set $Z=$ $=\{x\in M:T(x)^2\cap E$ is well-founded.$\}$

Z is an initial segment of M. For if $x\in Z$ and yEx then $\underline{M}\models y\subseteq\cup x$ so for $n\in\omega$ $\underline{M}\models\cup^n y\subseteq\cup^{n+1}x$, hence $T(y)\subseteq T(x)$, so $T(y)^2\cap E$ is well-founded, i.e. $y\in Z$.

$Z^2\cap E$ is well-founded. Suppose $A\subseteq Z, A\neq\phi$. Take some $x\in A$. Let $A'=T(x)\cap A$. If $A'=\phi$ then $yEx \Rightarrow y\in T(x) \Rightarrow y\notin A$ so x is E-minimal in A. If $A'\neq\phi$ let y be E-minimal in A'. If $t\in A, tEy$ then since $y\in T(x)$, $t\in T(x)$; so $t\in A'$; contradiction! Hence y is E-minimal in A.

$\langle Z,E\cap Z^2\rangle$ is extensional. For if $x,y\in Z$ and $x\neq y$ then $\underline{M}\models\exists t(t\in x\wedge t\notin y)$ say. So tEx, hence $t\in Z$. So by Mostowski's collapsing lemma there is a transitive Z' and π such that

$$\pi:\langle Z',\in\cap Z'^2\rangle\cong\langle Z,E\cap Z^2\rangle. \tag{4.1}$$

If $(M\smallsetminus Z)\cap Z'\neq\phi$ replace $M\smallsetminus Z$ by an isomorphic copy disjoint from Z'. Set $M'=Z'\cup(M\smallsetminus Z)$. Let $\pi'(x)=\pi(x)$ if $x\in Z'$, x otherwise. Say

$$xE'y \Leftrightarrow (x,y\in Z'\wedge x\in y)\vee(x\in Z'\wedge y\notin Z')\vee(x,y\notin Z'\wedge xEy).$$

We complete the proof by showing that $\langle M',E'\rangle$ is standard. Suppose N is an initial segment of M' and N is well-founded. Let $N'=\pi'"N$.

N' is a well-founded initial segment of M': for if $x\in N$, $zE\pi'(x)$ and $z=\pi'(z')$ then $z'E'x$ so $z'\in N$; hence $z\in N$. Suppose $x\in N'$. Then $zET(x) \Rightarrow z\in N'$; so $T(x)^2\cap E$ is well-founded, so $x\in Z$. Thus $N'\subseteq Z$, so $N\subseteq Z'$, giving the desired result.

Finally we are told that $E\cap z^2$ is well-founded, so $\pi:\omega\cong z$. If $\underline{M}\models$ " ω exists" then $\underline{M}'\models$"ω exists". But $E'\cap(\omega\cup\{\omega\})^2$ is well-founded so ω is ω in the sense of M'. Thus $\omega+1$ is an initial segment of M'.□

Next some terminology. Generally speaking $t_M(y_1\ldots y_n)$ denotes that y such that $\underline{M}\models y=t(y_1\ldots y_n)$. The M may be a subscript or a superscript as convenient.

For classes there is a small problem. Suppose η is a class of $\underline{M}$ defined by ϕ. The natural choice for η_M would be $\{x:\underline{M}\models\phi(x)\}$. Admittedly inside $\underline{M}$ we have written $x\in\eta$ - which should be read $xE_M\eta$- and now we are writing $x\in\eta$ in the real world. But $x\in\eta$ is only a convention with classes, whereas classes of $\underline{M}$ are sets in the real world, so that there is no ambiguity. Ambiguity does arise, though, when a class of $\underline{M}$ happens to be a set of $\underline{M}$. In $\underline{M}$ we identified η with a set; in the universe the corresponding set is $\{x:xE\eta\}$. Really this causes no trouble and we only mention it to account for some apparently bizarre uses later on.

Three conventions: firstly the use of $\in_M$ is restricted to cases where $\underline{M}$ is standard. This is handy when both a collapsed and an uncollapsed model are around together. If $\underline{M}$ is not standard we must explicitly give a name to its membership relation: it will usually be E. Secondly, we do not write V_N^{np} but simply $\underline{N}^{np}$. If n=0 this is unambiguous as $V_N^0=\underline{N}$. Finally, to avoid cumbersome expressions we do not underline structures that occur as subscripts and superscripts so that, for example, ρ_N denotes $\rho_{\underline{N}}$.

Lemma 4.4. (The condensation lemma) *Suppose $X\prec_{\Sigma_\Gamma}\underline{N}$ and $\underline{N}$ is a standard model of R. There is a standard model $\underline{M}$ of R and a map π with $\pi:\underline{M}\cong\underline{N}|X$. If X is a Σ_m substructure of $\underline{N}$ then in addition $\pi:\underline{M}\rightarrow_{\Sigma_m}\underline{N}$.*

Proof: R has a single Π_2 axiom so $\underline{N}|X\models R$. Also $\omega\subseteq X$; so the lemma may be proved by applying lemma 4.3 to $\underline{N}|X$. □

Lemma 4.5. *Suppose $\pi:\underline{M}\rightarrow_{\Sigma_\Gamma}\underline{N}$ with $\underline{M},\underline{N}$ standard. If $\underline{N}\models R^+$ (resp. RA) then $\underline{M}\models R^+$ (resp. RA).*

Proof: R^+ is Π_2 axiomatisable so $\underline{M}\models R^+$. Suppose $\underline{N}\models RA$. $\underline{M}\models R^+_\omega$ by lemma 4.2, and the axiom of acceptability is Π_2 so $\underline{M}\models RA$. □

There is also an upward form of lemma 4.5.

Definition 4.6. *$\pi:\underline{\bar{N}}\to_{\Sigma_0}\underline{N}$ cofinally provided $\pi:\underline{\bar{N}}\to_{\Sigma_0}\underline{N}$ and $\forall x\in N\exists y\in\bar{N}$ $\underline{N}\models x\in\pi(y)$.*

Lemma 4.7. *If $\pi:\underline{\bar{N}}\to_{\Sigma_0}\underline{N}$ cofinally then $\pi:\underline{\bar{N}}\to_{\Sigma_1}\underline{N}$.*
Proof: Suppose $\underline{N}\models\exists x\phi(x,\pi(z_1)\ldots\pi(z_n))$ with ϕ Σ_0. Suppose $\underline{N}\models\phi(x,\pi(z_1)\ldots\pi(z_n))$ and $\underline{N}\models x\in\pi(y)$. Then $\underline{N}\models\exists x\in\pi(y)\phi(x,\pi(z_1)\ldots\pi(z_n))$ so $\underline{\bar{N}}\models\exists x\in y\phi(x,z_1\ldots z_n)$. Thus $\underline{\bar{N}}\models\exists x\phi(x,z_1\ldots z_n)$. □

Lemma 4.8. *If $\pi:\underline{M}\to_{\Sigma_0}\underline{N}$ cofinally and $\underline{M}\models R$ then $\underline{N}\models R$.*

Proof: Use the finite axiomatisation of chapter 1. Extensionality and foundation are Π_1, so they hold in $\underline{N}$ by lemma 4.7. Suppose u,v are in N; we have to show that $\underline{N}\models F_i(u,v)$ exists for $0<i\leq 17+K$, where R is R^K. Suppose $\underline{N}\models u\in\pi(u'),v\in\pi(v')$. Then $\underline{M}\models F_i{}''u'\times v'$ exists by lemma 1.7. Say $\underline{M}\models w=F_i{}''u'\times v'$. Then $\underline{M}\models\forall x\in u'\forall y\in v'\exists t\in w\ t=F_i(x,y)$. So $\underline{N}\models\forall x\in\pi(u')\forall y\in\pi(v')\exists t\in\pi(w)t=F_i(x,y)$. In particular $\underline{N}\models\exists t\ t=F_i(u,v)$. □

Lemma 4.9. *If $\pi:\underline{M}\to_{\Sigma_0}\underline{N}$ cofinally and $\pi:\omega_M\to\omega_N$ onto then if $\underline{M}\models R$ there is a standard $\underline{N}'\models R$ with $\sigma:\underline{N}'\cong\underline{N}$ and $\sigma^{-1}\pi:\underline{M}\to_{\Sigma_0}\underline{N}'$ cofinally.*
Proof: By lemmas 4.3 and 4.8. □

Lemma 4.10. *If $\underline{M},\underline{N}$ are standard models of $R,\pi:\underline{M}\to_{\Sigma_0}\underline{N}$ cofinally and $\underline{M}\models R^+$ (resp. RA) then $\underline{N}\models R^+$ (resp. RA).*
Proof: $\underline{N}\models S_\nu$ exists for cofinal ν in On_N, hence for all $\nu\in On_N$. Given x in N, if $\underline{N}\models x\in\pi(y)$ and $\underline{M}\models y\in S_\nu$ then $\underline{N}\models\pi(y)\in S_{\pi(\nu)}$ (this is Σ_1) so $\underline{N}\models x\in S_{\pi(\nu)}$.

$\underline{N}\models R^+_\omega$ as $\underline{N}$ is standard. For $\underline{N}\models$RA observe that the axiom of acceptability restricted to $\lambda+\omega\in On$ is Π_1: so let λ be the largest limit ordinal. Let δ_n be least such that for some $a\subseteq\delta, a\in S_{\lambda+n}$, $a\notin S_\lambda$. Then $\delta_n\in rng(\pi)$; say $\delta_n=\pi(\bar{\delta}_n)$. But if f satisfies the acceptability axiom for $u=S_{\pi^{-1}(\lambda)+n}$, $\delta=\bar{\delta}_n$ in $\underline{M}$ then $\pi(f)$ satisfies it for $u=S_{\lambda+n}$, $\delta=\delta_n$ in $\underline{N}$. This suffices. □

Lemma 4.11. *If $\underline{M},\underline{N}$ are standard models of $R,\pi\,\underline{M}\to_{\Sigma_2}\underline{N}$ and $\underline{M}\models R^+$ (resp. RA) then $\underline{N}\models R^+$ (resp.RA).*
Proof: All axioms are Π_2. □

EXTENSION OF EMBEDDINGS

Lemma 4.12. *If $\pi:\underline{\bar{M}}\to_{\Sigma_1}\underline{N}^p$ and $\underline{N}\models RA_\rho$ is standard then there are $\underline{M},p'$ (unique up to isomorphism) with $\underline{M}^{p'}=\underline{\bar{M}}$ and $\underline{\bar{M}}\models V=h(\omega\rho\cup p')$.*
Proof: $\langle\underline{N}^p,T^p\rangle\models$RA+MC. Let $T=\{x\in\bar{M}:T^p(\pi(x))\}$; so $\langle\underline{\bar{M}},T\rangle\models$RA+MC and there are unique $\underline{M},p'$ with $\underline{M}^{p'}=\underline{\bar{M}}$, $T^{p'}=T$. T is unique by 3.24(iii).
Note: Σ_1 was only needed to ensure $\langle\underline{\bar{M}},T\rangle\models R^+$. □

The $\underline{M}$ in lemma 4.12 may be taken as standard. It will be a model of RA_ρ.

Lemma 4.13. *Suppose $\underline{M},\underline{N}$ are standard models of RA_ρ. Suppose $\pi:\underline{M}^{p'}\to_{\Sigma_0}\underline{N}^p$ and $\underline{M}\models V=h(\omega\rho\cup p')$. Then there is a unique $\tilde{\pi}$ such that $\tilde{\pi}:\underline{M}\to_{\Sigma_0}\underline{N}$ and $\tilde{\pi}(p')=p$. Also $\pi:\underline{M}\to_{\Sigma_1}\underline{N}$.*

Proof: By lemma 3.24 $\pi:\langle\underline{M}^{p'},T^{p'}\rangle\to_{\Sigma_0}\langle\underline{N}^p,T^p\rangle$, so for $x_1\ldots x_n\in M^p$ $i_1\ldots i_n\in\omega$, ϕ Σ_1

$$\underline{M}\models\phi(h(i_1,\langle x_1,p'\rangle)\ldots h(i_n,\langle x_n,p'\rangle)) \leftrightarrow \underline{N}\models\phi(h(i_1,\langle\pi(x_1),p\rangle)\ldots h(i_n,\langle\pi(x_n),p\rangle)) \quad (4.2)$$

and we may define

$$\tilde{\pi}(h_M(i,\langle x,p'\rangle))\simeq h_N(i,\langle\pi(x),p\rangle).$$

This is well-defined since $\underline{M}=V\models h(\omega\rho\cup p')$; and by (4.2) $\tilde{\pi}:\underline{M}\to_{\Sigma_1}\underline{N}$. We must show $\tilde{\pi}\supseteq\pi$. Let $s(j_0)$ be the formula $y=x$; then $h_M(j_0\langle x,p'\rangle)=x$, $h_N(j_0,\langle y,p\rangle)=y$. So

$$\tilde{\pi}(x)=\tilde{\pi}(h_M(j_0\langle x,p'\rangle)) =h_N(j_0,\langle\pi(x),p\rangle) =\pi(x).$$

Similarly $\tilde{\pi}(p')=p$.

It remains to prove uniqueness. Suppose $\pi':\underline{M}\to_{\Sigma_1}\underline{N},\pi'\supseteq\pi$ and $\pi'(p')=p$. Take $y\in M$ and suppose $\underline{M}\models y=h(i,\langle x,p'\rangle)$ with $x\in M^{p'}$. Then

$$\underline{N}\models\pi'(y)=h(i,\langle\pi'(x),\pi'(p')\rangle)=h(i,\langle\pi(x),p\rangle)=\tilde{\pi}(y). \quad \square$$

In fact we can improve this result if π is a stronger map.

Lemma 4.14. *Suppose $\pi:\underline{M}^{p'}\to_{\Sigma_h}\underline{N}^p$ where $\underline{M}\models V=h(\omega\rho\cup p')$, and $\underline{N}\models V=h(\omega\rho\cup p)$. Suppose $\tilde{\pi}\supseteq\pi$, $\tilde{\pi}:\underline{M}\to_{\Sigma_0}\underline{N}$ and $\tilde{\pi}(p')=p$. Then $\pi:\underline{M}\to_{\Sigma_{h+1}}\underline{N}$.*

Proof: Suppose ϕ is a Σ_{h+1} formula. For definiteness take h odd: so say $\phi\leftrightarrow\exists x_1\forall x_2\ldots\exists x_h\psi$ where ψ is Π_1. Let

$$D=\{\langle j,\gamma\rangle:\underline{M}\models j\in\omega \wedge \gamma\in M^{p'} \wedge h(j,\langle\gamma,p'\rangle) \text{ exists}\} \quad (4.3)$$

$$\bar{D}=\{\langle j,\gamma\rangle:\underline{N}\models j\in\omega \wedge \gamma\in N^{p} \wedge h(j,\langle\gamma,p\rangle) \text{ exists}\}$$

Say $R(\vec{x},\vec{z}) \leftrightarrow \underline{M}\models x_i=\langle j_i,x_i'\rangle\in D \wedge z_i=\langle k_i,z_i'\rangle\in D \wedge$

$$\wedge\ \psi(h(j_1,\langle x_1',p'\rangle)\ldots h(j_h,\langle x_h',p'\rangle),h(k_1\langle z_1',p'\rangle)\ldots \ldots h(k_n,\langle z_n',p'\rangle))$$

Define $\bar{R}$ similarly from $\underline{N},\bar{D},p$. Then

$$\underline{M}\models\phi(h(k_1,\langle z_1',p'\rangle)\ldots h(k_n,\langle z_n',p'\rangle) \leftrightarrow \exists x_1\in D\ldots\exists x_h\in D\ R(\vec{x},\vec{z}).$$

Similarly for $\underline{N}$.

R is rudimentary in $\underline{M}^{p'}$ so there is a Δ_1 formula χ such that

$$\underline{M}\models R(\vec{x},\vec{z}) \leftrightarrow \underline{M}^{p'}\models\chi(\vec{x},\vec{z}). \quad (4.4)$$

And there is a Δ_1 formula δ such that

$$\underline{M}\models x\in D \leftrightarrow \underline{M}^{p'}\models\delta(x). \quad (4.5)$$

Let $\chi'(\vec{z})$ be the formula

$$\exists x_1(\delta(x_1) \wedge \forall x_2(\delta(x_2) \Rightarrow \ldots \exists x_h(\delta(x_h) \wedge \chi(\vec{x},\vec{z})) \qquad (4.6)$$

(4.6) is Σ_h and

$$\underline{M} \models \phi(h(k_1,\langle z_1',p'\rangle)\ldots h(k_n,\langle z_n',p'\rangle)) \Leftrightarrow \underline{M}^{p'} \models \chi'(\vec{z}).$$

Similarly for $\underline{N}$. Hence

$$\underline{M} \models \phi(h(k_1,\langle z_1',p'\rangle)\ldots h(k_n,\langle z_n',p'\rangle)) \Leftrightarrow \underline{M}^{p'} \models \chi'(\vec{z})$$

$$\Leftrightarrow \underline{N}^{p} \models \chi'(\pi(\vec{z}))$$

$$\Leftrightarrow \underline{N} \models \phi(h(k_1,\langle \pi(z_1'),p\rangle)\ldots h(k_n,\langle \pi(z_n'),p\rangle)).$$

But $\tilde{\pi}(h(k_i,\langle z_i',p'\rangle)) = h(k_i\langle \pi(z_i'),p\rangle)$. □

<u>Lemma 4.15</u>. *Suppose $\underline{M}$, $\overline{N}$ are standard models of R and $\pi:\underline{M}^{p'} \to_{\Sigma_0} \overline{N}$ cofinally. Then there are $\underline{N}$, p, unique up to isomorphism, such that $\overline{N}=\underline{N}^p$ and $\underline{N} \models V=h(\omega\rho\cup p)$.*

Proof: By lemmas 4.10 ,3.26 and chapter 3 exercise 3. □

<u>Lemma 4.16</u>. *Suppose $\underline{M}$, $\overline{N}$ are standard models of R and $\pi:\underline{M}^{p'} \to_{\Sigma_2} \overline{N}$. Then the conclusion of Lemma 4.15 holds.*

Proof: By lemmas 4.11 , 3.26 and chapter 3 exercise 4. □

Lemmas 4.12-4.16 are called the extension of embeddings lemmas.

<u>SOUNDNESS</u>

In order to iterate the extension of embeddings lemmas we must first look at the notion of soundness.

<u>Definition 4.17</u>. *$\underline{M}$ is p-sound means $\underline{M} \models V=h(\omega\rho\cup p)$.*

Suppose p-sound is defined for all $p \in PA_n^M$. Let $\langle p_1 \ldots p_{n+1}\rangle \in PA_{n+1}^M$. Then $\underline{M}$ is p-sound provided $\underline{M}$ is $\langle p_1 \ldots p_n\rangle$-sound and $\underline{M}^{np_1\ldots p_n}$ is p_{n+1}-sound.

<u>Lemma 4.18</u> *Suppose $\underline{M} \models RA$ is p-sound. Then $\underline{M} \models RA_\rho$ and $p \in P_M$.*

Proof: Let $f \in \Sigma_1(\underline{M})$ with parameter p be such that $rng(f''\omega\rho_M) = P(On_M) \cap M$ Let $A=\{\xi:\xi \notin f(\xi)\}$; $A \in \Sigma_1(\underline{M})$ with parameter p. Suppose $A \cap \omega\rho_M = x \cap \omega\rho_M$ with $x \in M$. Assume $x \subseteq On_M$. Then if $x=f(\zeta)$, $\zeta \in \omega\rho_M$ we have

$$\zeta \in f(\zeta) \Leftrightarrow \zeta \in x$$
$$\Leftrightarrow \zeta \in A$$
$$\Leftrightarrow \zeta \notin f(\zeta).$$

Contradiction! □

A similar proof shows that if $V=h(\gamma \cup \{p\})$, any p, then $\omega\rho \subseteq \gamma$. This was used in lemma 3.26.

<u>Lemma 4.19</u>. *Suppose $\underline{M}$ is p-sound. Suppose $B \subseteq (\omega\rho)^{<\omega}$. Then B is $\Sigma_h(\underline{M}^p)$ if and only if there is $A \in \Sigma_{h+1}(\underline{M})$ with parameters from $p \cup (\omega\rho)^{<\omega}$ such that $B = A \cap M^p$ (h>0).*

Proof: (in RA) Firstly suppose $A \in \Sigma_{h+1}$ and $B = A \cap (\omega\rho)^{<\omega}$. By the argument of lemma 4.14 there is a Σ_h formula χ such that

$A(h(i,\langle x,p\rangle)) \Leftrightarrow V^p \models \chi(\langle i,x\rangle)$.

Let s(j) be the formula y=x. Then

$A(x) \Leftrightarrow A(h(j,\langle x,p\rangle))$

$\Leftrightarrow \chi(\langle j,x\rangle)$

for $x\in(\omega\rho)^{<\omega}$; so B is $\Sigma_h(V^p)$.

For the converse it is sufficient to consider the case h=1. Suppose, then, that B is $\Sigma_1(V^p)$.

Case 1 There is a Σ_1 map of some $\gamma\in\omega\rho$ cofinally into On. Call the map g. Suppose $x\in A^p \Leftrightarrow \exists y\psi(y,x,p)$ where ψ is Σ_0. Say $\phi(\nu,x) \Leftrightarrow (\exists y\in S_\nu)\psi(y,x,p)$.

Obviously it suffices to prove the result if B is $\Sigma_0(V^p)$. Furthermore, letting $A'=\{\langle\nu,x\rangle:\psi(\nu,x)\}$, since A^p is rudimentary in $\langle H_{\omega\rho},A'\rangle$ we may assume B is rudimentary in $\langle H_{\omega\rho},A'\rangle$. But there is a rudimentary function F such that

$$\langle H_{\omega\rho},A'\rangle \models x\in B \Leftrightarrow F(x)\neq\phi. \tag{4.7}$$

Hence it suffices to prove that the function $a(u)=A'\cap u$ is Σ_2. In fact it is Π_1, for

$$y=a(u) \Leftrightarrow \forall z(z\in y \Leftrightarrow A'(z) \wedge z\in u) \tag{4.8}$$

Case 2 There is no Σ_1 map of $\gamma\in\omega\rho$ cofinally into On.

Claim If H is Σ_0 and $u\in V^p$ then

$$\forall x\in u\exists yHxy \Rightarrow \exists v\forall x\in u\exists y\in vHxy. \tag{4.9}$$

Proof: Suppose $\forall x\in u\exists yHxy$. Suppose $u\in S_\gamma^{V^p}$, γ a limit, and suppose $f:\gamma\to S_\gamma^{V^p}$ is $\Sigma_1(\underline{S}_\gamma^{V^p})$ and onto. Define $g:\gamma\to On$ by

$g(\nu)=$ the least μ such that $\exists y\in S_\mu Hf(\nu)\mu$ if $f(\nu)\in u$

$= 0$ otherwise.

Then g is Σ_1. So rng(g) is bounded in On; say $g''\gamma\subseteq\gamma'\in On$. Then $v=S_{\gamma'}$ satisfies the claim. □(Claim)

Now as in case 1 we must show

$$y=a(u) \Leftrightarrow \forall z(z\in y \Leftrightarrow A^p(z)\wedge z\in u) \tag{4.10}$$

to be Σ_2. $(\forall z\in u)(A^p(z) \Rightarrow z\in y)$ is certainly Π_1. As for $(\forall z\in y)(A^p(z)\wedge z\in u)$, suppose $A^p(z) \Leftrightarrow \exists t\psi(t,z,p)$. By the claim

$(\forall z\in y)(\exists t)(\psi(t,z,p)\wedge z\in u) \Leftrightarrow (\exists v)(\forall z\in y)(\exists t\in v)(\psi(t,z,p)\wedge z\in u)$;

and this last is Σ_1. So y=a(u) is Σ_2. □

Now an easy induction gives

Lemma 4.20. *Suppose $\underline{M}$ is p-sound, $p\in PA_n^M$. Suppose $B\subseteq(\omega\rho_M^n)^{<\omega}$. Then B is $\Sigma_h(\underline{M}^{np})$ if and only if there is $A\in\Sigma_{h+n}(\underline{M})$ with parameters from $(\omega\rho_M^n)^{<\omega}\cup\{p\}$ such that $B=A\cap\underline{M}^{np}$. (h>0).*

Corollary 4.21. *Suppose $\underline{M}$ is p-sound with $p\in PA_n^M$. Then every x in $\underline{M}$ is Σ_n definable from parameters in $(\omega\rho_M^n)^{<\omega}\cup p$.*

Now we are in a position to iterate the extension of embeddings lemmas.

Lemma 4.22. *Suppose $\underline{N}$ is a standard model of R and $\pi:\overline{\underline{M}}\to_{\Sigma_1}\underline{N}^{np}$. Then there are $\underline{M},p'$ unique up to isomorphism with $\overline{\underline{M}}=\underline{M}^{np'}$ and $\underline{M}$ p'-sound.*

Proof: Iterate lemma 4.12. □

Lemma 4.23. *Suppose $\underline{M},\underline{N}$ are standard models of R and $\pi:\underline{M}^{np'}\to_{\Sigma_0}\underline{N}^{np}$, $\underline{M}$ p'-sound. Then there is a unique $\tilde{\pi}\supseteq\pi$ such that $\tilde{\pi}(p')=p$ and $\tilde{\pi}|M^{mp'_m}:\underline{M}^{mp'_m}\to_{\Sigma_0}\underline{N}^{mp_m}$ for all $m\leq n$. $(p'=\langle p'_m,p''_{m+1}\dots p''_n\rangle, p=\langle p_m,p^*_{m+1}\dots p^*_n\rangle)$.*

Lemma 4.24. *Suppose $\pi,\underline{M},\underline{N},\tilde{\pi},p,p'$ are as in lemma 4.23 and in addition $\underline{N}$ is p-sound. Then $\tilde{\pi}:\underline{M}\to_{\Sigma_{n+h}}\underline{N}$, provided that $\pi:\underline{M}^{np'}\to_{\Sigma_h}\underline{N}^{np}$.*

Lemma 4.25. *Suppose $\underline{M},\underline{N}$ are standard models of R and $\pi:\underline{M}^{np'}\to_{\Sigma_0}\overline{\underline{N}}$ cofinally. Then there are $\underline{N},p$ unique up to isomorphism such that $\overline{\underline{N}}=\underline{N}^{np}$ and $\underline{N}$ is p-sound (provided that $\underline{M}$ is p'-sound).*

Proof: Use lemma 4.15 to get the first step and lemma 4.16 thereafter. □

Definition 4.26. *Suppose $\pi:\underline{M}^{np}\to_{\Sigma_0}\overline{\underline{N}}$ cofinally and $\underline{M}$ is p-sound. Suppose $\tilde{\pi}:\underline{M}\to_{\Sigma_0}\underline{N}$ is the unique map of lemma 4.23. $\tilde{\pi}$ is called the n-completion of π to $\underline{M}$.*

In fact the n-completion depends only on $\underline{M}$ and π. For suppose $\tilde{\pi}:\underline{M}\to_{\Sigma_0}\underline{N}$ is the n-completion of $\pi:\underline{M}^{np}\to_{\Sigma_0}\underline{N}''$ and suppose $\underline{M}$ is p'-sound and $\pi:\underline{M}^{np'}\to_{\Sigma_0}\underline{N}'$ where $N'\models N''$. Then it suffices to show that $\underline{N}$ is $\tilde{\pi}(p')$- sound and $\underline{N}'=\underline{N}^{n\tilde{\pi}(p')}$. We may restrict ourselves to n=1. Then since $\underline{M}$ is p'-sound we may assume $p=h_M(i,\langle\xi,p'\rangle)$ $(i\in\omega,\xi\in\omega\rho_M)$ so $\tilde{\pi}(p)=h_N(i,\langle\pi(\xi),\tilde{\pi}(p')\rangle),\pi(\xi)\in\omega\rho_N$. But $\underline{N}$ is $\tilde{\pi}(p)$-sound so $\underline{N}$ is $\tilde{\pi}(p')$-sound.

To show that $\underline{N}'=\underline{N}^{\tilde{\pi}(p')}$ it suffices to show that $\pi:\langle M^{p'},A_M^{p'}\rangle\to_{\Sigma_0}$ $\to_{\Sigma_0}\langle N^{\tilde{\pi}(p')},A_N^{\tilde{\pi}(p')}\rangle$, that is, given $x\in M^{p'}$ $\pi(A_M^{p'}\cap x)=A_N^{\tilde{\pi}(p')}\cap\pi(x)$. Let $y=A_M^{p'}\cap x$. Then

$\forall i,z(\langle i,z\rangle\in y \leftrightarrow \underline{M}\models\phi_i(z,p')\wedge\langle i,z\rangle\in x)$

So $\forall i,z(\langle i,z\rangle\in\pi(y) \leftrightarrow \underline{N}'\models\phi_i(z,\tilde{\pi}(p'))\wedge\langle i,z\rangle\in\pi(x))$, as $\tilde{\pi}:\underline{M}\to_{\Sigma_2}\underline{N}'$. Thus $\pi(y)=A_N^{\tilde{\pi}(p')}\cap\pi(x)$.

In fact $\tilde{\pi}:\underline{M}\to\underline{N}$ is the unique $\tilde{\pi}':\underline{M}\to_{\Sigma_n}\underline{N}^*$ such that

(a) every y in N^* is Σ_n definable with parameters from $N^*\cup rng(\tilde{\pi}')$

(b) $\tilde{\pi}'"\omega\rho_M^m\leq\omega\rho_{N^*}^m$, all $m\leq n$.

THE CONSTRUCTIBLE UNIVERSE

$\underline{J}_\alpha$ is the unique transitive model $\underline{M}=\langle M,\in\rangle$ of R^+ with $On_M=\alpha$. We may deduce all the usual fine-structural properties of $\underline{J}_\alpha$ from our previous work; the following is the main result:

Lemma 4.27. *For all* α $\underline{J}_\alpha \models RA^+$.

Proof: By induction on α. Suppose it true for all $\beta<\alpha$. Suppose $a\in S_{\omega\alpha+\omega}\setminus S_{\omega\alpha}$, $a\subseteq\delta\leq\omega\alpha$. We have to show $\underline{S}_{\omega\alpha+\omega}\models\overline{\overline{\omega\alpha}}\leq\delta$. $\underline{S}_{\omega\alpha}\models RA^+$.

Since $a\subseteq\delta\leq\omega\alpha$ lemma 2.41 tells us that $a\in\Sigma_n(\underline{S}_{\omega\alpha})$ for some $n>0$. If $\omega\rho^m_{S_{\omega\alpha}}>\delta$ for all m then for some p $a\in\Sigma_1(\underline{S}^{n-1p}_{\omega\alpha})$ and $\omega\rho_{S^{n-1p}_{\omega\alpha}}>\delta$ so $a\in S^{n-1p}_{\omega\alpha}\subseteq S_{\omega\alpha}$. Thus $\omega\rho^m_{\underline{S}_{\omega\alpha}}\leq\delta$ for some m. So it suffices to show

Claim For all m there is $\langle p_1\ldots p_m\rangle$ such that $\underline{S}_{\omega\alpha}$ is $\langle p_1\ldots p_m\rangle$-sound.

Proof: Let m be the least for which this fails. m>0, of course. Now there is $A\in\Sigma_m$ such that $A \nsubseteq x\cap\omega\rho^m$, all $x\in S_{\omega\alpha}$, where ρ^m denotes $\rho^m_{S_{\omega\alpha}}$. Say $A\in\Sigma_1(\underline{S}^{m-1\langle p_1\ldots p_{m-1}\rangle}_{\omega\alpha})$ in parameter p_m, say, with p_m minimal in the canonical well-order (which really is a well-order in this case). Let $X=h_{\underline{S}^{m-1\vec{p}}_{\omega\alpha}}(\omega\rho^m\cup p_m)$. Then by lemma 4.4 let $\overline{M}$ be transitive, $\pi:\overline{\underline{M}}\cong\underline{S}^{m-1\vec{p}}_{\omega\alpha}|X$. Let $\pi:\underline{M}\to_{\Sigma_m}\underline{J}_\alpha$ with $\overline{\underline{M}}=\underline{M}^{m-1\vec{p}\,'}$ by lemma 4.22. Then $A\in\Sigma_m(\underline{M})$. But $\underline{M}=\underline{J}_\beta$ some $\beta\leq\omega\alpha$ and $a\notin S_{\omega\alpha}$ so $\beta=\omega\alpha$. $\overline{M}$ is $\pi^{-1}(p_m)$-sound; $\pi^{-1}(p_m)\leq p_m$ but by lemma 4.18 $\pi^{-1}(p_m)\in P_{\overline{M}}$ so $\pi^{-1}(p_m)=p_m$. Hence $\pi|\overline{M}=id|\overline{M}$. Hence $\underline{S}^{m-1\vec{p}}_{\omega\alpha}$ is p_m-sound. So $\underline{S}_{\omega\alpha}$ is $\langle p'_1\ldots p'_{m-1},p_m\rangle$-sound. □(Claim) □

Corollary 4.28. $L\models GCH$

Proof: $L\models$the power set axiom. Take α such that $P(\gamma)\in J_\alpha$. Then $\underline{J}_\alpha\models\overline{\overline{P(\gamma)}}=\gamma^+$ by corollary 3.4. So in L $\overline{\overline{P(\gamma)}}\leq\gamma^+$; so $\overline{\overline{P(\gamma)}}=\gamma^+$ by Cantor's theorem. □

Definition 4.29. $\underline{J}^{\vec{A}}_\alpha$ *denotes the unique model* $\langle M,\in,\vec{A}'\rangle$ *of* R^+ *which is transitive and has* $Qn_M=\omega\alpha$, $A'_k=A_k\cap M$ $(0<k\leq N)$.

$\underline{S}^{\vec{A}}_\gamma$ *denotes* $\underline{S}^{\vec{A}}_\gamma\alpha$ $(\omega\alpha>\gamma)$. *In particular if* $N=0$ *we write* $\underline{J}_\alpha$ *and* $\underline{S}_\gamma$.

These definitions are unambiguous provided it is clear that we are working in ZFC and not in some model of R. In a model with predicates, $\underline{M}=\langle M,\in_M,\vec{A}\rangle$ $\underline{J}_\alpha$ denotes a model with the same predicates. If there is any danger of confusion the latter must be indexed $\underline{J}^M_\alpha$.

Definition 4.30. *Models of RA are called acceptable; of* RA^+ *strongly acceptable.*

Exercises

1. Suppose $\pi:\underline{N}\to_{\Sigma_1}\underline{M}$ where $\underline{N},\underline{M}$ are standard models of RA. Do either of
   (a) $p\in P_N \Rightarrow \pi(p)\in P_M$;
   (b) $\pi(p)\in P_M \Rightarrow p\in P_N$

necessarily hold?

2.Show that there is a non-standard model of R.

3. Suppose $\underline{M},\underline{M}'$ are standard models of R. Suppose $\pi:\underline{M}\to_{\Sigma_0}\underline{M}'$ cofinally and $\langle M,A\rangle$ is amenable. Show that there is a unique A' such that $\pi:\langle \underline{M},A\rangle\to_{\Sigma_0}\langle \underline{M}',A'\rangle$.

4. Show that if $\underline{N}=\underline{J}_\alpha$ then $N^{np}= J_{\rho_N^n}$ for all $n\in\omega$.

5. Say $rud_{\vec{A}}(M)=$ the closure of $M\cup\{M\}$ under $F_1\ldots F_{15+N}$.
Show that if $\langle M,\vec{A}\rangle\models R^+$ and M is transitive then if $M'=rud_{\vec{A}}(M)$ and $\underline{M}'=\langle M',\in,A\rangle$, then $\underline{M}'\models R^+$ and $\underline{M}=\underline{S}^{M'}_{On_M}$.

PART TWO: NORMAL MEASURES

In this part we examine Kunen's theory of iterated ultrapowers. Most of the non-fine-structural apparatus is developed here; many important applications are deferred to part three, where fine-structure is reintroduced.

Much of our subsequent work could be done using the iterable premice developed in chapters 5 and 6. Indeed the core model will be seen to be the union of those iterable premice whose projectum lies at or below their measurable cardinal; and the existence of $0^{\#}$ is equivalent to the existence of an iterable premouse. Nevertheless we shall prefer to use "mice", which are indeed iterable premice, but which satisfy a weaker condition on the projectum than that stated above. Reasons will be given at the appropriate point.

Generally speaking the message of this part - especially of chapter 7 - is that if $\underline{N}$ is a premouse with critical point κ then nothing that affects $\mathcal{P}(\kappa)$ is altered by iterating $\underline{N}$. The reason why we might nevertheless want to iterate $\underline{N}$ is supplied by lemma 8.18; iterating restores to premice the L-like simplicity of levels of construction where $\alpha<\beta \Rightarrow J_\alpha = J_\alpha^{J_\beta}$. In part three we shall see that we are still living in a reasonably L-like universe.

5: NORMAL MEASURES AND ULTRAPOWERS

The dominant theme of studies of large cardinals has been the existence of embeddings of the universe into itself. Large cardinal axioms originally formulated in other terms have been translated into statements about embeddings; and these have revealed their connection with other axioms.

Measurability is one of the best examples. The original definition was motivated by classical measure theory. Early results proving measurables inaccesssible were purely set theoretic, with no "model theoretic" methods used. Scott used an embedding equivalent to obtain far stronger results - including the absence of measurables in L.

From our point of view, as in recent applications of forcing in large cardinal theory, the embeddings are central.

<u>Definition 5.1</u> *Suppose* $j:\underline{M}\to_{\Sigma_0}\underline{N}$ *where* $\underline{M}$, $\underline{N}$ *are standard models of* R. *Suppose that* $\in_N\cap(j''\kappa)^2$ *is an initial segment of* $\in_N$ *and* $\hat{\kappa}=\sup^N(j''\kappa)$ *exists. If* $\underline{N}\models\hat{\kappa}<j(\kappa)$, *then* κ *is called the critical point of* j. *(If* $\underline{M},\underline{N}$ *are transitive then* κ *is the smallest* κ *such that* $j(\kappa)>\kappa$*).*

<u>Definition 5.2</u>. κ *is measurable provided* κ *is the critical point of some* $j:V\to_e M$, *where* M *is an inner model.*

Note that definition 5.2 is made in ZFC; otherwise "measurable" is not defined. Also note that definition 5.2 is illegal as it quantifies over proper classes. A legal equivalent is provided later.

Until further notice κ is a fixed measurable cardinal and $j:V\to_e M$ with κ critical.

<u>Lemma 5.3</u>. $\kappa>\omega$

Proof: $j(\omega)=\omega$ and $j(n)=n$, all $n\in\omega$. □

<u>Lemma 5.4</u>. κ *is a strong limit*

Proof: Suppose $\gamma<\kappa\leq 2^\gamma$ and let $f:P(\gamma)\to\kappa$ onto. So $j(f):P_M(\gamma)\to j(\kappa)$ onto. But if $a\subseteq\gamma$ then $j(a)\subseteq j(\gamma)=\gamma$; and for $\beta<\gamma$

$$\beta\in a \Leftrightarrow j(\beta)\in j(a)$$
$$\Leftrightarrow \beta\in j(a)$$

so $j(a)=a$. Suppose $\kappa=j(f)(a)$ with $a\in P_M(\gamma)$. Suppose $\kappa'=f(a)$; $\kappa'<\kappa$ of course. So $j(\kappa')=j(f)(j(a))$, i.e. $\kappa'=j(f)(a)$. Contradiction! □

Corollary 5.5. *κ is a cardinal.*

Lemma 5.6. *κ is regular.*

Proof: Suppose $\gamma<\kappa$ and $f:\gamma\to\kappa$ cofinally. Then $j(f):\gamma\to j(\kappa)$ cofinally. Suppose $j(f)(\bar\beta)>\kappa$, $\bar\beta<\gamma$. Let $\beta'=f(\bar\beta)$. Then

$$\begin{aligned}\beta'=j(\beta')&=j(f)(j(\bar\beta))\\&=j(f)(\bar\beta)\\&>\kappa\\&>\beta'.\end{aligned}$$

Contradiction! □

Corollary 5.7. *κ is strongly inaccessible.*

For a long time it was an open problem whether κ might be the least such. The following lemma shows that it may not.

Lemma 5.8. *κ is not the smallest strong inaccessible.*

Proof: Since κ is strongly inaccessible it is strongly inaccessible in M. If κ were the smallest strong inaccessible then $M\models$"$j(\kappa)$ is the smallest strong inaccessible". But $\kappa<j(\kappa)$. Contradiction! □

There is a direct proof that $V\neq M$ (see corollary 24.4); but we will come upon a form of this result as we proceed with our alternative characterisation.

NORMAL MEASURES

Definition 5.9. *Suppose $\underline{M}\models R$. U is a normal measure on κ in $\underline{M}$ provided*

(i) $U\subseteq P_M(\kappa)$

(ii) $X\in U \wedge X\subseteq Y,\ Y\in P_M(\kappa) \Rightarrow Y\in U$

(iii) $X\in U \wedge Y\in U \Rightarrow X\cap Y\in U$

(i)-(iii) characterise a filter.

(iv) $X\in U \vee \kappa\setminus X\in U$, for all $X\in P_M(\kappa)$

(i)-(iv) characterise an ultrafilter.

(v) $\phi\notin U$

Such an ultrafilter is called proper.

(vi) for all $\gamma<\kappa$ $\{\gamma\}\notin U$

Such an ultrafilter is called non-principal

(vii) Suppose $\langle X_\gamma:\gamma<\delta\rangle$ is (in $\underline{M}$) a sequence with $X_\gamma\in U$ for all $\gamma<\delta<\kappa$; then $\bigcap_{\gamma<\delta} X_\gamma\in U$

Such an ultrafilter is called κ*-additive - the influence of classical measure theory shows through in the terminology - and a* κ*-additive non-principal ultrafilter is called a measure, or a two-valued measure.*

(viii) Suppose $\langle X_\gamma : \gamma<\kappa\rangle$ *is (in* $\underline{M}$*) a sequence with* $X_\gamma \in U$ *for all* $\gamma<\kappa$*; then* $\{\delta : (\forall\gamma<\delta)\delta\in X_\gamma\}\in U$.

$\{\delta : (\forall\gamma<\delta)\delta\in X_\gamma\}$ *is called the diagonal intersection of* $\langle X_\gamma : \delta<\kappa\rangle$.

There are redundancies in this list, of course. Some clauses are inserted merely for the intermediate definitions. Note that there is a Π_1 formula ϕ such that whenever $\langle \underline{M},U\rangle\models R$ we have:

$$U \text{ is a normal measure on } \kappa \text{ in } \underline{M} \Leftrightarrow \langle \underline{M},U\rangle\models\phi(\kappa). \qquad (5.1)$$

We do not insist on $\langle \underline{M},U\rangle\models R$ as part of the definition of normal measure, however.

Note that $\kappa\in U$ (since $\phi\notin U$); that $\gamma<\kappa \Rightarrow \gamma\notin U$ (since $\kappa\smallsetminus\{\delta\}\in U$ for all $\delta<\kappa$, by (vi), and $\kappa\smallsetminus\gamma = \bigcap_{\delta<\gamma} \kappa\smallsetminus\{\delta\}$); and (vii) fails if $\delta=\kappa$ (since $\phi = \bigcap_{\delta<\kappa} \kappa\smallsetminus\{\delta\}$).

Lemma 5.10. *Suppose* κ *is measurable. Then there is a normal measure on* κ *(in V)*

Proof: Let $U=\{X\in P(\kappa) : \kappa\in j(X)\}$.

(i) Immediate.

(ii) $\kappa\in j(X) \wedge X\subseteq Y \Rightarrow \kappa\in j(Y)$.

(iii) $j(X\cap Y)=j(X)\cap j(Y)$

(iv) $j(\kappa)=j(X)\cup j(\kappa\smallsetminus X)$

(v) $\kappa\notin j(\phi)$.

(vi) $\kappa\notin j(\{\gamma\})=\{\gamma\}$ $(\gamma<\kappa)$

(vii) Suppose $\kappa\in j(X_\gamma)$, all $\gamma<\delta$. Then $j(\bigcap_{\gamma<\delta} X_\gamma) = \bigcap_{\gamma<\delta} j(X_\gamma)$ as $\delta<\kappa$ so $\kappa\in j(\bigcap_{\gamma<\delta} X_\gamma)$.

(viii) Suppose $\kappa\in j(X_\gamma)$, all $\gamma<\kappa$. Then $j(X)_\gamma = j(X_\gamma)$ $(\gamma<\kappa)$ and

$j(\{\delta<\kappa : (\forall\gamma<\delta)\delta\in X_\gamma\}) = \{\delta<j(\kappa) : (\forall\gamma<\delta)\delta\in j(X)_\gamma\}$.

But $(\forall\gamma<\kappa)\kappa\in j(X_\gamma)$. □

The converse to lemma 5.10 is one of our essential tools. That is, we shall obtain from U an embedding $j:V\to_e M$. A word of warning, though: this procedure is not an inverse to that of lemma 5.10; if we start with j and let $U=\{X\in P(\kappa) : \kappa\in j(X)\}$, and then obtain an embedding from U we shall not necessarily return to j. Cases where this fails are encountered in part six.

Definition 5.11. *Suppose* $\underline{M}\models R+AC$ *is standard. Suppose* $j:\underline{M}\to_{\Sigma_0}\underline{N}$ *with* κ *critical. Suppose U is a normal measure on* κ *in* $\underline{M}$*. Then* $j:\underline{M}\to_U\underline{N}$ *means*

(i) For all $x\in N$*, there is* $f\in M$*,* $f:\kappa\to M$ *such that* $\underline{N}\models x=j(f)(\kappa)$ *where*

$\underline{N}\models\mathfrak{K}=\sup j''\kappa$.

(ii) For all $X\in P_M(\kappa)$ $X\in U \leftrightarrow \underline{N}\models\mathfrak{K}\in j(X)$.

$\underline{N}$ is called the ultrapower of $\underline{M}$ by U.

Suppose $j:\underline{M}\to_U\underline{N}$.

Lemma 5.12. *$j:\underline{M}\to_{\Sigma_0}\underline{N}$ cofinally.*

Proof: Suppose $x\in N$. Say $x=j(f)(\mathfrak{K})$ with $f\in M$, $f:\kappa\to M$. Suppose $\underline{M}\models y=\text{rng}(f)$. Then

$$\underline{N}\models(\forall\xi<j(\kappa))j(f)(\xi)\in j(y)$$

so in particular $\underline{N}\models x\in j(y)$. □

Corollary 5.13. *$j:\underline{M}\to_{\Sigma_1}\underline{N}$, $\underline{N}\models R+AC$*

Proof: By lemmas 4.7 and 4.8. □

Suppose from now on that $\underline{N}$ is standard. This we may safely do as $\underline{M}$ is standard, and $\kappa>\omega$ (since $\mathfrak{K}$ cannot be ω^N).

Lemma 5.14. *Suppose $j':\underline{M}\to_U\underline{N}'$. Then there is an isomorphism $\sigma:\underline{N}\tilde{=}\underline{N}'$ such that $\sigma j=j'$, $\sigma(\mathfrak{K})=\mathfrak{K}'$, where $\underline{N}'\models\mathfrak{K}'=\sup j''\kappa$.*

Proof: First of all observe that

$$\underline{N}\models j(f)(\mathfrak{K})=j(f')(\mathfrak{K}) \leftrightarrow \underline{N}'\models j'(f)(\mathfrak{K}')=j'(f')(\mathfrak{K}'). \tag{5.2}$$

For $\underline{N}\models j(f)(\mathfrak{K})=j(f')(\mathfrak{K}) \leftrightarrow \underline{N}\models\mathfrak{K}\in\{\xi<j(\kappa):j(f)(\xi)=j(f')(\xi)\}$

$\leftrightarrow \underline{N}\models\mathfrak{K}\in j(\{\xi<\kappa:\underline{M}\models f(\xi)=f'(\xi)\})$

$\leftrightarrow \{\xi<\kappa:\underline{M}\models f(\xi)=f'(\xi)\}\in U$

and similarly for $\underline{N}'$. By the same argument

$$\underline{N}\models j(f)(\mathfrak{K})\in j(f')(\mathfrak{K}) \leftrightarrow \underline{N}'\models j'(f)(\mathfrak{K}')\in j'(f')(\mathfrak{K}') \tag{5.3}$$

$$\underline{N}\models A_k(j(f)(\mathfrak{K})) \leftrightarrow \underline{N}'\models A_k(j'(f)(\mathfrak{K}')). \tag{5.4}$$

Define $\sigma:\underline{N}\to\underline{N}'$ by $\sigma(j(f)(\mathfrak{K}))=j'(f)(\mathfrak{K}')$. By (5.2)-(5.4) this is an isomorphism. Given $x\in\underline{M}$ let $c_x:\kappa\to M$ be defined by $c_x(\xi)=x$, all $\xi<\kappa$. Then $j(x)=j(c_x)(\mathfrak{K})$. So

$$\begin{aligned}\sigma(j(x))&=\sigma(j(c_x)(\mathfrak{K}))\\&=j'(c_x)(\mathfrak{K}')\\&=j'(x).\end{aligned}$$ □

Half of the above argument shows

Lemma 5.15. *If $j':\underline{M}\to\underline{N}'$ with κ critical and $X\in U \leftrightarrow \mathfrak{K}'\in j'(X)$ then there is $\sigma:\underline{N}\to_{\Sigma_0}\underline{N}'$ such that $\sigma j=j'$.*

This gives a "universal" property of $j:\underline{M}\to_U\underline{N}$.

The above results were uniqueness assertions: we also need an existence theorem.

<u>Lemma 5.16</u>. *Suppose U is a normal measure on κ in $\underline{M}\models R+AC$. Then there are $\underline{N}, j$ such that $j:\underline{M}\to_U\underline{N}$.*

Proof: Let $\overline{T}=\{f\in M: f:\kappa\to M\}$. Suppose $f,g\in\overline{T}$. We say

$$f\sim g \leftrightarrow \{\xi: f(\xi)=g(\xi)\}\in U \qquad (5.5)$$
$$f\overline{E}g \leftrightarrow \{\xi: f(\xi)\in g(\xi)\}\in U$$
$$\overline{A}_k f \leftrightarrow \{\xi: A_k f(\xi)\}\in U$$

Then $\sim$ is an equivalence relation on $\overline{T}$ and a congruence with respect to $\overline{E}$ and $\overline{A}_k$. Let $T=\overline{T}/\sim$ and define

$$[f]E[g] \leftrightarrow f\overline{E}G \qquad (5.6)$$
$$A_k[f] \leftrightarrow \overline{A}_k f$$

where $[f]$ is the equivalence class of f.

<u>Claim 1</u> For any Σ_0 formula ϕ

$$\langle T,E,\vec{A}\rangle\models\phi([f_1]...[f_n]) \leftrightarrow \{\xi:\phi(f_1(\xi)\dots f_n(\xi))\}\in U.$$

Proof: By induction on Σ_0 structure.

<u>Case 1</u> ϕ is atomic.

This is immediate.

<u>Case 2</u> ϕ is $-\psi$.

Then $\langle T,E,\vec{A}\rangle\models\phi([f_1]\dots[f_n]) \leftrightarrow \langle T,E,\vec{A}\rangle\not\models\psi([f_1]\dots[f_n])$
$\leftrightarrow \{\xi:\psi(f_1(\xi)\dots f_n(\xi))\}\notin U$
$\leftrightarrow \{\xi:\phi(f_1(\xi)\dots f_n(\xi))\}\in U.$

<u>Case 3</u> ϕ is $\psi\wedge\chi$.

Then $\langle T,E,\vec{A}\rangle\models\phi([f_1]\dots[f_n]) \leftrightarrow \langle T,E,\vec{A}\rangle\models\psi([f_1]\dots[f_n]),\chi([f_1]\dots[f_n])$
$\leftrightarrow \{\xi:\psi(f_1(\xi)\dots f_n(\xi))\},\{\xi:\chi(f_1(\xi)\dots f_n(\xi))\} \in U$
$\leftrightarrow \{\xi:\psi(f_1(\xi)\dots f_n(\xi))\}\cap\{\xi:\chi(f_1(\xi)\dots f_n(\xi))\} \in U$
$\leftrightarrow \{\xi:\psi\wedge\chi(f_1(\xi)\dots f_n(\xi))\}\in U.$

<u>Case 4</u> ϕ is $(\exists v_i\in v_j)\psi$

If $\langle T,E,\vec{A}\rangle\models\psi([g],[f_1]\dots[f_n]) \wedge [g]\in[f_j]$ then

$\{\xi:\psi(g(\xi),f_1(\xi)\dots f_n(\xi)) \wedge g(\xi)\in f_j(\xi)\}\in U$ so

$\{\xi:\exists v_i\in f_j(\xi)\psi(v_i,f_1(\xi)\dots f_n(\xi))\}\in U.$

Conversely suppose $\{\xi:\exists v_i\in f_j(\xi)\psi(v_i,f_1(\xi)\dots f_n(\xi))\}\in U$. Let $\tilde{g}(\xi)=\{x\in f_j(\xi):\psi(x,f_1(\xi)\dots f_n(\xi))\}$. Then $\tilde{g}$ is a set in $\underline{M}$. Since $\underline{M}\models AC$ there is h in $\underline{M}$ such that

$\underline{M}\models h(\xi)\in\tilde{g}(\xi)$ whenever $\tilde{g}(\xi)\neq\phi$.

$h:\kappa\to M$. But $\{\xi:\tilde{g}(\xi)\neq\phi\}\in U$ so $\{\xi:h(\xi)\in\tilde{g}(\xi)\}\in U$. Thus

$\{\xi:h(\xi)\in f_j(\xi) \wedge \psi(h(\xi),f_1(\xi)\dots f_n(\xi))\}\in U$ so

$\langle T,E,\vec{A}\rangle\models[h]\in[f_j] \wedge \psi([h],[f_1]\dots[f_n])$ □(Claim 1)

Note 1: If $\underline{M}\models ZFC$ then $\tilde{g}$ would exist even without the bound on x. Hence claim 1 would hold for all ϕ.

Note 2: Claim 1 only uses the fact that U is an ultrafilter.

Recall the definition of c_x from lemma 5.14. Let $j(x)=[c_x]$. Let $\underline{N}=\langle T,E,\vec{A}\rangle$.

So $j:\underline{M}\to_{\Sigma_0}\underline{N}$. For if ϕ is Σ_0 then

$\underline{N}\models\phi(j(x_1)\dots j(x_n)) \leftrightarrow \underline{N}\models\phi([c_{x_1}]\dots[c_{x_n}])$

$\Leftrightarrow \{\xi : \underline{M} \models \phi(x_1 \ldots x_n)\} \in U$ (by claim 1)

$\Leftrightarrow \underline{M} \models \phi(x_1 \ldots x_n)$.

If $\underline{M} \models ZFC$ then $j : \underline{M} \to_e \underline{N}$.

Claim 2 κ is the critical point of j.

Proof: First we must show that $j''\kappa$ is an initial segment of On_N. That is, if $\underline{N} \models x < j(\beta)$ and $\underline{M} \models \beta < \kappa$ then $x = j(\beta')$, some β' with $\underline{M} \models \beta' < \kappa$.

Suppose $x = [f]$. Then $\underline{N} \models [f] < [c_\beta]$ so $\{\xi : f(\xi) < \beta\} \in U$ by claim 1. Let $X_{\beta'} = \{\xi : f(\xi) = \beta'\}$. So $\bigcup_{\beta' < \beta} X_{\beta'} \in U$; hence for some $\beta' < \beta$ $X_{\beta'} \in U$. That is, $\underline{N} \models [f] = [c_{\beta'}]$, i.e. $x = j(\beta')$.

We shall now show that if ι is $id|\kappa$ then $[\iota] = \mathcal{R} = \sup j''\kappa$. Clearly if $\beta < \kappa$ then $\{\xi : \beta < \xi\} \in U$ so $[c_\beta] < [\iota]$. Suppose $\underline{N} \models [f] < [\iota]$. Then $\{\xi : f(\xi) < \xi\} \in U$. Let $X_\beta = \{\xi : f(\xi) \neq \beta\}$. So $\xi \in \bigcap_{\beta < \xi} X_\beta \Rightarrow f(\xi) \geq \xi$. So $\{\xi : \xi \in \bigcap_{\beta < \xi} X_\beta\} \notin U$, hence for some β $X_\beta \notin U$, thus $\{\xi : f(\xi) = \beta\} \in U$. So $[f] = [c_\beta]$. Hence $[\iota] = \mathcal{R}$.

But $[\iota] < [c_\kappa]$, as $\{\xi : \xi < \kappa\} \in U$. □(Claim 2)

Claim 3 $X \in U \Leftrightarrow \underline{N} \models [\iota] \in [c_X]$.

Proof: $X \in U \Leftrightarrow \{\xi : \xi \in X\} \in U$. □(Claim 3)

Claim 4 $\underline{N} \models [f] = [c_f]([\iota])$.

Proof: $\{\xi : f(\xi) = f(\xi)\} \in U$. □(Claim 4)

Hence $j : \underline{M} \to_U \underline{N}$. □

Note that we can always replace $\underline{N}$ by an isomorphic copy in which $\pi|\kappa = id|\kappa$ and $[id|\kappa] = \kappa$. Unless we specify otherwise this is assumed from now on.

Lemma 5.17. *Suppose U is a normal measure on κ. Then κ is measurable.*

Proof: Note first that $\kappa > \omega$; ω cannot carry a normal measure. Let $j : V \to_U M$. Then $j : V \to_e M$ and κ is critical. Take M standard. By the Mostowski collapsing lemma it is sufficient to show that M is well-founded (clearly for all $y \in M$ $\{x : x \in_M y\}$ is a set).

Suppose M were not well-founded and $\langle x_n : n \in \omega \rangle$ a sequence with $x_{n+1} \in_M x_n$, all $n \in \omega$. Then suppose $x_n = j(f_n)(\kappa)$. $\underline{M} \models j(f_{n+1})(\kappa) \in j(f_n)(\kappa)$ means

$X_n = \{\xi : f_{n+1}(\xi) \in f_n(\xi)\} \in U$.

Let $X = \bigcap_{n \in \omega} X_n$; then $X \in U$, as $\kappa > \omega$. Take $\xi \in X$. Then for all n $f_{n+1}(\xi) \in f_n(\xi)$; contradiction! □

Now we give some computations in ZFC that are needed later. Suppose $j : V \to_U M$.

Lemma 5.18. *$(2^\kappa)^M < j(\kappa) < (2^\kappa)^+$.*

Proof: $(2^\kappa)^M < j(\kappa)$ as $\underline{M} \models j(\kappa)$ is a strong limit. But $j(\kappa) = \{j(f)(\kappa) :$

$: f:\kappa\to\kappa\}$ so $\overline{\overline{j(\kappa)}}\leq\kappa^{\kappa}=2^{\kappa}$. Hence $j(\kappa)<(2^{\kappa})^{+}$. □

<u>Corollary 5.19</u>. *$V\neq M$*

<u>Corollary 5.20</u>. *$V\neq L$*

Proof: The only inner model of L is L. □

<u>Lemma 5.21</u>. *If $cf(\lambda)\neq\kappa$ then $j(\lambda)=\sup_{\alpha<\lambda} j(\alpha)$.*

Proof: Suppose $j(f)(\kappa)<j(\lambda)$. Then $\{\xi:f(\xi)<\lambda\}\in U$. Assume without loss of generality $f:\kappa\to\lambda$ and let $\mu=\sup f''\kappa$. If $cf(\lambda)>\kappa$ then $\mu<\lambda$ and $j(f)(\kappa)<j(\mu)$ so $j(\lambda)=\sup_{\alpha<\lambda} j(\alpha)$.

Suppose $cf(\lambda)<\kappa$. Say $\lambda=\sup_{i<\gamma}\lambda_i$ where $\gamma=cf(\lambda)$. Then $j(\lambda)= =\sup_{i<\gamma} j(\lambda_i)$ as $j(\gamma)=\gamma$. □

<u>Lemma 5.22</u>. *If $\lambda>\kappa$ and λ is a strong limit cardinal with $cf(\lambda)\neq\kappa$ then $j(\lambda)=\lambda$.*

Proof: By lemma 5.21 it suffices to show $j(\gamma)<\lambda$ whenever $\gamma<\lambda$. But $\overline{\overline{j(\gamma)}}\leq\overline{\overline{\{f: f:\kappa\to\gamma\}}}=\overline{\overline{\gamma}}^{\kappa}$ and $\overline{\overline{\gamma}}^{\kappa}\leq 2^{\gamma\cdot\kappa}<\lambda$, as λ is a strong limit cardinal. Hence $j(\gamma)<\lambda$. □

In particular if $\lambda>\kappa$ is strongly inaccessible then $j(\lambda)=\lambda$.

<u>Lemma 5.23</u>. *If $cf(\lambda)=\kappa$ then $j(\lambda)>\lambda$.*

Proof: Say $\lambda=\sup_{\alpha<\kappa}\hat{\lambda}_{\alpha}$ ($\hat{\lambda}_{\alpha}$ strictly increasing). Then $j(\lambda)=\sup j(\hat{\lambda})$.

But $\lambda=\sup_{\alpha<\kappa}\hat{\lambda}_{\alpha}\leq\sup_{\alpha<\kappa} j(\hat{\lambda}_{\alpha})\leq j(\hat{\lambda})_{\kappa}<j(\lambda)$. □

These results make essential use of ZFC; on the other hand lemmas 5.3-5.7 can be proved in R, giving

<u>Lemma 5.24</u>. *Suppose U is a normal measure on κ in $\underline{M}$. Then $\underline{M}\models\kappa$ is strongly inaccessible.*

See exercise 7.

<u>Lemma 5.25</u>. *Suppose $j:\underline{M}\to_U\underline{N}$ with U normal on κ in $\underline{M}$. Then $P_M(\kappa)\subseteq P_N(\kappa)$.*

Proof: This is loosely formulated, and involves an assumption about the relation $\in_N$; the reader should make the formulation precise. From now on we shall be less careful about these statements because they can become very cumbersome.

Suppose $\underline{M}\models a\subseteq\kappa$. Then for $\gamma<\kappa$

$\underline{M}\models\gamma\in a \leftrightarrow \underline{N}\models\gamma\in j(a)$

so $a=j(a)\cap\kappa\in N$. □

If $\underline{M} \models ZFC$ then equality holds. In general

Lemma 5.26. *Suppose $j:\underline{M} \to_U \underline{N}$ with U normal on κ in $\underline{M}$ and $\langle \underline{M},U \rangle \models R$. Then $P_M(\kappa)=P_N(\kappa)$.*

Proof: Say $a=j(f)(\kappa)$. Then

$$\beta \in a \leftrightarrow \beta \in j(f)(\kappa)$$
$$\leftrightarrow \{\xi:\beta \in f(\xi)\} \in U.$$

As $\langle \underline{M},U \rangle \models R$ $\{\beta:\{\xi:\beta \in f(\xi)\} \in U\}$ is a set of $\underline{M}$. □

Definition 5.27. *If $\underline{M}=\langle M,\in_M,U,\vec{A} \rangle \models R^+$ is standard and U is normal on κ in $\underline{M}$ then $\underline{M}$ is called a premouse. κ is called the critical point of $\underline{M}$. If $\underline{M}=\langle M,\in_M,U \rangle$ then $\underline{M}$ is called pure.*

Lemma 5.28. *There is a Π_1 sentence σ such that for all standard $\underline{M} \models R^+$ $\underline{M} \models \sigma \leftrightarrow \underline{M}$ is a premouse.*

Proof: Use the formula of (5.1). □

Corollary 5.29. *If $\pi:\underline{M} \to_{\Sigma_0} \underline{N}$ cofinally, $\underline{M}$ is a premouse and $\underline{N}$ is standard then $\underline{N}$ is a premouse.*

Corollary 5.30. *If $j:\underline{M} \to_U \underline{N}$ and $\underline{M}$ is a premouse then $\underline{N}$ is a premouse.*

Therefore we can take an ultrapower of $\underline{N}$. This is explained in the next chapter.

Exercises

1. Suppose $j:V \to_U M$. Show that $U \notin M$.

2. Show that if κ is uncountable and carries a measure then κ carries a normal measure.

3. Suppose $j:V \to_U M$ with κ critical. Show $\kappa^+=(\kappa^+)^M$.

4. Show that if κ is measurable then κ is the κth strongly inaccessible cardinal.

5. Suppose κ is measurable. What large cardinal properties will κ have in L? (In 1-5 assume ZFC).

6. Suppose U is normal on κ. Suppose $f:\kappa \to \kappa$ and $\{\xi:f(\xi)<\kappa\} \in U$. Show that for some $\beta<\kappa$ $\{\xi:f(\xi)=\beta\} \in U$.

7. The usual definition of strong inaccessibility assumes the power set axiom. Amend it so that lemma 5.24 makes sense.

We have just observed that if $\underline{M}$ is a premouse and $j:\underline{M}\to_U\underline{N}$ then $\underline{N}$ is a premouse, so that if $\underline{N}=\langle N,U'\rangle$ we may repeat the construction. And so on. We should thus obtain premice $\underline{N}_n$ and maps $j_n:\underline{N}_n\to_{U_n}\underline{N}_{n+1}$ where $\underline{N}_n=\langle N_n,U_n\rangle$. The maps may be composed to give $j_{mn}:\underline{N}_m\to_{\Sigma_0}\underline{N}_n$ cofinally whenever $m\leq n$.

In fact we want to iterate this procedure transfinitely. This requires the direct limit construction at limit stages; this is now defined.

Definition 6.1. *A commutative system is a pair* $\langle\langle \underline{A}_\alpha\rangle_{\alpha<\lambda},\langle\pi_{\alpha\beta}\rangle_{\alpha\leq\beta<\lambda}\rangle$ *where* $\underline{A}_\alpha\models R$, $\lambda>0$ *and*

(i) $\pi_{\alpha\beta}:\underline{A}_\alpha\to_{\Sigma_0}\underline{A}_\beta$ $(\alpha\leq\beta<\lambda)$

(ii) $\pi_{\beta\gamma}\cdot\pi_{\alpha\beta}=\pi_{\alpha\gamma}$ $(\alpha\leq\beta\leq\gamma<\lambda)$

(iii) $\pi_{\alpha\alpha}=id|A_\alpha$ $(\alpha<\lambda)$

Definition 6.2. *A direct limit of a commutative system* $\langle\langle \underline{A}_\alpha\rangle_{\alpha<\lambda},\langle\pi_{\alpha\beta}\rangle_{\alpha\leq\beta<\lambda}\rangle$ *is a pair* $\langle \underline{A},\langle\pi_\alpha\rangle_{\alpha<\lambda}\rangle$ *such that*

(i) $\pi_\alpha:\underline{A}_\alpha\to_{\Sigma_0}\underline{A}$ $(\alpha<\lambda)$

(ii) $\pi_\beta\cdot\pi_{\alpha\beta}=\pi_\alpha$ $(\alpha\leq\beta<\lambda)$

(iii) *for all* $a\in A$ *there is* $\alpha<\lambda$ *and* $\bar{a}\in A_\alpha$ *such that* $a=\pi_\alpha(\bar{a})$.

Lemma 6.3. *Suppose* $\langle \underline{A},\langle\pi_\alpha\rangle_{\alpha<\lambda}\rangle$ *and* $\langle \underline{A}',\langle\pi_\alpha'\rangle_{\alpha<\lambda}\rangle$ *are direct limits of the commutative system* $\langle\langle \underline{A}_\alpha\rangle_{\alpha<\lambda},\langle\pi_{\alpha\beta}\rangle_{\alpha\leq\beta<\lambda}\rangle$. *Then there is an isomorphism* $\sigma:\underline{A}\cong\underline{A}'$ *such that* $\sigma\pi_\alpha=\pi_\alpha'$ *for all* α.

Proof: Let $\sigma(\pi_\alpha(\bar{a}))=\pi_\alpha'(\bar{a})$.

Claim σ is well-defined.

Proof: Suppose $\pi_\alpha(a_1)=\pi_\beta(a_2)$, $\alpha\leq\beta$. Then

$$\pi_\beta(\pi_{\alpha\beta}(a_1))=\pi_\alpha(a_1)$$
$$=\pi_\beta(a_2)$$

so $\pi_{\alpha\beta}(a_1)=a_2$. Hence $\pi_\beta'(\pi_{\alpha\beta}(a_1))=\pi_\beta'(a_2)$, so $\pi_\alpha'(a_1)=\pi_\beta'(a_2)$ □(Claim)

The claim in reverse shows that σ one-one; it is clearly surjective. If $\underline{A}=\langle A,E,A_1\ldots A_N\rangle$ and $\underline{A}'=\langle A,E',A_1'\ldots A_N'\rangle$ then

$$\pi_\alpha(a_1)E\pi_\beta(a_2)\leftrightarrow\pi_\alpha'(a_1)E'\pi_\beta'(a_2)$$

just as in the claim, and similarly for A_k,A_k'. So σ is an isomorphism. $\sigma\pi_\alpha=\pi_\alpha'$ by definition. □

<u>Lemma 6.4</u>. *Suppose $\langle\langle \underline{A}_\alpha\rangle_{\alpha<\lambda}, \langle \pi_{\alpha\beta}\rangle_{\alpha\leq\beta<\lambda}\rangle$ is a commutative system. Then it has a direct limit.*

Proof: If $\lambda=\gamma+1$ then let $\underline{A}=\underline{A}_\gamma$, $\pi_\alpha=\pi_{\alpha\gamma}$. So suppose λ is a limit ordinal. A function $f:\lambda\to \bigcup_{\alpha<\lambda} A_\alpha$ is called convergent provided there is $\gamma<\lambda$ such that for all δ with $\gamma\leq\delta<\lambda$ $f(\delta)=\pi_{\gamma\delta}(f(\gamma))$.

Let $A^*=\{f: f:\lambda\to \bigcup_{\alpha<\lambda} A_\alpha \wedge f \text{ is convergent}\}$. For $f,g\in A^*$ say $f\sim g$ if and only if there is $\gamma<\lambda$ such that $\gamma\leq\delta<\lambda \Rightarrow f(\delta)=g(\delta)$. Then $\sim$ is an equivalence relation. Let $A=A^*/\sim$.

Suppose $\underline{A}_\alpha=\langle A_\alpha, E_\alpha, A^1_\alpha \ldots A^N_\alpha\rangle$. Let

$$fE^*g \Leftrightarrow (\exists\gamma<\lambda)(\forall\delta)(\gamma\leq\delta<\lambda \Rightarrow f(\delta)E_\delta g(\delta)) \tag{6.1}$$

$$A^*_k g \Leftrightarrow (\exists\gamma<\lambda)(\forall\delta)(\gamma\leq\delta<\lambda \Rightarrow A^k_\delta g(\delta))$$

Then $f\sim f', g\sim g', fE^*g \Rightarrow f'E^*g'$; similarly for A^*_k. Let $[f]$ denote the $\sim$ equivalence class of f. Then define

$$[f]E[g] \Leftrightarrow fE^*g \tag{6.2}$$

$$A_k[f] \Leftrightarrow A^*_k f.$$

Let $\underline{A}=\langle A,E,A_1\ldots A_N\rangle$.

Define $t^\alpha_x:\lambda\to \bigcup_{\gamma<\lambda} A_\gamma$ by

$$t^\alpha_x(\gamma)=0 \text{ if } \gamma<\alpha$$

$$t^\alpha_x(\gamma)=\pi_{\alpha\gamma}(x) \quad \gamma\geq\alpha$$

where $\gamma<\lambda$, $x\in A_\alpha$. Define $\pi_\alpha:A_\alpha\to A$ by $\pi_\alpha(x)=[t^\alpha_x]$. We shall show that $\langle \underline{A}, \langle\pi_\alpha\rangle_{\alpha<\lambda}\rangle$ is a direct limit.

If $\alpha\leq\beta<\lambda$ and $\delta\geq\beta$ then $t^\alpha_x(\delta)=\pi_{\alpha\delta}(x)$ and $t^\beta_{\pi_{\alpha\beta}(x)}(\delta)=\pi_{\beta\delta}(\pi_{\alpha\beta}(x))= =\pi_{\alpha\delta}(x)$ so $t^\alpha_x(\delta)=t^\beta_{\pi_{\alpha\beta}(x)}(\delta)$ so $t^\alpha_x\sim t^\beta_{\pi_{\alpha\beta}(x)}$, i.e. $\pi_\beta(\pi_{\alpha\beta}(x))=\pi_\alpha(x)$.

Given $f\in A^*$ suppose $\delta\geq\gamma, \delta<\lambda \Rightarrow f(\delta)=\pi_{\gamma\delta}(f(\gamma))$. Let $x=f(\gamma)$. Then $f\sim t^\gamma_x$ so $[f]=\pi_\gamma(x)$.

It remains to show $\pi_\alpha:\underline{A}_\alpha\to_{\Sigma_0}\underline{A}$ for $\alpha<\lambda$. This is done by induction on Σ_0 structure.

<u>Case 1</u> ϕ is $x=y$.

$\underline{A}_\alpha\models x=y \Rightarrow \underline{A}\models\pi_\alpha(x)=\pi_\alpha(y)$. But if $\pi_\alpha(x)=\pi_\alpha(y)$ then $t^\alpha_x\sim t^\alpha_y$ so $\pi_{\alpha\gamma}(x)= =\pi_{\alpha\gamma}(y)$, some $\gamma\geq\alpha$. So $x=y$.

<u>Case 2</u> ϕ is $x\in y$.

$$\underline{A}_\alpha\models x\in y \Rightarrow \underline{A}_\gamma\models\pi_{\alpha\gamma}(x)\in\pi_{\alpha\gamma}(y) \text{ all } \gamma\geq\alpha,\gamma<\lambda$$
$$\Rightarrow t^\alpha_x E^* t^\alpha_y$$
$$\Rightarrow \pi_\alpha(x)E\pi_\alpha(y).$$

If $t^\alpha_x E^* t^\alpha_y$, conversely, then there is $\gamma\geq\alpha$ with $\underline{A}_\gamma\models\pi_{\alpha\gamma}(x)\in\pi_{\alpha\gamma}(y)$ so $x\in y$.

<u>Case 3</u> ϕ is $A_k x$.

Similar

<u>Case 4</u> ϕ is $-\psi$.

$\underline{A}_\alpha\models\phi \Leftrightarrow \underline{A}_\alpha\not\models\psi \Leftrightarrow \underline{A}\not\models\psi \Leftrightarrow \underline{A}\models\phi$.

<u>Case 5</u> ϕ is $\psi\wedge\chi$.

$\underline{A}_\alpha \models \phi \Leftrightarrow \underline{A}_\alpha \models \psi$ and $\underline{A}_\alpha \models \chi \Leftrightarrow \underline{A} \models \psi$ and $\underline{A} \models \chi \Leftrightarrow \underline{A} \models \phi$.

Case 6 ϕ is $\exists y \in z \psi$

If $\underline{A}_\alpha \models \psi(y, x_1 \ldots x_n) \wedge y \in z$ then $\underline{A} \models \psi(\pi_\alpha(y), \pi_\alpha(x_1) \ldots \pi_\alpha(x_n)) \wedge \pi_\alpha(y) \in \in \pi_\alpha(z)$.

Conversely suppose $\underline{A} \models \psi(\pi_\beta(y), \pi_\alpha(x_1) \ldots \pi_\alpha(x_n))$ and $\pi_\beta(y) \in \pi_\alpha(z)$, $\alpha \leq \beta$. Then $\underline{A}_\beta \models \psi(\pi_\beta(y), \pi_{\alpha\beta}(x_1) \ldots \pi_{\alpha\beta}(x_n))$ and $\pi_\beta(y) \in \pi_{\alpha\beta}(z)$ so

$\underline{A}_\beta \models \phi(\pi_{\alpha\beta}(x_1) \ldots \pi_{\alpha\beta}(x_n))$ so

$\underline{A} \models \phi(x_1 \ldots x_n)$ □

The proof of case 6 also shows

Lemma 6.5. *If $\pi_{\alpha\beta}: \underline{A}_\alpha \to_{\Sigma_1} \underline{A}_\beta$ for $\alpha \leq \beta < \lambda$ then $\pi_\alpha: \underline{A}_\alpha \to_{\Sigma_1} \underline{A}$ for all $\alpha < \lambda$.*

Lemma 6.6. *If $\pi_{\alpha\beta}: \underline{A}_\alpha \to_{\Sigma_0} \underline{A}_\beta$ cofinally for $\alpha \leq \beta < \lambda$ then $\pi_\alpha: \underline{A}_\alpha \to_{\Sigma_0} \underline{A}$ cofinally for $\alpha < \lambda$.*

Proof: Suppose $y \in A$. $y = \pi_\beta(\bar{y})$ for some $\beta < \lambda$, $\bar{y} \in A_\beta$, $\beta \geq \alpha$. There is $z \in \underline{A}_\alpha$ such that $\bar{y} \in \pi_{\alpha\beta}(z)$. So $y \in \pi_\alpha(z)$. □

Corollary 6.7. *If $\pi_{\alpha\beta}: \underline{A}_\alpha \to_{\Sigma_0} \underline{A}_\beta$ cofinally for $\alpha \leq \beta < \lambda$ then $\underline{A} \models R$.*

Lemma 6.8. *If $\pi_{\alpha\beta}: \underline{A}_\alpha \to_{\Sigma_n} \underline{A}$ for all $\alpha \leq \beta < \lambda$ then $\pi_\alpha: \underline{A}_\alpha \to_{\Sigma_n} \underline{A}$.*

Proof: By induction on n. n=0 is known. Suppose n=m+1. Then $\pi_\alpha: \underline{A}_\alpha \to_{\Sigma_m} \underline{A}$. Suppose $\underline{A} \models \exists x \psi(x, \pi_\alpha(y_1) \ldots \pi_\alpha(y_n))$ with ψ Π_m. Say $\underline{A} \models \psi(\pi_\beta(\bar{x}), \pi_\alpha(y_1) \ldots \pi_\alpha(y_n))$, $\beta \geq \alpha$. $\pi_\beta: \underline{A}_\beta \to_{\Sigma_m} \underline{A}$ so $\underline{A}_\beta \models \psi(\bar{x}, \pi_{\alpha\beta}(y_1) \ldots \ldots \pi_{\alpha\beta}(y_n))$. Hence $\underline{A}_\beta \models \exists x \psi(x, \pi_{\alpha\beta}(y_1) \ldots \pi_{\alpha\beta}(y_n))$ so $\underline{A} \models \exists x \psi(x, y_1 \ldots y_n)$. □

ITERATED ULTRAPOWERS

Definition 6.9. *Suppose $\langle \underline{N}, U \rangle \models R + AC$ where U is a normal measure on κ in $\underline{N}$. A pair $\langle \langle \underline{N}_\alpha \rangle_{\alpha \in On}, \langle \pi_{\alpha\beta} \rangle_{\alpha \leq \beta \in On} \rangle$ is an iterated ultrapower of $\underline{N}$ by U provided there are U_α such that*

(i) $\langle \langle \underline{N}_\alpha, U_\alpha \rangle_{\alpha \in On}, \langle \pi_{\alpha\beta} \rangle_{\alpha \leq \beta \in On} \rangle$ is a commutative system

(ii) $\underline{N}_0 = \underline{N}$, $U_0 = U$

(iii) $\langle \underline{N}_\alpha, U_\alpha \rangle \models R + AC$ for all $\alpha \in On$

(iv) $\pi_{\alpha\alpha+1}: \langle \underline{N}_\alpha, U_\alpha \rangle \to_{U_\alpha} \langle \underline{N}_{\alpha+1}, U_{\alpha+1} \rangle$ for all $\alpha \in On$

(v) $\langle \langle \underline{N}_\lambda, U_\lambda \rangle, \langle \pi_{\alpha\lambda} \rangle_{\alpha < \lambda} \rangle$ is a direct limit of $\langle \langle \langle \underline{N}_\alpha, U_\alpha \rangle \rangle_{\alpha < \lambda}, \langle \pi_{\alpha\beta} \rangle_{\alpha \leq \beta < \lambda} \rangle$ whenever λ is a limit ordinal.

Lemma 6.10. *Suppose $\langle \underline{N}, U \rangle \models R + AC$ where U is a normal measure on κ in $\underline{N}$. There exists an iterated ultrapower of $\underline{N}$ by U.*

Proof: Induction based on lemma 5.16 at successors and lemma 6.4 at limits; each U_α is a normal measure on $\pi_{0\alpha}(\kappa)$ by lemma 5.28. □

Lemma 6.11. *Suppose* $\langle\langle \underline{N}_\alpha\rangle_{\alpha\in On},\langle \pi_{\alpha\beta}\rangle_{\alpha\leq\beta\in On}\rangle$ *and* $\langle\langle \underline{N}'_\alpha\rangle_{\alpha\in On},\langle \pi'_{\alpha\beta}\rangle_{\alpha\leq\beta\in On}\rangle$ *are iterated ultrapowers of* $\underline{N}$ *by* U. *Then there are isomorphisms* $\sigma_\alpha : N_\alpha \cong N'_\alpha$ *such that*

(i) $\sigma_0 = id|N$

(ii) $\sigma_\beta\pi_{\alpha\beta} = \pi'_{\alpha\beta}\sigma_\alpha$ $(\alpha\leq\beta\in On)$

(iii) $\sigma_\beta(\pi_{0\alpha}(\kappa)) = \pi'_{0\alpha}(\kappa)$ $(\alpha\leq\beta\in On)$

Proof: Induction based on lemma 5.14 at successors and lemma 6.3 at limits. □

Definition 6.12. *Suppose* $\langle\langle \underline{N}_\alpha\rangle_{\alpha\in On},\langle \pi_{\alpha\beta}\rangle_{\alpha\leq\beta\in On}\rangle$ *is an iterated ultrapower of* $\underline{N}$ *by* U. *Each* $\underline{N}_\alpha$ *is called an iterate of* $\underline{N}$ *by* U; $\underline{N}_\alpha$ *is called an* α*th iterate of* $\underline{N}$ *by* U.

Definition 6.13. $\underline{N}$ *is iterable by* U *provided each iterate of* $\underline{N}$ *by* U *is well-founded.*

Definition 6.14. *Suppose* $\underline{N} = \langle N, \in_N, U, \vec{A}\rangle$ *is a premouse. An iterate of* $\underline{N}$ *by* U *is called an iterate of* $\underline{N}$ *provided it is standard; similarly for* α*th iterate, iterable.*

Hence if a premouse is iterable then its αth iterate is unique. By lemma 5.17 the 1st iterate of V is always well-founded. In chapter 8 we shall see that this applies to all iterates of $V \models ZFC$.

Lemma 6.15. *Suppose* $\langle\langle \underline{N}_\alpha\rangle_{\alpha\in On},\langle \pi_{\alpha\beta}\rangle_{\alpha\leq\beta\in On}\rangle$ *is an iterated ultrapower of* $\underline{N}$ *by* U. *Let* $\kappa_\alpha = \pi_{0\alpha}(\kappa)$ *where* U *is a normal measure on* κ *in* $\underline{N}$. *Then*

(i) κ_α *is the critical point of* $\pi_{\alpha\beta}$ $(\alpha<\beta\in On)$

(ii) $\underline{N}_\beta \models \kappa_\alpha < \kappa_\beta$ $(\alpha<\beta\in On)$

(iii) $\underline{N}_\lambda \models \kappa_\lambda = \sup_{\alpha<\lambda} \kappa_\alpha$ $(\lambda\in On$ *a limit*$)$

(iv) $P_{N_\alpha}(\kappa_\alpha) = P_{N_\beta}(\kappa_\alpha)$ $(\alpha\leq\beta\in On)$

Proof: (i) $\pi_{\alpha\beta}(\xi) = \pi_{\alpha+1\beta}\pi_{\alpha\alpha+1}(\xi)$ $(\xi<\kappa_\alpha)$

$= \pi_{\alpha+1\beta}(\xi)$.

But $\kappa_\gamma = \pi_{\alpha\gamma}(\kappa_\alpha) \geq \kappa_\alpha$ for all γ with $\alpha<\gamma<\beta$ so $\pi_{\alpha+1\beta}(\xi) = \xi$. On the other hand $\pi_{\alpha\beta}(\kappa_\alpha) = \pi_{\alpha+1\beta}(\pi_{\alpha\alpha+1}(\kappa_\alpha))$

$> \pi_{\alpha+1\beta}(\kappa_\alpha)$

$\geq \kappa_\alpha$.

(ii) Immediate from (i).

(iii) Suppose $\underline{N}_\lambda \models \beta < \kappa_\lambda$. Then $\beta = \pi_{\alpha\lambda}(\bar\beta)$ for some $\alpha<\lambda, \bar\beta$. If $\bar\beta \geq \kappa_\alpha$ then $\beta \geq \kappa_\lambda$; so $\bar\beta < \kappa_\alpha$. Hence $\pi_{\alpha\lambda}(\bar\beta) = \bar\beta$, so $\beta < \kappa_\alpha$.

(iv) From lemma 5.26. □

In the above, and generally, we write, for example, $\alpha<\beta$ to mean $\underline{N}_\gamma \models \alpha<\beta$ if there is no danger of ambiguity.

We should refer here to a situation that sometimes arises. It may be the case that $M\models ZFC+R^+, P_M(\kappa)\subseteq J^M_\lambda$ and $\langle \underline{J}^M_\lambda, U\rangle\models R$ but $U\notin M$; thus $\langle M,U\rangle\not\models R$. Let $\underline{N}=\underline{J}^M_\lambda$; then we can define an iterated ultrapower of $\underline{N}$ but not, apparently, of M. But actually we can cobble something up:

Definition 6.16. *Suppose* $\underline{N}=\underline{J}^M_\lambda$, $P_M(\kappa)\subseteq J^M_\lambda$, $M\models cf(\lambda)>\kappa \wedge H_\lambda=J_\lambda$; *and suppose that* $\langle\langle \underline{N}_\alpha\rangle_{\alpha\in On},\langle \pi_{\alpha\beta}\rangle_{\alpha\leq\beta\in On}\rangle$ *is an iterated ultrapower of* $\underline{N}$ *by* U.
$\langle\langle M_\alpha\rangle_{\alpha\in On},\langle \pi'_{\alpha\beta}\rangle_{\alpha\leq\beta\in On}\rangle$ *is a weak iterated ultrapower of* M *by* U *provided:*

(i) $\langle\langle M_\alpha\rangle_{\alpha\in On},\langle \pi'_{\alpha\beta}\rangle_{\alpha\leq\beta\in On}\rangle$ *is a commutative system*

(ii) $\underline{N}_\alpha=\underline{J}^{M_\alpha}_{\lambda_\alpha}$ *for some* λ_α; $M_\alpha\models cf(\lambda_\alpha)>\kappa_\alpha \quad H_{\lambda_\alpha}=J_{\lambda_\alpha} \quad (\kappa_\alpha=\pi_{0\alpha}(\kappa))$

(iii) $\pi'_{\alpha\alpha+1}:M_\alpha\to_{U_\alpha} M_{\alpha+1}$ $(\alpha\leq\beta\in On)$ $(U_\alpha$ *as in definition 6.9)*

(iv) $\langle M_\lambda,\langle \pi'_{\alpha}\rangle_{\alpha\leq\lambda}\rangle$ *is a direct limit of* $\langle\langle M_\alpha\rangle_{\alpha<\lambda},\langle \pi_{\alpha\beta}\rangle_{\alpha\leq\beta<\lambda}\rangle$ *when* λ *is a limit*

(v) $M_0=M$.

Lemma 6.17. *Suppose* $\underline{N}=\underline{J}^M_\lambda$, $P_M(\kappa)\subseteq J^M_\lambda$ *and* U *is a normal measure on* κ *in* $\underline{N}$. *Suppose* $M\models cf(\lambda)>\kappa \wedge H_\lambda=J_\lambda$. *Then* M *has a weak iterated ultrapower by* U.

Proof: Suppose $\pi':M\to_U M'$ and $\pi:\underline{N}\to_U\underline{N}'$. We claim $\underline{N}'\cong\underline{J}^{M'}_{\lambda'}$ for some λ'. Let $\lambda'=\pi'(\lambda)$. Suppose $\pi'(f)(\kappa)\in J^{M'}_{\lambda'}$. Then without loss of generality $f:\kappa\to J^M_\lambda$, so $f:\kappa\to J^M_\delta$, some $\delta<\lambda$, so $f\in N$. Let $\sigma(\pi'(f)(\kappa))=\pi(f)(\kappa)$. σ is the required isomorphism.

Clearly $cf(\lambda')>\pi(\kappa)$ in M' and $M'\models J_{\lambda'}=H_{\lambda'}$. Thus we may iterate at successor ordinals.

Now suppose γ is a limit ordinal, and suppose $M_\alpha,\pi_{\alpha\beta}$ are defined for $\alpha\leq\beta<\gamma$. $M_\gamma,\pi_{\alpha\gamma}$ are determined by (iv). We must show that if $\bar\lambda=\pi_{0\gamma}(\lambda)$ then $\underline{N}_\lambda=\underline{J}^M_{\bar\lambda}\gamma$. If $x\in\underline{J}^M_{\bar\lambda}$ and $x=\pi_{\bar\gamma\gamma}(\bar x)$ with $\bar x\in J^{M_{\bar\gamma}}_{\lambda_{\bar\gamma}}$ and $\sigma_{\bar\gamma}:\underline{J}^{M_{\bar\gamma}}_{\lambda_{\bar\gamma}}\cong\underline{N}_{\bar\gamma}$ let $\sigma(x)=\pi_{\bar\gamma\gamma}(\sigma_{\bar\gamma}(x))$. The reader is left to check that this works. □

Lemma 6.18. *Lemma 6.11 applies to weak iterated ultrapowers.*

Exercises

1. Suppose $\langle\langle \underline{N}_\alpha\rangle_{\alpha\in On},\langle \pi_{\alpha\beta}\rangle_{\alpha\leq\beta\in On}\rangle$ is an iterated ultrapower of V by U, a normal measure on κ, and each $\underline{N}_\alpha$ transitive. Find a bound on the size of κ_α.

2. Let $\langle\langle \underline{N}_\alpha\rangle_{\alpha\in On},\langle \pi_{\alpha\beta}\rangle_{\alpha\leq\beta\in On}\rangle$ be the iterated ultrapower of an iterable premouse $\underline{N}$. Suppose $\underline{N}_\omega=\langle N_\omega,\in_{N_\omega},U_\omega\rangle$. Show that

$$X\in U_\omega \leftrightarrow \exists n\forall k>n\,\kappa_k\in X$$

where $X\in P_{N_\omega}(\kappa_\omega)$ and κ_α is the critical point of $\underline{N}_\alpha$.

7: INDISCERNIBLES

Our aim in this chapter is to show that $\{\kappa_\beta:\beta<\alpha\}$ are order Σ_1 indiscernibles for the structure $\langle \underline{N}_\alpha, \pi_{0\alpha}(x)\rangle_{x\in N}$. In other words given a Σ_1 formula ϕ and $x\in N$ and any two increasing sequences $i_1<\ldots<i_n<\alpha, j_1<\ldots<j_n<\alpha$

$$\underline{N}_\alpha\models\phi(\kappa_{i_1}\ldots\kappa_{i_n},\pi_{0\alpha}(x)) \Leftrightarrow \phi(\kappa_{j_1}\ldots\kappa_{j_n},\pi_{0\alpha}(x)). \tag{7.1}$$

An argument of Solovay is then used to extract Σ_n indiscernibles from these $\{\kappa_\beta:\beta<\alpha\}$ for any n.

For the time being fix a premouse $\underline{N}$ with iterated ultrapower $\langle\langle \underline{N}_\alpha\rangle_{\alpha\in On}, \langle\pi_{\alpha\beta}\rangle_{\alpha\leq\beta\in On}\rangle$; let κ_α be the critical point of $\underline{N}_\alpha$.

Lemma 7.1. *For all $x\in N_\beta$ there is $f\in N_\alpha$, $\underline{N}_\alpha\models f:\kappa^n\to V$, and there are $\gamma_1\ldots\gamma_n$, where $\alpha\leq\gamma_1<\ldots<\gamma_n<\beta$ such that*

$$\underline{N}_\beta\models x=\pi_{\alpha\beta}(f)(\kappa_{\gamma_1}\ldots\kappa_{\gamma_n})$$

Proof: Assume $\alpha=0$ without loss of generality. The result is proved by induction on β. Suppose $\beta=\gamma+1$, and the result is known for γ. Say $\pi=\pi_{\gamma\beta}$. Take some $x\in N_\beta$. Then $\underline{N}_\beta\models x=\pi(f)(\kappa_\gamma)$ for some $f\in N_\gamma$, $\underline{N}_\gamma\models f:\kappa_\gamma\to V$. By the induction hypothesis $\underline{N}_\gamma\models f=\pi_{0\gamma}(g)(\kappa_{\gamma_1}\ldots\kappa_{\gamma_m})$, say. Define a function $g'\in N$ with $g':\kappa^{m+1}\to V$ by

$$g'(\alpha_1\ldots\alpha_{m+1})=g(\alpha_1\ldots\alpha_m)(\alpha_{m+1}). \tag{7.2}$$

$$\begin{aligned}\text{Now } \pi_{0\beta}(g')(\kappa_{\gamma_1}\ldots\kappa_{\gamma_m},\kappa_\gamma) &= \pi_{0\beta}(g)(\kappa_{\gamma_1}\ldots\kappa_{\gamma_m})(\kappa_\gamma)\\ &= \pi(\pi_{0\gamma}(g)(\kappa_{\gamma_1}\ldots\kappa_{\gamma_m}))(\kappa_\gamma)\\ &= \pi(f)(\kappa_\gamma)\\ &= x \qquad\qquad (\text{in } \underline{N}_\beta)\end{aligned}$$

Suppose λ is a limit ordinal and the result is known for $\gamma<\lambda$. Given $x\in N_\lambda$, $x=\pi_{\gamma\lambda}(\bar{x})$ for some $\gamma<\lambda$; say $\underline{N}_\gamma\models\bar{x}=\pi_{0\gamma}(f)(\kappa_{\gamma_1}\ldots\kappa_{\gamma_m})$. Then $\underline{N}_\lambda\models x=\pi_{0\lambda}(f)(\kappa_{\gamma_1}\ldots\kappa_{\gamma_m})$. □

Lemma 7.2. *Suppose ϕ is a Σ_1 formula. Then there is a Σ_1 formula ϕ^* such that for all α*

$$\underline{N}_\alpha\models\phi(\kappa_{i_1}\ldots\kappa_{i_n},\pi_{0\alpha}(x)) \Leftrightarrow \underline{N}\models\phi^*(x)$$

whenever $i_1<\ldots<i_n<\alpha$.

Proof: ϕ^* is constructed by induction on n. If n=0 let ϕ^* be ϕ. Suppose n=m+1. Suppose $\phi(y_1\ldots y_n,x)$ is Σ_0. Say

$$\psi(y_1\ldots y_m,x) \leftrightarrow \{\xi:\phi(y_1\ldots y_m,\xi,x)\}\in U \tag{7.3}$$

and let ψ^* be such that

$$\underline{N}_\alpha\models\psi(\kappa_{i_1}\ldots\kappa_{i_m},\pi_{O\alpha}(x)) \leftrightarrow \underline{N}\models\psi^*(x) \tag{7.4}$$

for all α, $i_1<\ldots<i_m<\alpha$; this exists by induction hypothesis. Then

$$\begin{aligned}\underline{N}_\alpha\models\phi(\kappa_{i_1}\ldots\kappa_{i_n},\pi_{O\alpha}(x)) &\leftrightarrow \underline{N}_{i_n+1}\models\phi(\kappa_{i_1}\ldots\kappa_{i_n},\pi_{Oi_n+1}(x)) \\ &\leftrightarrow \underline{N}_{i_n}\models\{\xi:\phi(\kappa_{i_1}\ldots\kappa_{i_m},\xi,\pi_{Oi_n}(x))\}\in U_{i_n} \\ &\leftrightarrow \underline{N}_{i_n}\models\psi(\kappa_{i_1}\ldots\kappa_{i_m},\pi_{Oi_n}(x)) \\ &\leftrightarrow \underline{N}\models\psi^*(x)\end{aligned} \tag{7.5}$$

for all α, $i_1<\ldots<i_n<\alpha$. So we may let ϕ^* be ψ^*.

Suppose, then, that $\phi(y_1\ldots y_n,x) \leftrightarrow \exists y\psi(y,y_1\ldots y_n,x)$ where ψ is Σ_0. Let χ be the formula $\exists y\in x'\psi(y,y_1\ldots y_n,x)$ and let χ^* be such that

$$\underline{N}_\alpha\models\chi(\kappa_{i_1}\ldots\kappa_{i_n},\pi_{O\alpha}(x),\pi_{O\alpha}(x')) \leftrightarrow \underline{N}\models\chi^*(x,x') \tag{7.6}$$

Then

$$\begin{aligned}\underline{N}_\alpha\models\exists y\psi(\kappa_{i_1}\ldots\kappa_{i_n},\pi_{O\alpha}(x)) &\leftrightarrow \exists x'\in N\ \underline{N}_\alpha\models\exists y\in\pi_{O\alpha}(x')\psi(\kappa_{i_1}\ldots \\ &\qquad\ldots\kappa_{i_n},\pi_{O\alpha}(x)) \\ &\leftrightarrow \exists x'\in N\ \underline{N}_\alpha\models\chi(\kappa_{i_1}\ldots\kappa_{i_n},\pi_{O\alpha}(x),\pi_{O\alpha}(x')) \\ &\leftrightarrow \underline{N}\models\exists x'\chi^*(x,x').\end{aligned} \tag{7.7}$$

So we may let ϕ^* be $\exists x'\chi^*(x,x')$. □

Corollary 7.3. *$\{\kappa_i:i<\alpha\}$ are Σ_1 indiscernibles for $\langle\underline{N}_\alpha,\pi_{O\alpha}(x)\rangle_{x\in N}$.*

Proof: Suppose $i_1<\ldots<i_n<\alpha$, $j_1<\ldots<j_n<\alpha$, with ϕ Σ_1 and $x\in N$. Then

$$\begin{aligned}\underline{N}_\alpha\models\phi(\kappa_{i_1}\ldots\kappa_{i_n},\pi_{O\alpha}(x)) &\leftrightarrow \underline{N}\models\phi^*(x) \\ &\leftrightarrow \underline{N}_\alpha\models\phi(\kappa_{j_1}\ldots\kappa_{j_n},\pi_{O\alpha}(x)).\end{aligned} \tag{7.8}$$

□

Corollary 7.4. *$\Sigma_1(\underline{N}_\alpha)\cap P(\kappa)=\Sigma_1(\underline{N})\cap P(\kappa)$*

Proof: Suppose $X\in\Sigma_1(\underline{N})$, $X\subseteq\kappa$. Suppose ϕ is Σ_1 and $\xi\in X \leftrightarrow \underline{N}\models\phi(\xi,p)$. Then let $\overline{X}=\{\xi<\kappa:\underline{N}_\alpha\models\phi(\xi,\pi_{O\alpha}(p))\}$. Clearly $\overline{X}=X$; and $\overline{X}$ is $\Sigma_1(\underline{N}_\alpha)$.

Conversely suppose $X\in\Sigma_1(\underline{N}_\alpha)$. Suppose $\xi\in X \leftrightarrow \underline{N}_\alpha\models\phi(\xi,p)$. By lemma 7.1 there is $f\in N$ such that $\underline{N}_\alpha\models p=\pi_{O\alpha}(f)(\kappa_{i_1}\ldots\kappa_{i_n})$. Say $\phi'(\xi,y_1\ldots y_n,f) \leftrightarrow \phi(\xi,f(y_1\ldots y_n))$. Then ϕ' is Σ_1. And

$$\begin{aligned}\underline{N}_\alpha\models\phi(\xi,p) &\leftrightarrow \underline{N}_\alpha\models\phi'(\xi,\kappa_{i_1}\ldots\kappa_{i_n},\pi_{O\alpha}(f)) \\ &\leftrightarrow \underline{N}\models\phi'^*(\xi,f).\end{aligned} \tag{7.9}$$

Hence $X=\{\xi<\kappa:\underline{N}\models\phi'^*(\xi,f)\}$; so $X\in\Sigma_1(\underline{N})$. □

Occasionally a sharper form of lemma 7.2 is needed.

Definition 7.5. *Suppose U is a normal measure on κ in $\underline{M}$. Then $U^n\subseteq P(\kappa^n)$ is defined by: $U^1=U$*

$$X\in U^{n+1} \leftrightarrow \{\langle\xi_1\ldots\xi_n\rangle:\{\zeta:\langle\xi_1\ldots\xi_n,\zeta\rangle\in X\}\in U\}\in U^n.$$

Lemma 7.6. *$X\in U^n$ is a Σ_1 property of X,n (over $\underline{M}$, where U is a normal measure on κ in $\underline{M}$).*

Proof: $X\in U^n \leftrightarrow \exists f(\mathrm{dom}(f)=n \wedge f(0)=X \wedge (\forall m+1\in n)(f(m+1)=$
$=\{\langle \xi_1\ldots\xi_{n-m}\rangle:\{\zeta:\langle \xi_1\ldots\xi_{n-m},\zeta\rangle\in f(m)\}\in U\}) \wedge f(n-1)\in U)$ □

Lemma 7.7. *Suppose ϕ is a Σ_1 formula. Then there is a Σ_1 formula ϕ^* such that*
$$\underline{N}_\alpha\models\phi(\langle\kappa_{i_1}\ldots\kappa_{i_n}\rangle,\pi_{0\alpha}(x)) \leftrightarrow \underline{M}\models\phi^*(n,x) \quad (i_1<\ldots<i_n<\alpha)$$

Proof: If ϕ is Σ_0 then ϕ^* is
$$(n=0 \wedge \phi(x)) \vee (n>0 \wedge \{\langle\xi_1\ldots\xi_n\rangle:\phi(\langle\xi_1\ldots\xi_n\rangle,x)\}\in U^n). \quad (7.10)$$
The adaptation to ϕ Σ_1 is as before. □

GENERALISED INDISCERNIBLES

Solovay's generalised indiscernibles employ a combinatorial argument to obtain Σ_n indiscernibles. The central notion is that of a full sequence of indiscernibles.

Definition 7.8. *Every ordinal is 0-good.*

α is n+1-good provided α is a limit of n-good points (and hence n-good itself).

Note: α is n-good if and only if α is a multiple of ω^n, of course. (We count 0 as n-good). Solovay's proof was formulated in ordinal arithmetic: but we feel that the present approach gives more idea what is going on, and, more importantly, may be generalised to models with many measurables, where the ordinal arithmetic becomes hopelessly confused.

Definition 7.9. *Every increasing sequence $\alpha_1\ldots\alpha_k$ with $\alpha_1=0$ is 0-full.*

$\langle\alpha_1\ldots\alpha_k\rangle$ is n+1-full provided

(i) $\langle\alpha_1\ldots\alpha_k\rangle$ is n-full

(ii) if $\alpha_h<\beta<\alpha_{h+1}$ and α_{h+1} is not n+1-good then β is not n-good.

Definition 7.10. *$ch_n(\alpha)=\max(m\leq n$: α is m-good).*

Lemma 7.11. *Suppose $\alpha_h<\beta<\alpha_{h+1}$ and $\langle\alpha_1\ldots\alpha_k\rangle$ is n+1-full. Then $ch_n(\beta)<$ $<ch_{n+1}(\alpha_{h+1})$.*

Proof: Suppose $ch_n(\beta)\geq ch_{n+1}(\alpha_{h+1})=m$, say. Then $m\leq n$, so α_{h+1} is not m+1-good. So β is not m-good. But $ch_n(\beta)\geq m$. □

Lemma 7.12. *Suppose $\langle\alpha_1\ldots\alpha_k\rangle$ is an increasing sequence of ordinals. There is an n-full sequence $\langle\beta_1\ldots\beta_h\rangle$ with $\{\alpha_1\ldots\alpha_k\}\subseteq\{\beta_1\ldots\beta_h\}$ and $\alpha_k=\beta_h$.*

Proof: Suppose not. Let k be minimal such that the result fails for some $\langle\alpha_1\ldots\alpha_k\rangle$ and let $\langle\beta_1\ldots\beta_{h'}\rangle$ be n-full where $\{\beta_1\ldots\beta_{h'}\}\supseteq$

$\supseteq\{\alpha_1 \ldots \alpha_{k-1}\}$ and $\beta_{h'} = \alpha_{k-1}$ (provided $k>1$). So there are no $\{\beta_{h'+1} \ldots \beta_h\}$ with $\beta_h = \alpha_k$ and $\langle \beta_1 \ldots \beta_h \rangle$ n-full. Assume that α_k is minimal for the previous sentence. Then α_k is not n-good, otherwise $\langle \beta_1 \ldots \beta_{h'}, \alpha_k \rangle$ is n-full. Let $\alpha = \sup\{\beta < \alpha_k$: for some $k' \leq n$ α_k is not k'-good and β is k'-1-good$\}$. Then there are $\{\beta_{h'+1} \ldots \beta_m\}$ such that $\beta_m = \alpha$ and $\langle \beta_1 \ldots \beta_m \rangle$ are n-full as $\alpha < \alpha_k$. We claim $\langle \beta_1 \ldots \beta_m, \alpha_k \rangle$ is n-full. For suppose α_k is not k'-good, $k' \leq n$ and $\alpha < \beta < \alpha_k$. Then β is not k'-1-good. Contradiction! □

A few examples may help clarify this. A sequence is 1-full provided that whenever i_h is a successor ordinal $i_{h-1} = i_h - 1$. Hence $\langle 0, \omega 2+1, \omega^2+3 \rangle$ is not 1-full, but $\langle 0, \omega 2, \omega 2+1, \omega^2, \omega^2+1, \omega^2+2, \omega^2+3 \rangle$ is. The latter is not 2-full, but would be if ω were added.

The essential fact about n+1-full sequences is that from their ch_{n+1} values we may determine the ch_n values of their n-full extensions.

<u>Definition 7.13</u>. *Suppose* $\langle n_1 \ldots n_k \rangle \in (n+2)^k$, $\langle m_1 \ldots m_p \rangle \in (n+1)^p$. *Then* $\langle m_1 \ldots m_p \rangle \rightarrowtail_{n,X} \langle n_1 \ldots n_k \rangle$ *means*

(i) $X = \{u_1 \ldots u_k\}$, $0 < u_1 < \ldots < u_k = p$, *say*

(ii) $m_{u_h} = \max(n_h, n)$ $\quad (0 < h \leq k)$

(iii) $u_h < j < u_{h+1} \Rightarrow m_j < n_{h+1}$ $\quad (0 < h < k)$

<u>Definition 7.14</u>. $\Delta^{nX}_{\langle n_1 \ldots n_k \rangle} = \{\langle m_1 \ldots m_p \rangle : \langle m_1 \ldots m_p \rangle \rightarrowtail_{n,X} \langle n_1 \ldots n_k \rangle\}$.

<u>Lemma 7.15</u>. *Suppose* $\langle \beta_1 \ldots \beta_p \rangle$ *is n-full and* $\langle \alpha_1 \ldots \alpha_k \rangle$ *is n+1-full. Suppose* $\{\alpha_1 \ldots \alpha_k\} \subseteq \{\beta_1 \ldots \beta_p\}$ *and* $\beta_p = \alpha_k$. *Then, letting* $\alpha_h = \beta_{u_h}$ $(0 < h \leq k)$ *and* $X = \{u_1 \ldots u_k\}$

$$\langle ch_n(\beta_1), \ldots, ch_n(\beta_p) \rangle \rightarrowtail_{n,X} \langle ch_{n+1}(\alpha_1), \ldots, ch_{n+1}(\alpha_k) \rangle$$

Proof: Immediate from lemma 7.11. □

Conversely

<u>Lemma 7.16</u>. *Suppose* $\langle \alpha_1 \ldots \alpha_k \rangle$ *is n+1-full and* $\langle m_1 \ldots m_p \rangle \rightarrowtail_{n,X} \langle ch_{n+1}(\alpha_1) \ldots \ldots ch_{n+1}(\alpha_k) \rangle$. *Suppose* $X = \{u_1 \ldots u_k\}, u_1 < \ldots < u_k$. *Then there is an n-full sequence* $\langle \beta_1 \ldots \beta_p \rangle$ *with* $\beta_{u_h} = \alpha_h$ $(0 < h \leq k)$ *and* $ch_n(\beta_{h'}) = m_{h'}$, $(0 < h' \leq p)$.

Proof: $\beta_{h'}$ is defined by induction on h'. Suppose it is defined for h"<h'.

<u>Case 1</u> $h' = u_h$.

Then let $\beta_{h'} = \alpha_h$. $ch_n(\beta_{h'}) = \max(n, ch_{n+1}(\alpha_h))$
$= m_{h'}$.

<u>Case 2</u> $u_h < h' < u_{h+1}$

Let h"=h'-1. Let $m = m_{h'}$. Then $m < ch_{n+1}(\alpha_{h+1})$. Hence α_{h+1} is m+1-good.

So $\sup\{\beta<\alpha_{h+1}:\beta$ is m-good$\}=\alpha_{h+1}$. Let β be least such that β is m-good and $\beta>\beta_{h''}$. Then $ch_n(\beta)\geq m$. Suppose $ch_n(\beta)>m$; then $\sup\{\beta'<\beta:\beta'$ is m-good$\}=\beta$, contradicting the minimal choice of β. So $ch_n(\beta)=m$. Let $\beta_{h'}=\beta$.

It remains to check that $\langle\beta_1\ldots\beta_p\rangle$ is n-full. So suppose $\beta_{h'}$ is not n-good and $\beta_{h''}<\beta<\beta_{h'}$ $(h'=h''+1)$

Case 1 $h'=u_h$

Then $\alpha_{h-1}<\beta<\alpha_h$ and α_h is not n-good, so β is not n-1-good.

Case 2 $u_h<h'<u_{h+1}$

Since $\beta_{h'}$ is not n-good $ch_n(\beta_{h'})<n$; that is, $m_{h'}<n$. By the minimal choice of $\beta_{h'}$ $ch_n(\beta)<m_{h'}$, so β is not n-1-good. □

That concludes the disguised ordinal arithmetic. We are at last in a position to prove a result about indiscernibles.

Lemma 7.17. *Suppose ϕ is a Σ_{n+1} formula. Then there is a Σ_{n+1} formula ϕ^* such that whenever $\langle\alpha_1\ldots\alpha_k,\alpha\rangle$ is n-full then*

$$\underline{N}_\alpha\models\phi(\langle\kappa_{\alpha_1}\ldots\kappa_{\alpha_k}\rangle,\pi_{0\alpha}(x))\Leftrightarrow\underline{N}\models\phi^*(\langle ch_n(\alpha_1)\ldots ch_n(\alpha_k)\rangle,ch_n(\alpha),x).$$

Note: As in lemmas 7.2 and 7.7 ϕ^* depends only on ϕ and not on $\underline{N}$.

Proof: By induction on n. n=0 is lemma 7.7. So suppose ϕ is $\exists y\psi(y,z,x)$ and ψ is Π_n. Then the formula

$$\psi'(\langle\gamma_1\ldots\gamma_p\rangle,X,x,f)\Leftrightarrow\psi(f(\gamma_1\ldots\gamma_p),\langle\gamma_{u_1}\ldots\gamma_{u_k}\rangle,x)\wedge\wedge X=\{u_1\ldots u_k,p+1\}\wedge u_1<\ldots<u_k\leq p \tag{7.11}$$

is Π_n. By induction hypothesisthere is Π_n ψ'^* such that

$$\underline{N}_\alpha\models\psi'(\langle\kappa_{\beta_1}\ldots\kappa_{\beta_p}\rangle,X,\pi_{0\alpha}(x),\pi_{0\alpha}(f))\Leftrightarrow\Leftrightarrow\underline{N}\models\psi'^*(\langle ch_{n-1}(\beta_1)\ldots ch_{n-1}(\beta_p)\rangle,ch_{n-1}(\alpha),X,x,f) \tag{7.12}$$

provided $\beta_1\ldots\beta_p,\alpha$ is n-1-full. Set:

$$\phi^*(\langle n_1\ldots n_k\rangle,n_{k+1},x)\Leftrightarrow\exists f\exists X\exists\langle m_1\ldots m_{p+1}\rangle(\langle m_1\ldots m_{p+1}\rangle\in\in\Delta^{n-1X}_{\langle n_1\ldots n_{k+1}\rangle}\wedge\psi'^*(\langle m_1\ldots m_p\rangle,m_{p+1},X,x,f)). \tag{7.13}$$

First of all, suppose $\underline{N}_\alpha\models\phi(\langle\kappa_{\alpha_1}\ldots\kappa_{\alpha_k}\rangle,\pi_{0\alpha}(x))$. That is, there is y such that $\underline{N}_\alpha\models\psi(y,\langle\kappa_{\alpha_1}\ldots\kappa_{\alpha_k}\rangle,\pi_{0\alpha}(x))$. By lemma 7.1 there are $f\in N$ and and $\langle\kappa_{\beta'_1}\ldots\kappa_{\beta'_r}\rangle$ such that $\underline{N}_\alpha\models y=\pi_{0\alpha}(f)(\kappa_{\beta'_1}\ldots\kappa_{\beta'_r})$. By lemma 7.12 there are $\{\beta_1\ldots\beta_p,\alpha\}\supseteq\{\beta'_1\ldots\beta'_r\}\cup\{\alpha_1\ldots\alpha_k,\alpha\}$ such that $\langle\beta_1\ldots\beta_p,\alpha\rangle$ is n-1-full. So letting $\pi_{0\alpha}(f)(\kappa_{\beta'_1}\ldots\kappa_{\beta'_r})=\pi_{0\alpha}(f')(\kappa_{\beta_1}\ldots\kappa_{\beta_p})$

$$\underline{N}_\alpha\models\psi(\pi_{0\alpha}(f')(\kappa_{\beta_1}\ldots\kappa_{\beta_p}),\langle\kappa_{\alpha_1}\ldots\kappa_{\alpha_k}\rangle,\pi_{0\alpha}(x)). \tag{7.14}$$

Let $\alpha_h=\beta_{u_h}$ $(0<h\leq k)$ and $X=\{u_1\ldots u_k,p+1\}$. Then

$$\underline{N}_\alpha\models\psi'(\langle\kappa_{\beta_1}\ldots\kappa_{\beta_p}\rangle,X,\pi_{0\alpha}(x),\pi_{0\alpha}(f')) \tag{7.15}$$

so $\underline{N}\models\psi'^*(\langle ch_{n-1}(\beta_1)\ldots ch_{n-1}(\beta_p)\rangle,ch_{n-1}(\alpha),X,x,f')$. Let $m_h=ch_{n-1}(\beta_h)$ $(0<h\leq p)$, $m_{p+1}=ch_{n-1}(\alpha)$.

By lemma 7.15 $\langle m_1 \ldots m_{p+1} \rangle \to_{n-1,X} \langle ch_n(\alpha_1) \ldots ch_n(\alpha_k), ch_n(\alpha) \rangle$. So

$\underline{N} \models \phi^*(\langle ch_n(\alpha_1) \ldots ch_n(\alpha_k) \rangle, ch_n(\alpha), x)$. (7.16)

Conversely suppose $\underline{N} \models \phi^*(\langle ch_n(\alpha_1) \ldots ch_n(\alpha_k) \rangle, ch_n(\alpha), x)$.
Let $n_h = ch_n(\alpha_h)$ $(0<h\leq k)$, $n_{k+1} = ch_n(\alpha)$. Let $f, X, \langle m_1 \ldots m_{p+1} \rangle$ be such that $\langle m_1 \ldots m_{p+1} \rangle \in \Delta^{n-1,X}_{n_1 \ldots n_{k+1}}$ and $\underline{N} \models \psi'^*(\langle m_1 \ldots m_p \rangle, m_{p+1}, X, x, f)$. By lemma 7.16 there is an n-1-full sequence $\langle \beta_1 \ldots \beta_p \rangle$ with $\alpha_h = \beta_{u_h}$ where $X = \{u_1 \ldots u_{k+1}\}, u_1 < \ldots < u_{k+1}$ and $ch_{n-1}(\beta_{h'}) = m_{h'}$, $0<h'\leq p$. So

$\underline{N}_\alpha \models \psi'(\langle \kappa_{\beta_1} \ldots \kappa_{\beta_p} \rangle, X, \pi_{0\alpha}(x), \pi_{0\alpha}(f))$, so (7.17)

$\underline{N}_\alpha \models \psi(\pi_{0\alpha}(f)(\kappa_{\beta_1} \ldots \kappa_{\beta_p}), \langle \kappa_{\alpha_1} \ldots \kappa_{\alpha_k} \rangle, \pi_{0\alpha}(x))$, so

$\underline{N}_\alpha \models \phi(\langle \kappa_{\alpha_1} \ldots \kappa_{\alpha_k} \rangle, \pi_{0\alpha}(x))$ □

Lemma 7.17 is one of the major technical results we need. We can immediately derive some corollaries.

<u>Corollary 7.18</u>. $P(\kappa) \cap \Sigma_{n+1}(\underline{N}_\alpha) \subseteq P(\kappa) \cap \Sigma_{n+1}(\underline{N})$
Proof: Just like half of corollary 7.4. □

<u>Corollary 7.19</u>. *Suppose* ϕ *is a* Σ_{n+1} *formula and* $\langle \kappa_{\alpha_1} \ldots \kappa_{\alpha_p} \rangle, \langle \kappa_{\beta_1} \ldots \kappa_{\beta_p} \rangle$ *are sequences such that:*

(i) $\langle \alpha_1 \ldots \alpha_p, \alpha \rangle, \langle \beta_1 \ldots \beta_p, \alpha \rangle$ *are n-full*

(ii) $ch_n(\alpha_h) = ch_n(\beta_h)$ $(0<h\leq p)$

then $\underline{N}_\alpha \models \phi(\kappa_{\alpha_1} \ldots \kappa_{\alpha_p}, \pi_{0\alpha}(x)) \leftrightarrow \phi(\kappa_{\beta_1} \ldots \kappa_{\beta_p}, \pi_{0\alpha}(x))$.

<u>Corollary 7.20</u>. *Suppose* α *is n-good and* $C = \{\kappa_\beta : \beta<\alpha$ *is n-good*$\}$. *Then C are* Σ_{n+1}*-indisernibles for* $\langle \underline{N}_\alpha, \pi_{0\alpha}(x) \rangle_{x \in N}$.

<u>Exercises</u>

1. Suppose $0<\alpha<\beta$ and α, β are multiples of ω^ω. Show that $\pi_{\alpha\beta}$ is an elementary embedding.

2. Show that $\{\kappa_\beta : \beta<\alpha\}$ are remarkable, that is, that for all $f \in rng(\pi_{0\alpha})$ $\alpha_1 < \ldots < \alpha_m < \beta_1 < \ldots < \beta_n < \gamma_1 < \ldots < \gamma_n$ and $f(\kappa_{\alpha_1} \ldots \kappa_{\alpha_m}, \kappa_{\beta_1} \ldots \kappa_{\beta_n}) < \kappa_{\beta_1}$ imply $f(\kappa_{\alpha_1} \ldots \kappa_{\alpha_m}, \kappa_{\beta_1} \ldots \kappa_{\beta_n}) = f(\kappa_{\alpha_1} \ldots \kappa_{\alpha_m}, \kappa_{\gamma_1} \ldots \kappa_{\gamma_n})$.

8: ITERABILITY

This chapter begins with an embedding theorem which yields relative criteria of iterability; continues to some absolute criteria; and concludes with some important consequences of iterability.

Review the proof of lemma 5.17. We used the fact that M=V together with κ-additivity to obtain well-foundedness of the ultrapower. The reason that we needed M=V was that otherwise $\langle X_i : i\in\omega \rangle$ might not have been a member of M, even if each X_i were. Observe that we only used the ω_1-additivity of U (i.e. $X_i \in U$ for all $i\in\omega \Rightarrow \bigcap_{i\in\omega} X_i \in U$). We may formulate a weaker condition:

<u>Definition 8.1</u>. *U is countably complete provided that whenever $X_i \in U$ for all $i\in\omega$ then $\bigcap_{i\in\omega} X_i \neq \phi$.*

In the case that M=V this is actually equivalent to ω_1-additivity.

<u>Lemma 8.2</u>. *Suppose U is an ultrafilter on κ. Then if U is countably complete then U is ω_1-additive.*

Proof: Suppose $X_i \in U, i\in\omega$, but $\bigcap_{i\in\omega} X_i \notin U$. Let $Y_0 = \kappa \setminus \bigcap_{i\in\omega} X_i$; so $Y_0 \in U$. Let $Y_{i+1} = X_i$, $i\in\omega$; so $\bigcap_{i\in\omega} Y_i = \phi$. □

Looking at the proof of lemma 5.17 it is only the fact that the intersection of X_i is non-empty that is used to prove well-foundedness; so it might be hoped that if $\underline{N} = \underline{J}^U_\alpha$ is a premouse and U is countably complete then, letting $j : \underline{N} \to_U \underline{M}$, $\underline{M}$ is well-founded. But we want to go further and prove that $\underline{N}$ is iterable.

Unfortunately the obvious inductive proof fails. Let $\langle \langle \underline{N}_\alpha \rangle_{\alpha\in On}, \langle \pi_{\alpha\beta} \rangle_{\alpha\leq\beta\in On} \rangle$ be an iterated ultrapower of $\underline{N}$; suppose κ is the critical point of $\underline{N}$. Let $\kappa_\alpha = \pi_{0\alpha}(\kappa)$, for $\alpha\in On$. We know (lemma 6.15(iii)) that $\kappa_\omega = \sup\{\kappa_i : i\in\omega\}$; let $X_i = \kappa_\omega \setminus \kappa_i$, for $i\in\omega$. $\underline{N}_\omega = \langle N_\omega, \in_{N_\omega}, U_\omega \rangle$. Then $X_i \in U_\omega$ for $i\in\omega$ but $\bigcap_{i\in\omega} X_i = \phi$; so U_ω could not possibly be countably complete. This also shows that countable completeness cannot be necessary for iterability, for $\underline{N}_\omega$ is iterable if $\underline{N}$ is.

The proof of iterability from countable completeness uses the following very versatile embedding lemma.

<u>Lemma 8.3</u>. *Suppose $\sigma: \underline{N} \to_{\Sigma_0} \underline{M}$ where $\underline{N}$ and $\underline{M}$ are premice. Suppose $f: On \to On$ is order-preserving. Suppose $\underline{N} = \langle N, \in_N, U, A_1 \ldots A_K \rangle$ and $\underline{M} = \langle M, \in_M, U', A'_1 \ldots A'_K \rangle$. Let $\langle \langle \underline{N}_\alpha \rangle_{\alpha \in On}, \langle \pi_{\alpha\beta} \rangle_{\alpha \leq \beta \in On} \rangle, \langle \langle \underline{M}_\alpha \rangle_{\alpha \in On}, \langle \pi'_{\alpha\beta} \rangle_{\alpha \leq \beta \in On} \rangle$ be iterated ultrapowers of $\underline{N}, \underline{M}$ respectively. Let $\kappa_\alpha, \kappa'_\alpha$ be the critical points of $\underline{N}_\alpha, \underline{M}_\alpha$ respectively. Then there are unique maps σ_α such that*

(i) $\sigma_\alpha : \underline{N}_\alpha \to_{\Sigma_0} \underline{M}_{f(\alpha)}$

(ii) $\sigma_\alpha \pi_{0\alpha} = \pi'_{0f(\alpha)} \sigma$

(iii) $\sigma_\alpha(\kappa_\beta) = \kappa'_{f(\beta)}$ $\quad (\beta \leq \alpha)$

If $\sigma: \underline{N} \to_{\Sigma_1} \underline{M}$ then $\sigma_\alpha : \underline{N}_\alpha \to_{\Sigma_1} \underline{M}_{f(\alpha)}$; and if $\sigma: \underline{N} \to_{\Sigma_0} \underline{M}$ cofinally then $\sigma_\alpha : \underline{N}_\alpha \to_{\Sigma_0} \underline{M}_{f(\alpha)}$ cofinally.

Proof: By induction on α. Let $\sigma_0 = \pi'_{0f(0)}$.

Suppose σ_α is defined. Define $\sigma_{\alpha+1} : \underline{N}_{\alpha+1} \to_{\Sigma_0} \underline{M}_{f(\alpha+1)}$ by

$$\sigma_{\alpha+1}(\pi_{\alpha\alpha+1}(f')(\kappa_\alpha)) = \pi'_{f(\alpha)f(\alpha+1)}(\sigma_\alpha(f'))(\kappa'_{f(\alpha)}) \tag{8.1}$$

<u>Claim 1</u> $\sigma_{\alpha+1}$ is well-defined, $\sigma_{\alpha+1} : \underline{N}_{\alpha+1} \to_{\Sigma_0} \underline{M}_{f(\alpha+1)}$

Proof: Suppose ϕ is Σ_0. Then

$$\underline{M}_{f(\alpha+1)} \models \phi(\pi'_{f(\alpha)f(\alpha+1)}(\sigma_\alpha(f_1))(\kappa'_{f(\alpha)}) \ldots \tag{8.2}$$
$$\ldots \pi'_{f(\alpha)f(\alpha+1)}(\sigma_\alpha(f_n))(\kappa'_{f(\alpha)})) \Leftrightarrow$$
$$\Leftrightarrow \quad \underline{M}_{f(\alpha)+1} \models \phi(\pi'_{f(\alpha)f(\alpha)+1}(\sigma_\alpha(f_1))(\kappa'_{f(\alpha)}) \ldots$$
$$\ldots \pi'_{f(\alpha)f(\alpha)+1}(\sigma_\alpha(f_n))(\kappa'_{f(\alpha)})) \Leftrightarrow$$
$$\Leftrightarrow \quad \{\xi : \underline{M}_{f(\alpha)} \models \phi(\sigma_\alpha(f_1)(\xi) \ldots \sigma_\alpha(f_n)(\xi))\} \in U'_{f(\alpha)} \Leftrightarrow$$
$$\Leftrightarrow \quad \{\xi : \underline{N}_\alpha \models \phi(f_1(\xi) \ldots f_n(\xi))\} \in U_\alpha \Leftrightarrow$$
$$\Leftrightarrow \quad \underline{N}_{\alpha+1} \models \phi(\pi_{\alpha\alpha+1}(f_1)(\kappa_\alpha) \ldots \pi_{\alpha\alpha+1}(f_n)(\kappa_\alpha))$$

where $\underline{N}_\alpha = \langle N_\alpha, \in_{N_\alpha}, U_\alpha, \vec{A}_\alpha \rangle$, $\underline{M}_\alpha = \langle M_\alpha, \in_{M_\alpha}, U'_\alpha, \vec{A}'_\alpha \rangle$. □(Claim 1)

<u>Claim 2</u> $\sigma_{\alpha+1}\pi_{0\alpha+1} = \pi'_{0f(\alpha+1)}\sigma$

Proof:
$$\begin{aligned} \sigma_{\alpha+1}\pi_{0\alpha+1} &= \sigma_{\alpha+1}\pi_{\alpha\alpha+1}\pi_{0\alpha} \\ &= \pi'_{f(\alpha)f(\alpha+1)}\sigma_\alpha\pi_{0\alpha} \\ &= \pi'_{f(\alpha)f(\alpha+1)}\pi'_{0f(\alpha)}\sigma \\ &= \pi'_{0f(\alpha+1)}\sigma \end{aligned}$$
□(Claim 2)

<u>Claim 3</u> $\sigma_{\alpha+1}(\kappa_\beta) = \kappa'_{f(\beta)}$ $\quad (\beta \leq \alpha+1)$

Proof: Suppose $\beta < \alpha$. Then $\pi_{\alpha\alpha+1}(\kappa_\beta) = \kappa_\beta$ and $\pi'_{f(\alpha)f(\alpha+1)}(\kappa'_{f(\beta)}) = \kappa'_{f(\beta)}$.

Hence
$$\begin{aligned} \sigma_{\alpha+1}(\kappa_\beta) &= \sigma_{\alpha+1}(\pi_{\alpha\alpha+1}(\kappa_\beta)) \\ &= \pi'_{f(\alpha)f(\alpha+1)}\sigma_\alpha(\kappa_\beta) \\ &= \pi'_{f(\alpha)f(\alpha+1)}(\kappa'_{f(\beta)}) \\ &= \kappa'_{f(\beta)}. \end{aligned}$$

Suppose $\beta = \alpha$. $\sigma_{\alpha+1}(\kappa_\alpha) = \kappa'_{f(\alpha)}$ by definition.

Suppose $\beta = \alpha+1$. Then
$$\begin{aligned} \sigma_{\alpha+1}(\kappa_{\alpha+1}) &= \sigma_{\alpha+1}(\pi_{0\alpha+1}(\kappa)) \\ &= \pi'_{0f(\alpha+1)}\sigma(\kappa) \\ &= \pi'_{0f(\alpha+1)}(\kappa') \\ &= \kappa'_{f(\alpha+1)}. \end{aligned}$$
□(Claim 3)

<u>Claim 4</u> $\sigma_{\alpha+1}$ is the unique map satisfying claims 1-3.

Proof: Suppose $\hat{\sigma} : \underline{N}_{\alpha+1} \to_{\Sigma_0} \underline{M}_{f(\alpha+1)}$, $\hat{\sigma}\pi_{0\alpha+1} = \pi'_{0f(\alpha+1)}\sigma$ and $\hat{\sigma}(\kappa_\beta) = \kappa'_{f(\beta)}$ $(\beta \leq \alpha)$. Given $x \in N_{\alpha+1}$ there are $f', \kappa_{\alpha_1} \ldots \kappa_{\alpha_n}$ such that $\underline{N}_{\alpha+1} \models x =$

$=\pi_{0\alpha+1}(f')(\kappa_{\alpha_1}\ldots\kappa_{\alpha_n})$ $(\alpha_1<\ldots<\alpha_n\leq\alpha)$. Without loss of generality $\alpha_n=\alpha$. Then

$$\begin{aligned}\hat{\sigma}(x)&=\hat{\sigma}\pi_{0\alpha+1}(f')(\hat{\sigma}(\kappa_{\alpha_1})\ldots\hat{\sigma}(\kappa_{\alpha_n}))\\ &=\pi'_{0f(\alpha+1)}\sigma(f')(\kappa'_{f(\alpha_1)}\ldots\kappa'_{f(\alpha_n)})\\ &=\pi'_{f(\alpha)f(\alpha+1)}\pi'_{0f(\alpha)}\sigma(f')(\kappa'_{f(\alpha_1)}\ldots\kappa'_{f(\alpha_n)})\\ &=\pi'_{f(\alpha)f(\alpha+1)}\sigma_\alpha\pi_{0\alpha}(f')(\kappa'_{f(\alpha_1)}\ldots\kappa'_{f(\alpha_n)})\\ &=\pi'_{f(\alpha)f(\alpha+1)}(\sigma_\alpha(\pi_{0\alpha}(\hat{f})(\kappa_{\alpha_1}\ldots\kappa_{\alpha_{n-1}})))(\kappa'_{f(\alpha_n)})\\ &=\sigma_{\alpha+1}(\pi_{\alpha\alpha+1}(\pi_{0\alpha}(\hat{f})(\kappa_{\alpha_1}\ldots\kappa_{\alpha_{n-1}}))(\kappa_{\alpha_n}))\\ &=\sigma_{\alpha+1}(\pi_{0\alpha+1}(\hat{f})(\kappa_{\alpha_1}\ldots\kappa_{\alpha_{n-1}})(\kappa_{\alpha_n}))\\ &=\sigma_{\alpha+1}(\pi_{0\alpha+1}(f')(\kappa_{\alpha_1}\ldots\kappa_{\alpha_n}))\\ &=\sigma_{\alpha+1}(x),\end{aligned}$$

where $\hat{f}(\gamma_1\ldots\gamma_{n-1})(\gamma_n)=f'(\gamma_1\ldots\gamma_n)$. □(Claim 4)

<u>Claim 5</u> If $\sigma_\alpha:\underline{N}_\alpha\to_{\Sigma_1}\underline{M}_{f(\alpha)}$ then $\sigma_\alpha:\underline{N}_{\alpha+1}\to_{\Sigma_1}\underline{M}_{f(\alpha+1)}$.

Proof: Suppose ϕ is Σ_1, $\phi\leftrightarrow\exists y\psi$ where ψ is Σ_0. Then

$$\begin{aligned}&\underline{M}_{f(\alpha+1)}\models\phi(\sigma_{\alpha+1}(x_1)\ldots\sigma_{\alpha+1}(x_n))\leftrightarrow\underline{M}_{f(\alpha+1)}\models\exists y\psi(y,\sigma_{\alpha+1}(x_1)\ldots\\ &\qquad\ldots\sigma_{\alpha+1}(x_n))\leftrightarrow\\ &\rightarrow\ \underline{M}_{f(\alpha+1)}\models\exists y\in\pi'_{f(\alpha)f(\alpha+1)}(z)\psi(y,\sigma_{\alpha+1}(x_1)\ldots\sigma_{\alpha+1}(x_n)),\text{say, }z\in M_{f(\alpha)}\\ &\rightarrow\ \underline{M}_{f(\alpha+1)}\models\exists y\in\pi'_{f(\alpha)f(\alpha+1)}(z)\psi(y,\pi'_{f(\alpha)f(\alpha+1)}(\sigma_\alpha(f_1))(\kappa'_{f(\alpha)})\ldots\\ &\qquad\ldots\pi'_{f(\alpha)f(\alpha+1)}(\sigma_\alpha(f_n))(\kappa'_{f(\alpha)}))\\ &\rightarrow\ \underline{M}_{f(\alpha)}\models\exists z\{\xi:(\exists y\in z)\psi(y,\sigma_\alpha(f_1)(\xi)\ldots\sigma_\alpha(f_n)(\xi))\}\in U'_{f(\alpha)}\\ &\rightarrow\ \underline{N}_\alpha\models\exists z\{\xi:(\exists y\in z)\psi(y,f_1(\xi)\ldots f_n(\xi))\}\in U_\alpha\\ &\rightarrow\ \underline{N}_{\alpha+1}\models\phi(x_1\ldots x_n)\end{aligned}$$

where $x_k=\pi_{\alpha\alpha+1}(f_k)(\kappa_\alpha)$ $(0<k\leq n)$ □(Claim 5)

<u>Claim 6</u> If $\sigma_\alpha:\underline{N}_\alpha\to_{\Sigma_0}\underline{M}_{f(\alpha)}$ cofinally then $\sigma_{\alpha+1}:\underline{N}_{\alpha+1}\to_{\Sigma_0}\underline{M}_{f(\alpha+1)}$ cofinally.

Proof: $\pi'_{f(\alpha)f(\alpha+1)}\sigma_\alpha=\sigma_{\alpha+1}\pi_{\alpha\alpha+1}$, so $\text{rng}(\sigma_{\alpha+1})\supseteq\text{rng}(\pi'_{f(\alpha)f(\alpha+1)}\sigma_\alpha)$.
□(Claim 6)

This concludes the successor case.

In the limit case, where σ_α is defined for $\alpha<\lambda$, we may define $\sigma_\lambda(\pi_{\alpha\lambda}(x))=\pi'_{f(\alpha)f(\lambda)}\sigma_\alpha(x)$. For if $\pi_{\alpha\lambda}(x)=\pi_{\beta\lambda}(y)$, $\alpha\leq\beta$, then $\pi_{\alpha\beta}(x)=y$ so $\sigma_\beta\pi_{\alpha\beta}(x)=\sigma_\beta(y)$. But $\sigma_\beta\pi_{\alpha\beta}(x)=\pi'_{f(\alpha)f(\beta)}\sigma_\alpha(x)$ so $\sigma_\beta(y)=$ $=\pi'_{f(\alpha)f(\beta)}\sigma_\alpha(x)$; hence $\pi'_{f(\beta)f(\lambda)}\sigma_\beta(y)=\pi'_{f(\beta)f(\lambda)}\pi'_{f(\alpha)f(\beta)}\sigma_\alpha(x)=$ $=\pi'_{f(\alpha)f(\lambda)}\sigma_\alpha(x)$.

(i) $\sigma_\lambda:\underline{N}_\lambda\to_{\Sigma_0}\underline{M}_{f(\lambda)}$ since $\sigma_\alpha:\underline{N}_\alpha\to_{\Sigma_0}\underline{M}_{f(\alpha)}$.

(ii) $\sigma_\lambda\pi_{0\lambda}=\pi'_{f(0)f(\lambda)}\sigma_0=\pi'_{f(0)f(\lambda)}\pi'_{0f(0)}\sigma$
$\qquad=\pi'_{0f(\lambda)}\sigma$.

(iii) $\sigma_\lambda(\kappa_\alpha)=\sigma_\lambda\pi_{\alpha+1\lambda}(\kappa_\alpha)$ $(\alpha<\lambda)$
$\qquad=\pi'_{f(\alpha+1)f(\lambda)}\sigma_{\alpha+1}(\kappa_\alpha)$
$\qquad=\pi'_{f(\alpha+1)f(\lambda)}(\kappa'_\alpha)$
$\qquad=\kappa'_\alpha$.

(iv) $\sigma_\lambda(\kappa_\lambda)=\sigma_\lambda(\pi_{O\lambda}(\kappa))$
$=\pi'_{Of(\lambda)}(\kappa')$
$=\kappa'_{f(\lambda)}$.

(v) σ_λ is the unique map with properties (i)-(iii). For suppose $\sigma':\underline{N}_\lambda\to\underline{M}_{f(\lambda)}$ satisfies (i)-(iii); given $x\in N_\lambda$ suppose $\hat{f}\in N$ and $\alpha_1<\ldots<\alpha_n$ such that $x=\pi_{O\lambda}(\hat{f})(\kappa_{\alpha_1}\ldots\kappa_{\alpha_n})$. Then

$$\begin{aligned}\sigma'(x)&=\sigma'(\pi_{O\lambda}(\hat{f})(\kappa_{\alpha_1}\ldots\kappa_{\alpha_n}))\\ &=\pi'_{Of(\lambda)}\sigma(\hat{f})(\sigma(\kappa_{\alpha_1})\ldots\sigma(\kappa_{\alpha_n}))\\ &=\pi'_{Of(\lambda)}\sigma(\hat{f})(\kappa'_{f(\alpha_1)}\ldots\kappa'_{f(\alpha_n)})\\ &=\pi'_{f(\alpha_n+1)f(\lambda)}(\pi'_{Of(\alpha_n+1)}\sigma(\hat{f})(\kappa'_{f(\alpha_1)}\ldots\kappa'_{f(\alpha_n)}))\\ &=\pi'_{f(\alpha_n+1)f(\lambda)}(\sigma_{\alpha_n+1}(\pi_{O\alpha_n+1}(\hat{f})(\kappa_{\alpha_1}\ldots\kappa_{\alpha_n})))\\ &=\sigma_\lambda\pi_{\alpha_n+1\lambda}(\pi_{O\alpha_n+1}(\hat{f})(\kappa_{\alpha_1}\ldots\kappa_{\alpha_n}))\\ &=\sigma_\lambda(\pi_{O\lambda}(\hat{f})(\kappa_{\alpha_1}\ldots\kappa_{\alpha_n}))\\ &=\sigma_\lambda(x).\end{aligned}$$

(vi) If each $\sigma_\alpha:\underline{N}_\alpha\to_{\Sigma_1}\underline{M}_{f(\alpha)}$ then $\sigma_\lambda:\underline{N}_\lambda\to_{\Sigma_1}\underline{M}_{f(\lambda)}$.

(vii) If $\sigma_\alpha:\underline{N}_\alpha\to_{\Sigma_0}\underline{M}_{f(\alpha)}$ cofinally for $\alpha<\lambda$ then $\sigma_\lambda:\underline{N}_\lambda\to_{\Sigma_0}\underline{M}_{f(\alpha)}$ cofinally. For $\mathrm{rng}(\sigma_\lambda)\supseteq\mathrm{rng}(\pi'_{Of(\lambda)}\sigma)$. □

The lemma was expressed for premice but it is easy to see that it works too for weak iterated ultrapowers. Lemma 8.3 is very useful: see exercise 1, for example. It immediately proves:

Lemma 8.4. *Suppose $\underline{M}$ is an iterable premouse and $\underline{N}$ is a premouse, $\sigma:\underline{N}\to_{\Sigma_0}\underline{M}$. Then $\underline{N}$ is iterable.*

Proof: Suppose $\langle\langle\underline{N}_\alpha\rangle_{\alpha\in On},\langle\pi_{\alpha\beta}\rangle_{\alpha\leq\beta\in On}\rangle$, $\langle\langle\underline{M}_\alpha\rangle_{\alpha\in On},\langle\pi'_{\alpha\beta}\rangle_{\alpha\leq\beta\in On}\rangle$ are iterated ultrapowers of $\underline{N},\underline{M}$ respectively. Suppose $\underline{N}_\alpha$ is not well-founded; let $f:On\to On$ be the identity and let $\sigma_\alpha:\underline{N}_\alpha\to_{\Sigma_0}\underline{M}_\alpha$ be as given by lemma 8.3. Suppose $\langle x_i:i\in\omega\rangle$ is such that $x_{i+1}\in_{N_\alpha}x_i$, all $i\in\omega$; then $\sigma_\alpha(x_{i+1})\in_{M_\alpha}\sigma_\alpha(x_i)$, all $i\in\omega$. So $\underline{M}_\alpha$ is not well-founded. Contradiction! □

Corollary 8.5. *If $\underline{J}^U_\alpha$ is an iterable premouse with critical point κ and if $\kappa<\beta<\alpha$ then $\underline{J}^U_\beta$ is an iterable premouse.*

Like lemma 8.3, lemma 8.4 is true for weak iterated ultrapowers.

Lemma 8.6. *If $\underline{N}$ is a premouse and $\langle\langle\underline{N}_\alpha\rangle_{\alpha\in On},\langle\pi_{\alpha\beta}\rangle_{\alpha\leq\beta\in On}\rangle$ is an iterated ultrapower of $\underline{N}$, and $\underline{N}_\alpha$ is well-founded for $\alpha<\omega_1$ then $\underline{N}$ is iterable.*

Proof: Suppose $\underline{N}_\alpha$ is not well-founded. Suppose $\langle x_i:i\in\omega\rangle$ is a sequence such that $x_{i+1} \in_{N_\alpha} x_i$, all $i \in \omega$. Suppose

$$x_i = \pi_{0\alpha}(f_i)(\kappa_{\alpha_1^i} \ldots \kappa_{\alpha_{p_i}^i}) \tag{8.3}$$

where κ_α is the critical point of $\underline{N}_\alpha$, $\alpha_1^i < \ldots < \alpha_{p_i}^i < \alpha$. Let $S = \{\alpha_j^i : i \in \omega \wedge 0 < j \leq p_i\}$. Let β be the order-type of $\langle S, \in \rangle$; so β is countable. Let $f' : \beta \to S$ be order preserving and surjective. Define $f : On \to On$ by

$$f(\gamma) = f'(\gamma) \qquad (\gamma < \beta) \tag{8.4}$$

$$f(\beta + \delta) = \alpha + \delta \qquad (\text{all } \delta)$$

Then f is monotone. Apply lemma 8.3 with $\underline{M} = \underline{N}$ and $\sigma = \mathrm{id}|N$. Then $\sigma_\beta : \underline{N}_\beta \to_{\Sigma_0} \underline{N}_\alpha$. Let $\bar{\alpha}_j^i = f^{-1}(\alpha_j^i)$ $(i < \omega, 0 < j \leq p_i)$. So $\sigma_\beta(\kappa_{\bar{\alpha}_j^i}) = \kappa_{\alpha_j^i}$. Now let

$$\bar{x}_i = \pi_{0\beta}(f_i)(\kappa_{\bar{\alpha}_1^i} \ldots \kappa_{\bar{\alpha}_{p_i}^i}). \tag{8.5}$$

Then $\sigma_\beta(\bar{x}_i) = \pi_{0\alpha}(f_i)(\kappa_{\alpha_1^i} \ldots \kappa_{\alpha_{p_i}^i}) = x_i$. So for $i \in \omega$ $\bar{x}_{i+1} \in_{N_\beta} \bar{x}_i$; hence $\underline{N}_\beta$ is not well-founded. But $\beta < \omega_1$. □

We may modify this proof slightly to get

Lemma 8.7. *$\underline{N}$ is iterable if and only if whenever $\sigma : \underline{M} \to_{\Sigma_1} \underline{N}$ and M is countable then $\underline{M}$ is iterable.*

Proof: Suppose not and let f_i, x_i be as in the previous lemma. Let $X \prec_{\Sigma_1} \underline{N}$ with $\{f_i : i \in \omega\} \subseteq X$. Let $\sigma : \underline{M} \cong \underline{N}|X$ where X is countable and $\underline{M}$ is standard. Let $\bar{f}_i = \sigma^{-1}(f_i)$. Let $\langle \langle \underline{M}_\alpha \rangle_{\alpha \in On}, \langle \pi'_{\alpha\beta} \rangle_{\alpha \leq \beta \in On} \rangle$ be an iterated ultrapower of $\underline{M}$ with κ'_α the critical point of $\underline{M}_\alpha$. We may now repeat the previous proof, getting countable β and $\bar{x}_i \in M_\beta$ such that $\sigma_\alpha(\bar{x}_i) = x_i$, $i \in \omega$. So $\underline{M}_\beta$ is not well-founded. □

Lemma 8.8. *Suppose $\underline{N} = \langle N, \in_N, U, \vec{A} \rangle$ is a premouse with critical point κ. U is countably complete if and only if whenever $\sigma : \underline{M} \to_{\Sigma_0} \underline{N}$ and $\underline{M}$ is a countable premouse $\langle M, \in_M, U', \vec{A}' \rangle$ and $j : \underline{M} \to_{U'} \underline{M}'$ then there is $\sigma' : \underline{M}' \to_{\Sigma_0} \underline{N}$ such that $\sigma' j = \sigma$.*

Proof: Firstly suppose U is countably complete. Suppose σ, $\underline{M}$, $\underline{M}'$ and j are as in the statement and let $\bar{\kappa}$ be the critical point of $\underline{M}$. Let $\langle \bar{X}_i \rangle_{i < \omega}$ enumerate the $\bar{X} \in P_M(\bar{\kappa})$ such that $X \in U'$. Let $X_i = \sigma(\bar{X}_i)$ $(i < \omega)$. Then $X_i \in U$. Hence $\bigcap_{i<\omega} X_i \neq \phi$. Say $\tau \in \bigcap_{i<\omega} X_i$. Define $\sigma' : \underline{M}' \to \underline{N}$ by $\sigma'(j(f)(\bar{\kappa})) = \sigma(f)(\tau)$. Then for Σ_0 ϕ

$$\begin{aligned} \underline{M}' \models \phi(j(f_1)(\bar{\kappa}) \ldots j(f_n)(\bar{\kappa})) &\Rightarrow \{\xi : \phi(f_1(\xi) \ldots f_n(\xi))\} \in U' \\ &\Rightarrow \{\xi : \phi(\sigma(f_1)(\xi) \ldots \sigma(f_n)(\xi))\} \in U \\ &\Rightarrow \tau \in \{\xi : \phi(\sigma(f_1)(\xi) \ldots \sigma(f_n)(\xi))\} \\ &\Rightarrow \underline{N} \models \phi(\sigma(f_1)(\tau) \ldots \sigma(f_n)(\tau)). \end{aligned} \tag{8.6}$$

Conversely suppose the stated condition holds. Let $\langle X_i : i \in \omega \rangle$ be a sequence with $X_i \in U$, all $i < \omega$. Let $X \prec_{\Sigma_1} \underline{N}$ with $\{X_i : i \in \omega\} \subseteq X$, X countable.

Let $\sigma:\underline{M}\tilde{=}\underline{N}|X$, $\underline{M}$ standard. Let $\underline{X}_i=\sigma^{-1}(X_i)$. Then $\underline{M}$ is a premouse; say $\underline{M}=\langle M,\in_M,U',\vec{A}'\rangle$, $\bar{\kappa}$ the critical point of $\underline{M}$. Let $j:\underline{M}\to_{U'}\underline{M}'$ and let $\sigma':\underline{M}'\to_{\Sigma_0}\underline{N}$ where $\sigma'j=\sigma$. Let $\tau=\sigma'(\bar{\kappa})$. Then $\tau=\sigma'(\bar{\kappa})<\sigma'(j(\bar{\kappa}))=\sigma(\bar{\kappa})=\kappa$.

And $\tau=\sigma'(\bar{\kappa})\in\sigma'(j(\underline{X}_i))$ (as $X_i\in U$)

$=\sigma(\underline{X}_i)$

$=X_i$.

So $\bigcap_{i<\omega}X_i\neq\phi$. □

Lemma 8.9. *If $\underline{N}$ is a premouse, $\underline{N}=\langle N,\in_N,U,\vec{A}\rangle$, $\underline{N}$ is well-founded and U is countably complete then $\underline{N}$ is iterable.*

Proof: By lemmas 8.6 and 8.7 we need only show that if $\sigma:M\to_{\Sigma_1}\underline{N}$ and $\underline{M}$ is countable with iterated ultrapower $\langle\langle\underline{M}_\alpha\rangle_{\alpha\in On},\langle\pi_{\alpha\beta}\rangle_{\alpha\leq\beta\in On}\rangle$ then $\underline{M}_\alpha$ is well-founded for all $\alpha<\omega_1$. We shall do this by obtaining a map $\sigma':\underline{M}_{\omega_1}\to_{\Sigma_0}\underline{N}$; in fact by induction we define $\sigma_\alpha:\underline{M}_\alpha\to_{\Sigma_0}\underline{N}$ such that $\sigma_\alpha\pi_{\beta\alpha}=\sigma_\beta$ (all $\beta<\alpha$). σ_0 is σ. If σ_α is defined $\sigma_{\alpha+1}$ is given by lemma 8.8. If σ_α is defined for all $\alpha<\lambda$ and λ is a limit ordinal then define $\sigma_\lambda(\pi_{\alpha\lambda}(x))=\sigma_\alpha(x)$. Then if $\pi_{\alpha\lambda}(x)=\pi_{\beta\lambda}(y)$ $(\alpha\leq\beta)$, $\pi_{\alpha\beta}(x)=y$ so $\sigma_\beta(y)=\sigma_\beta\pi_{\alpha\beta}(x)=\sigma_\alpha(x)$; so σ_λ is well-defined. Let $\sigma'=\sigma_{\omega_1}$.

It is necessary to check that $\underline{M}_\alpha$ is countable when $\alpha<\omega_1$. This is easily done. □

All of the above applies to weak iterated ultrapowers. Also it applies to arbitrary $\underline{N}\models ZFC$ with $U\in N$. Hence

Lemma 8.10. *If U is a normal measure on κ then V is iterable by U.*

Unfortunately there is no obvious way of applying this proof to premice without using the "external" property of countable completeness. We should like to be able to code some of the apparatus internally. There is a difficulty here; if $\underline{N}$ is a premouse and $j:\underline{N}\to_U\underline{N}'$ but $On_{N'}>On_N$ we shall not even have $N'\subseteq N$. This difficulty could be got around if we lifted the requirement of standardness; the proof of lemma 5.16, for example, could then be done purely internally. But that only gives one stage; it is not likely that with a set of ordinals we can represent a proper class of iterates. But for well-foundedness we only need ω_1 iterates; so if $On_N\geq\omega_1$ we might hope to code the first ω_1 iterates and appeal to lemma 8.6.

All the coding has already been done in chapter 7.

Lemma 8.11. *Suppose ϕ is a Σ_1 formula. There is a Σ_1 formula ϕ^+ such that whenever $\underline{N}$ is a premouse with iterated ultrapower $\langle\langle\underline{N}_\alpha\rangle_{\alpha\in On},\langle\pi_{\alpha\beta}\rangle_{\alpha\leq\beta\in On}\rangle$ then* *the relation* $R=\{\langle x,y\rangle : \underline{N}_\alpha \models \phi(x,y)\}$ *is well-founded if and only if* $R^+= =\{\langle x,y\rangle : \underline{N}\models\phi^+(x,y,\alpha)\}$ *is, provided* $\alpha\cup\{\alpha\}$ *is an initial segment of* On_N.

Proof: Let $T_\alpha=\{\langle f,u\rangle\in N:\ f:(\alpha)^m\to N$ and $u\in[\alpha]^m$ some $m<\omega\}$. Given $v\in[\alpha]^n$, $m\le n$ we define a function f^{uv} by

$$f^{uv}(\gamma_1\ldots\gamma_n)=f(\gamma_{i_1}\ldots\gamma_{i_m}) \tag{8.7}$$

where $v=\{\alpha_1\ldots\alpha_n\}$, $\alpha_1<\ldots<\alpha_n$ and $u=\{\alpha_{i_1}\ldots\alpha_{i_m}\}(i_1<\ldots<i_m)$. Now let ψ be the formula

$$\psi(\langle\gamma_1\ldots\gamma_n\rangle,f,g)\leftrightarrow\phi(f(\gamma_1\ldots\gamma_n),g(\gamma_1\ldots\gamma_n)). \tag{8.8}$$

By lemma 7.7 there is a formula ψ^* such that

$$\underline{N}\models\psi^*(f,g,n)\leftrightarrow\underline{N}_\alpha\models\psi(\langle\kappa_{\alpha_1}\ldots\kappa_{\alpha_n}\rangle,\pi_{0\alpha}(f),\pi_{0\alpha}(g)) \tag{8.9}$$

where $\alpha_1<\ldots<\alpha_n<\alpha$, κ_β is the critical point of $\underline{N}_\beta$. Then say

$$\phi^+(\langle f,u\rangle,\langle g,v\rangle,\alpha)\leftrightarrow\langle f,u\rangle,\langle g,v\rangle\in T_\alpha\wedge\psi^*(f^{uw},g^{vw},n) \tag{8.10}$$

where $w=u\cup v$.

Suppose R is well-founded. Suppose $\langle f_i,u_i\rangle\in T_\alpha$, all $i\in\omega$ and and $R^+(\langle f_{i+1},u_{i+1}\rangle,\langle f_i,u_i\rangle)$ $(i\in\omega)$. If $u_i=\{\alpha^i_1\ldots\alpha^i_{p_i}\}$ $\alpha^i_1<\ldots<\alpha^i_{p_i}$ let $x_i=\pi_{0\alpha}(f_i)(\kappa_{\alpha^i_1}\ldots\kappa_{\alpha^i_{p_i}})$. Let $w_i=u_i\cup u_{i+1}$ and let $w_i=\{\gamma^i_1\ldots\gamma^i_{n_i}\}$where $\gamma^i_1<\ldots<\gamma^i_{n_i}$. Then

$$\underline{N}\models\psi^*(f^{u_{i+1}w_i}_{i+1},f^{u_iw_i}_i,n_i)$$

so $\quad \underline{N}_\alpha\models\psi(\langle\kappa_{\gamma^i_1}\ldots\kappa_{\gamma^i_{n_i}}\rangle,\pi_{0\alpha}(f^{u_{i+1}w_i}_{i+1}),\pi_{0\alpha}(f^{u_iw_i}_i))$

so $\quad \underline{N}_\alpha\models\phi(\pi_{0\alpha}(f^{u_{i+1}w_i}_{i+1})(\kappa_{\gamma^i_1}\ldots\kappa_{\gamma^i_{n_i}}),\pi_{0\alpha}(f^{u_iw_i}_i)(\kappa_{\gamma^i_1}\ldots\kappa_{\gamma^i_{n_i}}))$

so $\quad \underline{N}_\alpha\models\phi(\pi_{0\alpha}(f_{i+1})(\kappa_{\alpha^{i+1}_1}\ldots\kappa_{\alpha^{i+1}_{p_{i+1}}}),\pi_{0\alpha}(f_i)(\kappa_{\alpha^i_1}\ldots\kappa_{\alpha^i_{p_i}}))$

so $\quad \underline{N}_\alpha\models\phi(x_{i+1},x_i)$ (all $i<\omega$).

But R is well-founded.

Suppose, conversely, that R^+ is well-founded, but that there is a sequence $\langle x_i : i<\omega\rangle$ with $R(x_{i+1},x_i)$, all $i<\omega$. Let

$$x_i=\pi_{0\alpha}(f_i)(\kappa_{\alpha^i_1}\ldots\kappa_{\alpha^i_{p_i}}) \qquad (\alpha^i_1<\ldots<\alpha^i_{p_i}). \tag{8.11}$$

Let $u_i=\{\alpha^i_1\ldots\alpha^i_{p_i}\}$. Then $R^+(\langle f_{i+1},u_{i+1}\rangle,\langle f_i,u_i\rangle)$ all $i\in\omega$. □

Corollary 8.12. *There is a relation R uniformly* $\Sigma_1(\underline{N})$, *where* $\underline{N}$ *is a premouse and* $\omega_1\cup\{\omega_1\}$ *is an initial segment of* $\underline{N}$, *such that* $\underline{N}$ *is iterable if and only if R is well-founded.*

Proof: By lemma 8.6, letting $\phi(x,y)\leftrightarrow x\in y$ in lemma 8.11. □

Corollary 8.13. *If* $\underline{M}\models ZF^-$ *and* $\omega_1\in M$ *then*

$M\models$"$\underline{N}$ *is an iterable premouse"* $\leftrightarrow$ $\underline{N}$ *is an iterable premouse.*

ITERABLE PREMICE

Two important results on iterable premice follow. The first is the comparability property. Although two iterable premice $\underline{J}^U_\alpha, \underline{J}^{U'}_\beta$ are not necessarily directly comparable, in the sense that $\alpha<\beta \Rightarrow J^U_\alpha \in J^{U'}_\beta$, yet they may be iterated to premice with this property.

Lemma 8.14. *Suppose $\underline{N}$ is an iterable premouse with iterated ultrapower $\langle\langle \underline{N}_\alpha\rangle_{\alpha\in On}, \langle \pi_{\alpha\beta}\rangle_{\alpha\leq\beta\in On}\rangle$. Suppose $\underline{N}_\alpha = \langle N_\alpha, \in, U_\alpha, \vec{A}_\alpha\rangle$ and κ_α is the critical point of $\underline{N}_\alpha$. If λ is a limit ordinal then for all $x\in P(\kappa_\lambda)\cap N_\lambda$*

$$x\in U_\lambda \Leftrightarrow \exists\gamma<\lambda\forall\delta(\gamma<\delta<\lambda \Rightarrow \kappa_\delta\in x).$$

Proof: Suppose $x\in U_\lambda$; say $x=\pi_{\alpha\lambda}(\bar{x})$, so $\bar{x}\in U_\alpha$. Hence $\kappa_\alpha\in\pi_{\alpha\alpha+1}(\bar{x})$, so $\kappa_\alpha\in\pi_{\alpha\lambda}(\bar{x})=x$. That is, $x\in rng(\pi_{\alpha\lambda}) \Rightarrow \kappa_\alpha\in x$. Let γ be least such that $x\in rng(\pi_{\gamma\lambda})$.

If $x\notin U_\lambda$ then $\kappa_\lambda\setminus x\in U_\lambda$ so $\exists\gamma<\lambda\forall\delta(\gamma<\delta<\lambda \Rightarrow \kappa_\delta\notin x)$. □

Lemma 8.15. *Suppose θ is a cardinal and $\theta>\overline{\overline{P(\kappa)\cap N}}$, where $\underline{N}$ is an iterable premouse with critical point κ. If $\langle\langle \underline{N}_\alpha\rangle_{\alpha\in On}, \langle \pi_{\alpha\beta}\rangle_{\alpha\leq\beta\in On}\rangle$ is the iterated ultrapower of $\underline{N}$ then θ is the critical point of $\underline{N}_\theta$.*

Proof: Let κ_α be the critical point of $\underline{N}_\alpha$. An easy induction on $\gamma<\theta$ shows $\overline{\overline{P(\kappa_\gamma)\cap N_\gamma}}<\theta$, so $\kappa_\gamma<\theta$. And $\kappa_\theta=\sup_{\gamma<\theta}\kappa_\gamma$ so $\kappa_\theta\leq\theta$. But κ_α is a normal sequence so $\kappa_\theta=\theta$. □

Lemma 8.16. *Suppose $\underline{N}$ is an iterable premouse with iterated ultrapower $\langle\langle \underline{N}_\alpha\rangle_{\alpha\in On}, \langle \pi_{\alpha\beta}\rangle_{\alpha\leq\beta\in On}\rangle$. Suppose $\theta>\overline{\overline{N}}$, θ regular. Then for all $x\in P(\theta)\cap N_\theta$*

$x\in U_\theta \Leftrightarrow x$ contains a closed unbounded set

where $\underline{N}_\theta = \langle N_\theta, \in, U_\theta, \vec{A}_\theta\rangle$.

Proof: By lemma 8.14 if $x\in U_\theta$ then x contains a closed unbounded set $\{\kappa_\alpha : \gamma<\alpha<\theta\}$ for some $\gamma<\theta$. If $x\notin U_\theta$ then $\theta\setminus x$ contains a closed unbounded set so x does not. □

Definition 8.17. *Iterable premice $\underline{M},\underline{N}$ are comparable if and only if there is α such that either $\underline{M}=\underline{J}^N_\alpha$ or $\underline{N}=\underline{J}^M_\alpha$.*

Lemma 8.18. *Suppose $\underline{M},\underline{N}$ are iterable pure premice with iterated ultrapowers $\langle\langle \underline{M}_\alpha\rangle_{\alpha\in On}, \langle \pi_{\alpha\beta}\rangle_{\alpha\leq\beta\in On}\rangle$, $\langle\langle \underline{N}_\alpha\rangle_{\alpha\in On}, \langle \pi'_{\alpha\beta}\rangle_{\alpha\leq\beta\in On}\rangle$ respectively. Suppose θ is regular, $\theta>\overline{\overline{M}},\overline{\overline{N}}$. Then $\underline{M}_\theta,\underline{N}_\theta$ are comparable.*

Proof: θ is the critical point of both $\underline{M}_\theta$ and $\underline{N}_\theta$ by lemma 8.15. So the result follows immediately from lemma 8.16. □

Of course, lemma 8.18 does not hold if $\underline{M},\underline{N}$ have additional predicates A_i. Does it follow from the fact that $\underline{M}_\theta=\underline{N}_\theta$ that either $\underline{M}=\underline{N}$ or $\underline{M}$ is an iterate of $\underline{N}$ or $\underline{N}$ is an iterate of $\underline{M}$? Not in general: but we shall show in the next part that it does in the cases that interest us.

The second property of iterable premice that we need is of less obvious significance.

Lemma 8.19. *Suppose $\underline{N}$ is an iterable premouse and that $\underline{M}$ is an iterate of $\underline{N}$. Suppose $\sigma:\underline{N}\to_{\Sigma_0}\underline{M}$; suppose $\langle\langle \underline{N}_\alpha\rangle_{\alpha\in On},\langle\pi_{\alpha\beta}\rangle_{\alpha\leq\beta\in On}\rangle$ is the iterated ultrapower of $\underline{N}$; say $\underline{M}=\underline{N}_\alpha$. Then for all $\xi\in On_N$, $\sigma(\xi)\geq\pi_{0\alpha}(\xi)$.*

Proof: Suppose not. Say $\sigma(\xi)<\pi_{0\alpha}(\xi)$. By lemma 8.3 there are $\sigma_\beta:\underline{N}_\beta\to_{\Sigma_0}\underline{N}_{\alpha+\beta}$ such that $\pi_{\alpha+\beta,\alpha+\gamma}\sigma_\beta=\sigma_\gamma\pi_{\beta\gamma}$, all $\beta\leq\gamma$. Define inductively

$\xi_0=\xi$

$\xi_{n+1}=\sigma_{\alpha.n}(\xi_n)$.

So $\xi_n\in N_{\alpha.n}$. Let $\zeta_n=\pi_{\alpha.n,\alpha.\omega}(\xi_n)$.

We claim $\zeta_n>\zeta_{n+1}$. For $\zeta_0>\zeta_1$, because $\sigma(\xi)<\pi_{0\alpha}(\xi)$. And

$$\begin{aligned}\zeta_{n+2}&=\pi_{\alpha(n+2),\alpha.\omega}(\xi_{n+2})\\&=\pi_{\alpha(n+2),\alpha.\omega}(\sigma_{\alpha(n+1)}(\xi_{n+1}))\\&=\sigma_{\alpha.\omega}\pi_{\alpha(n+1),\alpha.\omega}(\xi_{n+1})\\&=\sigma_{\alpha.\omega}(\zeta_{n+1})\\&<\sigma_{\alpha.\omega}(\zeta_n)\qquad\text{(induction hypothesis)}\\&=\zeta_{n+1}.\end{aligned}$$

So $\underline{N}_{\alpha.\omega}$ is not well-founded; contradiction! □

Corollary 8.20. *Suppose $\underline{M}$ is an iterate of $\underline{N}$, an iterable premouse. Suppose $\pi:\underline{N}\to_{\Sigma_0}\underline{M}$. Then $\pi:\underline{N}\to_{\Sigma_0}\underline{M}$ cofinally.*

Exercises

1. Derive corollary 7.3 from lemma 8.3.

2. Suppose U is a normal measure on κ and $\langle\langle M_\alpha\rangle_{\alpha\in On},\langle\pi_{\alpha\beta}\rangle_{\alpha\leq\beta\in On}\rangle$ is the iterated ultrapower of V by U. Suppose $\theta>\kappa$ is a strong limit cardinal with $cf(\theta)>\kappa$. Show that for all $\alpha<\theta$ $\pi_{0\alpha}(\theta)=\theta$.

3. Suppose $\underline{M},\underline{M}'$ are comparable iterable premice, $\underline{N},\underline{N}'$ iterates of $\underline{M},\underline{M}'$ respectively and $\underline{N},\underline{N}'$ comparable. Show that

$M\in M' \Leftrightarrow N\in N'$.

4. Suppose $\underline{N}_\alpha,\underline{M}_\alpha,\pi_{\alpha\beta},\pi'_{\alpha\beta},\sigma_\alpha,f$ are as in lemma 8.3. Show that for all $\beta\leq\gamma$ $\sigma_\gamma\pi_{\beta\gamma}=\pi_{f(\beta)f(\gamma)}\sigma_\beta$ (this fact was used in lemma 8.19).

PART THREE: MICE

Part three brings together fine-structure and the theory of iterated ultrapowers. Chapter 9 will introduce mice, the main technical device of this study. The development here is rather different from that in [10]. We are concerned to show the parallel with the results of part two more closely because this is important for generalisations. To eliminate the operator $\to_U^n$ makes the proof of the covering lemma harder.

Chapter 10 reveals a simple structure theorem for mice: every mouse is an iterate of its core. So given two mice with a common mouse iterate one is a mouse iterate of the other. This fails to generalise even to mice with two measurable cardinals.

Chapter 11 proves acceptability for iterable premice. This would seem to make the specification of acceptability in the definition of mouse redundant; but actually it is much simpler to put it in. Mice are used essentially in the proof of corollary 11.29. Chapter 11 uses many techniques that will be needed for the construction of the core model.

Chapters 12 and 13 stand a little aside from the main development, relating to class models of ZFC rather than to mice. None of the results in these chapters require mice for their proofs, but the results are essential for the statement of later theorems, which is why they are included here. Chapter 12 gives a first example of a mouse and chapter 13 a very weak generalisation of mice. This is, moreover, only sketched; but the theory of double mice is not very different from the theory of single mice presented in this book. The descriptive set theory at the end of chapter 13 is included to show how mice may be used to prove results that predated their invention. It is not necessary for anything that comes later in the book.

Consider an iterable premouse $\underline{N}$. Let us assume for the moment that the critical point of $\underline{N}$ is κ and that $\omega\rho_{\underline{N}} \leq \kappa$. Then it is clear that if $\langle\langle \underline{N}_\alpha \rangle_{\alpha \in On}, \langle \pi_{\alpha\beta} \rangle_{\alpha<\beta \in On} \rangle$ is the iterated ultrapower of $\underline{N}$ then for all α $\rho_{N_\alpha} = \rho_N$, since $\overline{P}(\kappa) \cap \Sigma_1(\underline{N}_\alpha) = P(\kappa) \cap \Sigma_1(\underline{N})$ by lemma 7.4, and $P(\kappa) \cap \underline{N}_\alpha = P(\kappa) \cap \underline{N}$. In fact we shall show that $A_N^{\overline{p}} = A_{N_\alpha}^{p_{N_\alpha}}$ and $\pi_{0\alpha}(p_N) = p_{N_\alpha}$, where p_N is a standard member of P_N to be defined.

Not so, however, if $\omega\rho_N > \kappa$. Suppose $\omega\rho_{\underline{N}}^n \leq \kappa$. In general we do not have $\pi_{0\alpha} : \underline{N} \to_{\Sigma_n} \underline{N}_\alpha$ and we cannot conclude that $\rho_{N_\alpha}^n = \rho_N^n$ or that $A_{N_\alpha}^n = A_N^n$. For reasons we explained in the introduction we are not interested in premice with $\omega\rho_N^\omega > \kappa$. But we shall need to consider the cases where $\omega\rho_N > \kappa$ but $\omega\rho_{\underline{N}}^n \leq \kappa$ for some n; and we shall need a form of iteration - mouse iteration, it will be called - that treats fine structure more gently than ordinary iteration.

Actually the real work was done in chapter 4 in a very general framework. If $\omega\rho_{\underline{N}}^{n+1} \leq \kappa$ but $\omega\rho_N^n > \kappa$ we are going to iterate $\underline{N}^{np}$ rather than $\underline{N}$ to get $\pi : \underline{N}^{np} \to_U \underline{M}$, say. Then the mouse iteration map will be the n completion of π, $\pi' : \underline{N} \to_{\Sigma_{n+1}} \underline{M}'$, and $\underline{M}'^{np'}$ will be $\underline{M}$. In general $\underline{M}'$ will be bigger than the usual iterate of $\underline{N}$.

Fine structural preliminaries precede the main definition.

Definition 9.1. *Suppose* $p,q \in [On]^{<\omega}$. *Then*

$$p <_* q \leftrightarrow \exists\gamma (p \setminus \gamma = q \setminus \gamma \wedge q \cap \gamma \neq \phi \wedge (p \cap \gamma \neq \phi \rightarrow \max(p \cap \gamma) < \max(q \cap \gamma))).$$

Lemma 9.2. $<_*$ *is a well-order of* $[On]^{<\omega}$.

Proof: If $p <_* q$ and $q <_* r$, with γ, γ' repectively as in the definition, let $\overline{\gamma} = \max(\gamma, \gamma')$. Then $p \setminus \overline{\gamma} = r \setminus \overline{\gamma}$. $r \cap \gamma' \neq \phi$ so $r \cap \overline{\gamma} \neq \phi$.

Case 1 $\overline{\gamma} = \gamma$.

Then either $p \cap \overline{\gamma} = \phi$, in which case $p <_* r$, or $\max(p \cap \overline{\gamma}) < \max(q \cap \overline{\gamma})$. If $\max(q \cap \overline{\gamma}) = \max(r \cap \overline{\gamma})$ then $p <_* r$. Otherwise $q \cap \overline{\gamma} = q \cap \gamma'$, $r \cap \overline{\gamma} = r \cap \gamma'$ so $\max(q \cap \overline{\gamma}) < \max(r \cap \overline{\gamma})$. Hence $p <_* r$.

Case 2 $\overline{\gamma} = \gamma'$.

If $p \cap \overline{\gamma} = \phi$ $p <_* r$. Otherwise $q \cap \gamma \neq \phi$ so $q \cap \overline{\gamma} \neq \phi$. Hence $\max(q \cap \overline{\gamma}) < \max(r \cap \overline{\gamma})$. The proof goes as in case 1. So $<_*$ is transitive.

Given $p,q \in [On]^{<\omega}$, $p \neq q$ let $a = \max(p \setminus q)$ (undefined if $p \subseteq q$) and $b = \max(q \setminus p)$ (undefined if $q \subseteq p$). Then $a \neq b$. $p <_* q$ if $a < b$ or a undefined;

$q <_* p$ otherwise.

Finally, $<_*$ is a well-order. For suppose $p_{i+1} <_* p_i$, all $i < \omega$. Suppose γ_i is such that $p_{i+1} \setminus \gamma_i = p_i \setminus \gamma_i$, $p_i \cap \gamma_i \neq \phi$ and $\max(p_{i+1} \cap \gamma_i) < \max(p_i \cap \gamma_i)$ or $p_{i+1} \cap \gamma_i = \phi$. We may assume $i \le j \Rightarrow \gamma_i \le \gamma_j$ by taking a final segment. And we may assume that for all i $\max(p_{i+1} \cap \gamma_i) \neq \max(p_{i+1} \cap \gamma_{i+1})$ But then p_0 is infinite. Contradiction! □

<u>Lemma 9.3</u>. *Suppose $\sigma: \underline{M} \to_{\Sigma_0} \underline{N}$, $\underline{M}, \underline{N}$ transitive. Then $\sigma(p) \ge_* p$, all $p \in [On_M]^{<\omega}$.*

Proof: Let $p = \{\alpha_1 \ldots \alpha_n\}$, $\alpha_1 < \ldots < \alpha_n$. Then $\sigma(p) = \{\sigma(\alpha_1) \ldots \sigma(\alpha_n)\}$. Suppose for $i > k$ $\alpha_i = \sigma(\alpha_i)$ $(0 \le k \le n$, k minimal). If $k = 0$ $p = \sigma(p)$. If $k > 0$ $\alpha_k \neq \sigma(\alpha_k)$ so $\alpha_k < \sigma(\alpha_k)$. Thus $\max(p \cap \tilde{\alpha}) = \alpha_k < \sigma(\alpha_k) = \max(\sigma(p) \cap \tilde{\alpha})$, where $\tilde{\alpha} = \sigma(\alpha_k) + 1$. □

We can use $<_*$ to simplify definition 3.23.

<u>Definition 9.4</u>. *Suppose $\underline{N}$ is a transitive model of RA (hence a model of RA_ρ). Then we define by induction:*

$$p_N^0 = \phi;\ \underline{N}^0 = \underline{N};\ A_N^0 = \phi.$$
$$p_N^{n+1} = \text{the } <_*\text{-least member of } P_N n$$
$$\underline{N}^{n+1} = \underline{N}^{n+1p} \text{ where } p = \langle p_N^1 \ldots p_N^{n+1} \rangle$$
$$A_N^{n+1} = A_N^{n+1p};\ T_N^{n+1} = T_N^{n+1p}.$$

<u>Definition 9.5</u>. *$\underline{N}$ is n-sound if and only if $\underline{N}$ is $\langle p_N^1 \ldots p_N^n \rangle$-sound.*

<u>Lemma 9.6</u>. *Suppose $\underline{N}$ is p-sound, $p = \langle p_1 \ldots p_n \rangle$. Then $\rho_N^{n+1} = \rho_N np$.*

Proof: $\omega\rho_N^{n+1} = \bigcap\{\omega\rho_N nq : q \in PA_n^N\}$. A simple induction shows that $A_N^{mq_1 \ldots q_m}$ is rudimentary in $\underline{N}^{mp_1 \ldots p_m}$ for $m \le n$. But then $A \in \Sigma_1(\underline{N}^{nq}) \Rightarrow A \in \Sigma_1(\underline{N}^{np})$, giving the result. □

<u>Lemma 9.7</u>. *Suppose $\underline{N}$ is a transitive acceptable premouse of the form J_α^U and $\omega\rho_N^{n+1} \ge \kappa$, where κ is the critical point of $\underline{N}$. Then $\underline{N}$ is n+1-sound.*

Proof: By induction on n. Suppose the result holds for all $m < n$. Let $\underline{M} = \underline{N}^n$. Let $X = h_M(\omega\rho_N^{n+1} \cup p_N^{n+1})$. Let $\sigma: \overline{\underline{M}} \cong \underline{M} | X$ where $\overline{\underline{M}}$ is transitive. Then $\sigma: \overline{\underline{M}} \to_{\Sigma_1} \underline{M}$. By lemma 4.22 there are $\overline{\underline{N}}, p'$ with $\overline{\underline{N}}$ p'-sound and $\overline{\underline{N}}^{np'} = \overline{\underline{M}}$. By lemma 4.23 there is $\tilde{\sigma} \supseteq \sigma$ such that $\tilde{\sigma}: \overline{\underline{N}} \to_{\Sigma_0} \underline{N}$ and $\tilde{\sigma}(p') = \langle p_N^1 \ldots p_N^n \rangle$. By lemma 4.24 $\tilde{\sigma}: \overline{\underline{N}} \to_{\Sigma_{n+1}} \underline{N}$. Now $\kappa + 1 \subseteq X$; so $\tilde{\sigma} | \kappa+1 = id | \kappa+1$. Hence $\tilde{\sigma}(X) = X$ for $X \in \overline{\underline{N}} \cap P(\kappa)$. Say $\overline{\underline{N}} = J_{\bar{\alpha}}^{\bar{U}}$. Then $X \in \bar{U} \cap \overline{N} \Rightarrow X \in U$. Thus $\overline{\underline{N}} = J_{\bar{\alpha}}^{U}$.

By lemma 9.6 there is $A \in \Sigma_1(\underline{N}^n)$ such that $A \cap \omega\rho_N^{n+1} \neq x$, all $x \in N^n$, and A is definable using parameter p_N^{n+1}. Hence letting B have the same Σ_1 definition over $\overline{\underline{M}}$ from $\sigma^{-1}(p_N^{n+1})$, $B \cap \omega\rho_N^{n+1} = A \cap \omega\rho_N^{n+1}$. But by lemma 4.20 $B \cap \omega\rho_N^{n+1}$ is definable over $\overline{\underline{N}}$. Thus $\alpha = \bar{\alpha}$, $\overline{\underline{N}} = \underline{N}$. But then B is $\Sigma_1(\overline{\underline{N}}^n)$ so $\sigma^{-1}(p_N^{n+1}) \in P_{\overline{N}} n = P_N n$. $\sigma^{-1}(p_N^{n+1}) \le_* p_N^{n+1}$ by lemma 9.3

so $\sigma^{-1}(p_N^{n+1})=p_N^{n+1}$. But then σ is the identity map so $X=\bar{M}=N^n$. □

Compare the proof of lemma 4.27. A similar argument yields

Lemma 9.8. *Suppose $\underline{N}$ is a transitive acceptable premouse and $\omega\rho_N^n>\kappa$ but $\omega\rho_N^{n+1}\leq\kappa$, where κ is the critical point of $\underline{N}$. Then $\underline{N}^n\models V=h(\kappa\cup p_N^{n+1})$.*

Definition 9.9. *Suppose $\underline{N}\models RA$. $\underline{N}$ is strongly acceptable above γ provided that whenever λ is a limit ordinal $\geq\gamma$ and $a\subseteq\delta$, $a\in S_{\lambda+\omega}^N\setminus S_\lambda^N$ then $\underline{S}_{\lambda+\omega}^N\models\bar{\lambda}\leq max(\gamma,\delta)$.*

Lemma 9.10. *Suppose $\underline{N}$ is strongly acceptable above γ. Then whenever $\gamma\in\omega\rho_N^n$ $N^n=J_{\rho_N^n}^N$.*

Proof: We have to show $H_{\omega\rho_N}^N n=J_{\rho_N}^N n$. We know that $J_{\rho_N}^N n\subseteq H_{\omega\rho_N}^N n$. Suppose $a\in H_{\omega\rho_N}^N n$. We may assume that $a\subseteq\mu<\omega\rho_N^n$, for $H_{\omega\rho_N}^N n=\cup\{J_{\rho_N}^a n: a\in N \wedge a\subseteq\mu<\omega\rho_N^n\}$ and $a\in J_{\rho_N}^N n \Rightarrow J_{\rho_N}^{aN} n\subseteq J_{\rho_N}^N n$ since $\omega\rho_N^n$ is closed under addition (unless n=0 when the result is trivial).

Suppose λ is least such that λ is a limit ordinal and $a\in S_{\lambda+\omega}^N$. Then $\underline{S}_{\lambda+\omega}^N\models\lambda\leq max(\gamma,\mu)$. Since $\omega\rho_N^n$ is a Σ_1 cardinal in $\underline{N}$ and $\gamma,\mu\in\omega\rho_N^n$, therefore $\lambda\in\omega\rho_N^n$. Thus $a\in J_{\rho_N}^N n$. □

Lemma 9.11. *Suppose $\underline{N}$ is a transitive acceptable premouse, with κ the critical point. Then $\underline{N}$ is strongly acceptable above κ. ($\underline{N}$ must be of the form $\underline{J}_\alpha^U$).*

Proof: Suppose $a\subseteq\delta$, $a\in S_{\lambda+\omega}^N\setminus S_\lambda^N$, λ a limit.

Case 1 $P(\kappa)\cap S_{\lambda+\omega}^N=P(\kappa)\cap S_\lambda^N$.

Then by lemma 2.41 $a\in\Sigma_\omega(\underline{S}_\lambda^N)$. Say $a\in\Sigma_n(\underline{S}_\lambda^N)$; then $\omega\rho_{S_\lambda^N}^n\leq max(\kappa,\delta)$. But we may apply lemmas 9.7 and 9.8 to $\underline{S}_\lambda^N$ to get an $\underline{S}_\lambda^N$ definable map of $max(\delta,\kappa)$ onto λ.

Case 2 $P(\kappa)\cap S_{\lambda+\omega}^N\neq P(\kappa)\cap S_\lambda^N$.

Suppose $\Sigma_\omega(\underline{S}_\lambda^N)\cap P(\kappa)\subseteq S_\lambda^N$. Let M be $rud_{U\cap S_\lambda^N}(S_\lambda^N)$. Then by lemma 2.41 $P(\kappa)\cap M=P(\kappa)\cap\Sigma_\omega(\underline{S}_\lambda^N)\subseteq S_\lambda^N$. Hence $\langle M,\in,U\rangle\models R^+$. But then $S_{\lambda+\omega}^N\subseteq M$. Contradiction!

So $\Sigma_\omega(\underline{S}_\lambda^N)\cap P(\kappa)\not\subseteq S_\lambda^N$, so for some m $\omega\rho_{S_\lambda^N}^m\leq\kappa$ and we may argue as in case 1. □

THE OPERATION $\eta:N\to_U^n N'$

Definition 9.12. *A premouse $\underline{N}$ is n-suitable provided:*

(i) $\underline{N}$ is acceptable.

(ii) $\underline{N}$ is strongly acceptable above κ, where κ is the critical point of $\underline{N}$.

(iii) $\kappa\in\omega\rho_N^n$.

(iv) $\underline{N}$ is p-sound for some $p\in PA_n^N$.

(ii) is not necessary but may always be assumed in the cases we are considering. Of course, (iv) means that $\underline{N}^m \models RA_\rho$ for $m<n$, by lemma 4.18. If $\underline{N}=\underline{J}_\alpha^U$ then by lemmas 9.7 and 9.11 (ii) and (iv) are redundant.

Definition 9.13. *Suppose $\underline{N}$ is an n-suitable premouse, $\underline{N}=\langle N,\in_N,U,\vec{A}\rangle$. Then $\eta:\underline{N}\to_U^n\underline{N}'$ means that for some $p\in PA_n^N$ $\underline{N}$ is p-sound, $\eta|N^{np}:\underline{N}^{np}\to_U \underline{N}'^{n\eta(p)}$ and η is the n-completion of $\eta|N^{np}$ to $\underline{N}$.*

So $\eta:\underline{N}\to_U^0\underline{N}'$ means $\eta:\underline{N}\to_U\underline{N}'$.

Lemma 9.14. *Suppose $\underline{N}$ is an n-suitable premouse. If $\eta:\underline{N}\to_U^n\underline{N}'$, $\eta':\underline{N}\to_U^n\underline{N}''$ then there is an isomorphism $\sigma:\underline{N}'\cong\underline{N}''$ such that $\sigma\eta=\eta'$.*

Proof: By the remark after definition 4.26 we may assume that each of $\eta|N^{np},\eta'|N^{np}:\underline{N}^{np}\to_U\underline{N}'^{n\eta(p)}$, $\underline{N}''^{n\eta'(p)}$. So by lemma 5.14 there is $\hat{\sigma}$ such that $\hat{\sigma}:\underline{N}'^{n\eta(p)}\cong N''^{n\eta'(p)}$ and $\hat{\sigma}\eta|N^{np}=\eta'|N^{np}$.
By lemma 4.25, then, there is $\sigma\supseteq\hat{\sigma}$ such that $\sigma:\underline{N}'\cong\underline{N}''$. But then $\sigma\eta$ satisfies the definition of the n-completion of $\eta'|N^{np}$ to $\underline{N}$, so by lemma 4.23 $\sigma\eta=\eta'$. □

Lemma 9.15. *Suppose $\underline{N}$ is an n-suitable premouse. Then if $\underline{N}=\langle N,\in_N,U,\vec{A}\rangle$ there are $\eta,\underline{N}'$ such that $\eta:\underline{N}\to_U^n\underline{N}'$.*

Proof: By lemmas 5.16, 4.25 and 4.23. Note that $\eta:\underline{N}\to_{\Sigma_{n+1}}\underline{N}'$. □

Lemma 9.16. *If $\eta:\underline{N}\to_U^n\underline{N}'$ then $\underline{N}'$ is n-suitable.*

Proof: $\underline{N}'$ is acceptable by lemma 4.10 or 4.11. By the same argument $\underline{N}'$ is strongly acceptable above $\eta(\kappa)$. $\underline{N}'$ is $\eta(\vec{p})$-sound, provided $\underline{N}$ is $\vec{p}$-sound, by definition, and clearly $\eta(\kappa)\in\omega\rho_{N'}^n$. □

Lemma 9.16 gives us a basis for iterating the operation $\eta:\underline{N}\to_U^n\underline{N}'$. But first we need a couple of general results about n-completions.

Lemma 9.17. *If $\eta:\underline{N}\to\underline{N}'$ is the n-completion of $\hat{\eta}:\underline{N}^{np}\to\underline{N}'^{n\eta(p)}$ and η' is the n-completion of $\hat{\eta}':\underline{N}'^{n\eta(p)}\to\underline{N}''^{n\eta'\eta(p)}$ then $\eta'\eta$ is the n-completion of $\hat{\eta}'\hat{\eta}$.*

Proof: $\underline{N}''$ is $\eta'\eta(p)$-sound and $\eta'\eta:\underline{N}\to_{\Sigma_0}\underline{N}''$, $\eta'\eta\supseteq\hat{\eta}'\hat{\eta}$. □

Lemma 9.18. *Suppose $\langle\langle\underline{N}_\alpha\rangle_{\alpha<\lambda},\langle\pi_{\alpha\beta}\rangle_{\alpha\leq\beta<\lambda}\rangle$ is a commutative system, λ a limit ordinal, $\langle\underline{N},\langle\pi_\alpha\rangle_{\alpha<\lambda}\rangle$ a direct limit of the system and for each $\alpha\leq\beta<\lambda$ $\pi_{\alpha\beta}:\underline{N}_\alpha\to\underline{N}_\beta$ is the n-completion of $\pi'_{\alpha\beta}=\pi_{\alpha\beta}|N^{n\pi_{0\alpha}(p)}$ (some fixed p such that $p\in PA_n^{N_0}$, N_0 is p-sound). Let $\pi'_\alpha=\pi_\alpha|N^{n\pi_{0\alpha}(p)}$; then $\langle\underline{N}^{n\pi_0(p)},\langle\pi'_\alpha\rangle_{\alpha<\lambda}\rangle$ is a direct limit of $\langle\langle\underline{N}_\alpha^{n\pi_{0\alpha}(p)}\rangle_{\alpha<\lambda},\langle\pi'_{\alpha\beta}\rangle_{\alpha<\beta<\lambda}\rangle$ and π_α is the n-completion of π'_α.*

Proof: Suppose first of all that $x\in N^{n\pi_0(p)}$. Then $x=\pi_\alpha(\bar{x})$ for some $\alpha<\lambda$, $\bar{x}\in N$. $\underline{N}\models\overline{\overline{TC(x)}}\in\omega\rho_N^n$.

It suffices to prove the result for n=1. Let $\gamma=\{\nu\in N:\underline{N}\models\nu\in\pi_\alpha(\delta)$, some $\alpha<\lambda$, $\delta\in\omega\rho_{N_\alpha}\}$. Then $\underline{N}\models V=h(\gamma\cup\pi_0(p))$: for given $y=\pi_\alpha(\bar{y})$ $\underline{N}_\alpha\models\bar{y}=h(i,\langle\vec{\delta},\pi_{0\alpha}^{-1}(p)\rangle)$, $\vec{\delta}\in\omega\rho_{N_\alpha}$, then $\underline{N}\models y=h(i,\langle\pi_\alpha(\vec{\delta}),\pi_0(p)\rangle)$. So $\omega\rho_N\subseteq\gamma$. Hence $\underline{N}\models TC(x)\in\gamma$, so $\underline{N}_\alpha\models TC(\bar{x})\in\omega\rho_{N_\alpha}$, without loss of generality. So $\bar{x}\in N^{\pi_{0\alpha}(p)}$.

For the second part it suffices to show that $\underline{N}$ is p-sound. If it is not then $\omega\rho_N\neq\gamma$: say $\pi_\alpha(\delta)\in\gamma\setminus\omega\rho_N$, $\delta\in\omega\rho_{N_\alpha}$. Let A be $\Sigma_1(\underline{N})$ in parameter q such that $A\cap\pi_\alpha(\delta)\notin N$. We may assume $q=\pi_\alpha(\bar{q})$. Letting $\bar{A}$ have the same Σ_1 definition ϕ over $\underline{N}_\alpha$ from $\bar{q}$ $\bar{A}\cap\delta\in\underline{N}_\alpha$. But then

$$\underline{N}_\alpha\models\forall\mu\,\mu\in\bar{A}\cap\delta\leftrightarrow\phi(\mu,\bar{q})$$

which is Π_2; and $\pi_{\alpha\beta}:\underline{N}_\alpha\to_{\Sigma_2}\underline{N}_\beta$, as $\pi_{\alpha\beta}$ is the 1-completion of $\pi'_{\alpha\beta}$, so by lemma 6.8 $\pi_\alpha:\underline{N}_\alpha\to_{\Sigma_2}\underline{N}$. So $A\cap\pi_\alpha(\delta)=\pi_\alpha(\bar{A}\cap\delta)$; contradiction! □

MICE

Definition 9.19. *Suppose $\underline{N}$ is an n-suitable premouse. Then* $\langle\langle\underline{N}_\alpha\rangle_{\alpha\in On},\langle\pi_{\alpha\beta}\rangle_{\alpha\leq\beta\in On}\rangle$ *is an n-iterated ultrapower of $\underline{N}$ provided*

(i) $\langle\langle\underline{N}_\alpha\rangle_{\alpha\in On},\langle\pi_{\alpha\beta}\rangle_{\alpha\leq\beta\in On}\rangle$ *is a commutative system.*

(ii) each $\underline{N}_\alpha$ is n-suitable, $\underline{N}_\alpha=\langle N_\alpha,\in_{N_\alpha},U_\alpha,\vec{A}_\alpha\rangle$.

(iii) $\underline{N}_0=\underline{N}$.

(iv) $\pi_{\alpha\alpha+1}:\underline{N}_\alpha\to^n_{U_\alpha}\underline{N}_{\alpha+1}$ *all α.*

(v) $\langle\underline{N}_\lambda,\langle\pi_{\alpha\lambda}\rangle_{\alpha<\lambda}\rangle$ *is a direct limit of* $\langle\langle\underline{N}_\alpha\rangle_{\alpha<\lambda},\langle\pi_{\alpha\beta}\rangle_{\alpha\leq\beta<\lambda}\rangle$ *(λ a limit).*

Lemma 9.20. *If $\underline{N}$ is an n-suitable premouse then $\underline{N}$ has an n-iterated ultrapower.*

Proof: Construct $\underline{N}_\alpha$ by induction on α, proving that $\underline{N}_\alpha$ is n-suitable. Let $\underline{N}_0=\underline{N}$. In the successor case we use lemma 9.15 to get $\underline{N}_{\alpha+1}$. $\underline{N}_{\alpha+1}$ is n-suitable by lemma 9.16.

In the limit case $\underline{N}_\lambda$ exists by lemma 6.4. To show that $\underline{N}_\lambda$ is n-suitable we observe that by an inductive argument $\alpha<\beta<\lambda\to$ $\to\pi_{\alpha\beta}$ is the n-completion of $\pi_{\alpha\beta}|N_\alpha^{n\pi_{0\alpha}(p)}$ for some fixed p in PA_n^N such that $\underline{N}$ is p-sound. The induction uses lemmas 9.17 and 9.18. Hence by lemma 9.18 $\pi_{0\lambda}$ is the n-completion of $\pi_{0\lambda}|N^{np}$ so $\underline{N}_\lambda$ is $\pi_{0\lambda}(p)$-sound. The other clauses are easily checked. □

Lemma 9.21. *Suppose* $\langle\langle\underline{N}_\alpha\rangle_{\alpha\in On},\langle\pi_{\alpha\beta}\rangle_{\alpha\leq\beta\in On}\rangle$ *and* $\langle\langle\underline{N}'_\alpha\rangle_{\alpha\in On},\langle\pi'_{\alpha\beta}\rangle_{\alpha\leq\beta\in On}\rangle$ *are n-iterated ultrapowers of $\underline{N}$. Then there are isomorphisms* $\sigma_\alpha:\underline{N}_\alpha\cong\underline{N}'_\alpha$ *such that*

(i) $\sigma_\beta\pi_{\alpha\beta}=\pi'_{\alpha\beta}\sigma_\alpha$ *for* $\alpha\leq\beta$.

(ii) $\sigma_\beta(\kappa_\alpha)=\kappa'_\alpha$ *($\kappa_\alpha,\kappa'_\alpha$ the critical points of $\underline{N}_\alpha,\underline{N}'_\alpha$ respectively).*

Definition 9.22. *An n-suitable premouse $\underline{N}$ is called critical if* $\kappa\notin\omega\rho_N^{n+1}$ *where κ is the critical point of $\underline{N}$. This n is written n($\underline{N}$).*

Definition 9.23. *If $\underline{N}$ is a critical premouse then an $n(\underline{N})$-iterated ultrapower of $\underline{N}$ is called a mouse iteration of $\underline{N}$.*

Definition 9.24. *Suppose $\langle\langle \underline{N}_\alpha\rangle_{\alpha\in On}, \langle \pi_{\alpha\beta}\rangle_{\alpha\leq\beta\in On}\rangle$ is an n-iterated ultrapower of $\underline{N}$. Then each $\underline{N}_\alpha$ is called an n-iterate of $\underline{N}$, or a mouse iterate if $n=n(\underline{N})$.*

Definition 9.25. *$\underline{N}$ is n-iterable provided each n-iterate of $\underline{N}$ is well-founded. If $\underline{N}$ is critical and $n(\underline{N})$-iterable then $\underline{N}$ is called a mouse (provided $\underline{N}$ is pure).*

The reason for introducing mice is quickly seen if we try to apply the ordinary iteration operation to critical premice with $n(\underline{N})>0$. Say $\eta:\underline{N}\to_U\underline{N}'$. Suppose $n(\underline{N})=1$; we do not know that $\omega\rho^2_{N'}\leq\kappa$, or even that $\omega\rho^\omega_{N'}\leq\kappa$ (indeed this will fail if $\omega\rho^2_{N'}>\kappa$ and $\underline{N}$ is a mouse.) But there are good reasons for excluding such premice; if they are admitted to the core model indiscriminately then we shall end up with measurable cardinals in the core model, and we saw in the introduction that this would be a bad thing.

We saw that if $\underline{N}$ is a mouse $\eta:\underline{N}\to^n_U\underline{N}'$ and $\eta':\underline{N}\to_U\underline{N}''$ then $\underline{N}''$ is an initial segment of $\underline{N}'$; and $P(\kappa)\cap N''=P(\kappa)\cap N'$ where κ is the critical point of $\underline{N}$. $\underline{N}'$ adds enough levels to $\underline{N}''$ to get a construction stage where $\omega\rho^{n(N)+1}_{N'}\leq\eta(\kappa)$. Of course it could be the case that $\underline{N}$ was an iterable premouse but not a mouse, and then these levels might not exist.

By the results of chapter 7 there will often be α,β such that $\pi_{\alpha\beta}:\underline{N}_\alpha\to_{\Sigma_n}\underline{N}_\beta$ when $\pi_{\alpha\beta}$ is the ordinary iteration map. In this case the maps may or may not coincide.

Exercises

1. Suppose V=L[U]. Assuming that L[U]is acceptable and U is a normal measure on κ construct a premouse $\underline{N}$ with $n(\underline{N})=1$.

2. Is there a sentence σ such that

$\underline{N}\models\sigma \leftrightarrow \underline{N}$ is a critical premouse?

Such that $\underline{N}\models\sigma \leftrightarrow \underline{N}$ is a critical premouse and $n(\underline{N})=0$?

3. Suppose $\eta:\underline{N}\to^n_U\underline{N}'$, $n=n(\underline{N})>0$. Do we necessarily have $\eta(\rho^n_N)=\rho^n_{N'}$?

4. Prove the equivalent of Los theorem for $\eta:\underline{N}\to^n_U\underline{N}'$.

10: PROPERTIES OF MICE

In this chapter we shall prove the basic structure theorem about mice. Roughly speaking this says that if $\underline{M}$ and $\underline{N}$ have a common mouse iterate then either $\underline{M}$ is a mouse iterate of $\underline{N}$ or $\underline{N}$ is a mouse iterate of $\underline{M}$. Indeed there is a smallest mouse $\overline{\underline{N}}$ of which $\underline{N}$ is a mouse iterate, called the core of $\underline{N}$; this mouse will be $n(\underline{N})+1$ sound (indeed, as it turns out it will be m-sound for all m). All manner of corollaries are proved as we go along.

PRESERVATION OF PARAMETERS

Suppose until further notice that $\underline{N}$ is a mouse, $n=n(\underline{N})$, and that $\langle\langle \underline{N}_\alpha\rangle_{\alpha\in On}, \langle \pi_{\alpha\beta}\rangle_{\alpha\leq\beta\in On}\rangle$ is its mouse iteration. Let κ_α be the critical point of $\underline{N}_\alpha$. Let $\kappa=\kappa_0$.

<u>Lemma 10.1</u>. *$\Sigma_{n+1}(\underline{N}_\alpha)\cap P(\kappa)=\Sigma_{n+1}(\underline{N})\cap P(\kappa)$.*

Proof: $\Sigma_1(\underline{N}_\alpha^n)\cap P(\kappa)=\Sigma_1(\underline{N}^n)\cap P(\kappa)$. □

<u>Lemma 10.2</u>. *Each $\underline{N}_\alpha$ is n-sound.*

Proof: By lemma 9.7. □

<u>Lemma 10.3</u>. *$\underline{N}^n\models V=h(\kappa_\alpha\cup p^{n+1}_{N_\alpha})$*

Proof: By lemma 9.8. □

<u>Lemma 10.4</u>. *$\pi_{0\alpha}(p^m_N)=p^m_{N_\alpha}$ $(m\leq n)$.*

Proof: By induction on m. Suppose $\pi_{0\alpha}(p_N^{m+1})>_* p^{m+1}_{N_\alpha}$. Then there are $i\in\omega$, $\vec{\gamma}\in(\omega\rho^{m+1}_{N_\alpha})^{<\omega}$ such that

$$\underline{N}_\alpha^m\models(\exists r<_*\pi_{0\alpha}(p_N^{m+1}))(\pi_{0\alpha}(p_N^{m+1})=h(i,\langle\vec{\gamma},r\rangle)). \qquad (10.1)$$

Hence

$$\underline{N}^m\models(\exists r<_* p_N^{m+1})(p_N^{m+1}=h(i,r)). \qquad (10.2)$$

This is because $r<_*\pi_{0\alpha}(p_N^{m+1}) \Rightarrow r\cup\vec{\gamma}<_*\pi_{0\alpha}(p_N^{m+1})$. It follows that $\underline{N}^m$ is r-sound, so $r\in P_N m$. Contradiction!

Suppose, then, that $\pi_{0\alpha}(p_N^{m+1})<_* p_N^{m+1}$. Then as $\pi_{0\alpha}$ is an n-completion of $\pi_{0\alpha}|N^n:\underline{N}^n\to\underline{N}^n_\alpha$ $\underline{N}_\alpha^{mp}$ is $\pi_{0\alpha}(p_N^{m+1})$-sound, where $p=$ $=\langle\pi_{0\alpha}(p^1_N)\ldots\pi_{0\alpha}(p^m_N)\rangle$. But by the induction hypothesis $p=\langle p^1_{N_\alpha}\ldots p^m_{N_\alpha}\rangle$, so $\pi_{0\alpha}(p_N^{m+1})\in P_{N_\alpha}m$. Contradiction! □

Note that the above is a general property of n-completions.

Lemma 10.5. $\rho_N^{n+1}=\rho_{N_\alpha}^{n+1}$; $\pi_{0\alpha}(p_N^{n+1})=p_{N_\alpha}^{n+1}$.

Proof: The first claim is immediate from lemma 10.1.

Let $q=p_N^{n+1}\cap\kappa$, $r=p_N^{n+1}\setminus\kappa$. Let $\bar{r}=p_{N_\alpha}^{n+1}\setminus\kappa_\alpha$. ($\kappa=\kappa_0$)

Claim 1 $\pi_{0\alpha}(r)=\bar{r}$.

Proof: Suppose $\pi_{0\alpha}(r)>_*\bar{r}$. Then suppose $i<\omega$, $\vec{\gamma}<\kappa$ such that

$$\underline{N}_\alpha^n\models(\exists\vec{\gamma}<\pi_{0\alpha}(\kappa))(\exists\bar{r}<_*\pi_{0\alpha}(r))(\pi_{0\alpha}(r)=h(i,\langle\vec{\gamma},\bar{r}\rangle)) \qquad (10.3)$$

and so

$$\underline{N}^n\models(\exists\vec{\gamma}<\kappa)(\exists\hat{r}<r)(r=h(i,\langle\vec{\gamma},\hat{r}\rangle)).$$

So $\underline{N}^n$ is $\hat{r}\cup q$-sound. But $\hat{r}\cup q<_* r\cup q$; and $\hat{r}\cup q\in P_{N}n$. Contradiction!

If A is $\Sigma_1(\underline{N}^n)$ with parameter p_N^{n+1} and A' has the same definition over $\underline{N}^n$ with parameter $\pi_{0\alpha}(p_N^{n+1})$ then $A'\cap\kappa=A\cap\kappa$; suppose $A\cap\kappa\notin N^n$, then $A\cap\kappa\notin N_\alpha^n$. So $\pi_{0\alpha}(p_N^{n+1})\in P_{N_\alpha}n$, so $\pi_{0\alpha}(p_N^{n+1})\geq_* p_{N_\alpha}^{n+1}$. Thus $\pi_{0\alpha}(p_N^{n+1})\setminus\kappa_\alpha\geq_*$ $\geq_* p_{N_\alpha}^{n+1}\setminus\kappa_\alpha$, that is, $\pi_{0\alpha}(r)\geq_*\bar{r}$. □(Claim 1)

Claim 2 $q=p_{N_\alpha}^{n+1}\cap\kappa$.

Proof: As $\pi_{0\alpha}(p_N^{n+1})\in P_{N_\alpha}n$ and $\pi_{0\alpha}(p_N^{n+1})\setminus\kappa_\alpha=p_N^{n+1}\setminus\kappa_\alpha$ we know $\pi_{0\alpha}(p_N^{n+1})\cap\kappa_\alpha\geq_* p_{N_\alpha}^{n+1}\cap\kappa_\alpha$. Hence as $\pi_{0\alpha}(p_N^{n+1})\cap\kappa_\alpha=\pi_{0\alpha}(p_N^{n+1}\cap\kappa)=\pi_{0\alpha}(q)=q\subseteq\kappa$, therefore $p_{N_\alpha}^{n+1}\cap\kappa_\alpha\subseteq\kappa$. Thus $p_{N_\alpha}^{n+1}\cap\kappa_\alpha=\pi_{0\alpha}(p_{N_\alpha}^{n+1}\cap\kappa_\alpha)$; so $p_{N_\alpha}^{n+1}\cap\kappa_\alpha\geq_* q$.

□(Claim 2)

□

Corollary 10.6. $\underline{N}^{n+1}=N_\alpha^{n+1}$.

CORE MICE

Let $X=h_{N}n(\omega\rho_N^{n+1}\cup p_N^{n+1})$. Let $\sigma:\underline{M}'\cong\underline{N}^n|X$ where $\underline{M}'$ is transitive; so $\sigma:\underline{M}'\to_{\Sigma_1}\underline{N}^n$. Let $\underline{M}$, p be unique such that $\underline{M}'=\underline{M}^{np}$, $\underline{M}$ is p-sound and $\tilde{\sigma}:\underline{M}\to_{\Sigma_{n+1}}\underline{N}$ with $\tilde{\sigma}\supseteq\sigma$ and $\tilde{\sigma}(p)=\langle p_N^1\ldots p_N^n\rangle$.

Lemma 10.7. *$\underline{M}$ is a mouse.*

Proof: $\underline{M}$ is an acceptable premouse, and is transitive. By the construction $\sigma^{-1}(\kappa)\in\omega\rho_M^n$ so $\underline{M}$ is n-suitable.

Suppose $\rho_M^{n+1}>\rho_N^{n+1}$. Suppose A is $\Sigma_1(\underline{N}^n)$ in parameter p_N^{n+1} and $A\cap\omega\rho_N^{n+1}\notin N^n$. Let A' have the same Σ_1 definition over $\underline{M}'$ from $\sigma^{-1}(p_N^{n+1})$ Then for $\gamma\in\omega\rho_N^{n+1}$ $\gamma\in A\leftrightarrow\gamma\in A'$ since $\sigma(\gamma)=\gamma$. Now $A'\cap\omega\rho_M^{n+1}\in M$. Let $B=\sigma(A'\cap\omega\rho_M^{n+1})$; so $B\cap\omega\rho_N^{n+1}=A\cap\omega\rho_N^{n+1}\in N$; contradiction! Hence $\rho_M^{n+1}\leq\rho_N^{n+1}$ so $\omega\rho_M^{n+1}<\sigma^{-1}(\kappa)$. Hence $\underline{M}$ is critical and $n(\underline{M})=n$.

It remains to show that $\underline{M}$ is n-iterable. Let $\langle\langle\underline{M}_\alpha\rangle_{\alpha\in On},\langle\pi'_{\alpha\beta}\rangle_{\alpha\leq\beta\in On}\rangle$ be its n-iterated ultrapower. $\underline{M}$ is n-sound because it is transitive. So each $\underline{M}_\alpha$ is $\pi'_{O\alpha}(\langle p^1_M \ldots p^n_M\rangle)$-sound. And $\langle\langle \underline{M}'_\alpha\rangle_{\alpha\in On}, \langle \pi'_{\alpha\beta}\rangle_{\alpha\leq\beta\in On}\rangle$, where $\underline{M}'_\alpha = \underline{M}_\alpha^{n\pi'_{O\alpha}(\langle p^1_M\ldots p^n_M\rangle)}$, is an iterated ultrapower of $\underline{M}^n$. By lemma 10.4 $\sigma:\underline{M}^n \to_{\Sigma_1} \underline{N}^n$; so by lemma 8.3 there are maps $\sigma_\alpha:\underline{M}'_\alpha \to_{\Sigma_1} \underline{N}^n_\alpha$ such that $\sigma_\beta\pi'_{\alpha\beta}|M' = \pi_{\alpha\beta}\sigma_\alpha$, when $\alpha\leq\beta$; and $\sigma_O=\sigma$. Let $\tilde{\sigma}_\alpha:\underline{M}_\alpha \to_{\Sigma_{n+1}} \underline{N}_\alpha$ be the n-completion of σ_α. $\underline{N}_\alpha$ is well-founded, so $\underline{M}_\alpha$ is. □

<u>Lemma 10.8</u>. *Suppose $\underline{Q}$ and $\underline{Q}'$ are mice. Suppose $n=n(\underline{Q})$ and $n'=n(\underline{Q}')$. Let $\langle\langle \underline{Q}_\alpha\rangle_{\alpha\in On}, \langle\sigma_{\alpha\beta}\rangle_{\alpha\leq\beta\in On}\rangle$, $\langle\langle \underline{Q}'_\alpha\rangle_{\alpha\in On}, \langle\sigma'_{\alpha\beta}\rangle_{\alpha\leq\beta\in On}\rangle$ be the mouse iterations of $\underline{Q}$, $\underline{Q}'$ respectively. Let θ be regular and $\theta>\overline{Q},\overline{Q}'$; let $\overline{\underline{Q}}_\alpha = \underline{J}^{Q_\alpha}_{\rho^n_{Q_\alpha}}$, $\overline{\underline{Q}}' = \underline{J}^{Q'}_{\rho^{n'}_{Q'_\alpha}}$. Then Q_θ, Q'_θ are comparable and*

$$\overline{\underline{Q}}_\theta \neq \underline{J}^{\overline{Q}'}_{\beta}\theta \quad (\textit{some } \omega\beta\in\overline{Q}'_\theta) \Leftrightarrow \underline{Q}_\theta = \underline{J}^{Q'}_{\beta}\theta \quad (\textit{some } \omega\beta\in\underline{Q}').$$

Proof: $\overline{\underline{Q}}_\theta = H^Q_{\omega\rho^n_{Q_\theta}}$, so $P(\theta)\cap\overline{Q}_\theta = P(\theta)\cap Q_\theta$; similarly $P(\theta)\cap\overline{Q}'_\theta = P(\theta)\cap Q'_\theta$. By lemma 8.18 $\overline{\underline{Q}}_\theta$, $\overline{\underline{Q}}'_\theta$ are comparable. So $\underline{Q}_\theta$, $\underline{Q}'_\theta$ are comparable.

Suppose $\overline{\underline{Q}}_\theta = \underline{J}^{\overline{Q}'}_{\beta}\theta$, $\omega\overline{\beta}\in\overline{Q}'_\theta$. $\overline{\underline{Q}}_\theta \neq \overline{\underline{Q}}'_\theta$ so $\underline{Q}_\theta \neq \underline{Q}'_\theta$. But if $\underline{Q}'_\theta = \underline{J}^{Q}_{\beta}\theta$, $\omega\beta\in Q_\theta$ then $P(\theta)\cap Q'_\theta \neq P(\theta)\cap Q_\theta$ as $\omega\rho^{n'+1}_{Q'_\theta}<\theta$. Hence $P(\theta)\cap\overline{Q}'_\theta \neq P(\theta)\cap Q_\theta$. Contradiction!

Suppose $\underline{Q}_\theta = \underline{J}^{Q'}_{\beta}\theta$, $\omega\beta\in Q'_\theta$. Then by the last paragraph $\overline{\underline{Q}}'_\theta \neq \underline{J}^{\overline{Q}}_{\beta}\theta$, all $\omega\overline{\beta}\in\overline{Q}_\theta$. Suppose $\overline{\underline{Q}}'_\theta = \overline{\underline{Q}}_\theta$. Then $P(\theta)\cap Q_\theta \neq P(\theta)\cap Q'_\theta$ but $P(\theta)\cap\overline{Q}_\theta = P(\theta)\cap\overline{Q}'_\theta$. Contradiction! □

<u>Lemma 10.9</u>. *For some θ $\underline{M}_\theta = \underline{N}_\theta$.*

Proof: Let $\underline{M}'_\alpha = \underline{J}^{M_\alpha}_{\rho^n_M}$, $\underline{N}'_\alpha = \underline{J}^{N_\alpha}_{\rho^n_N}$. Let θ be regular, $\theta>\overline{\overline{N}}$. By lemma 10.8 either $\underline{M}_\theta\in\underline{N}_\theta$ or $\underline{M}_\theta=\underline{N}_\theta$ or $\underline{N}_\theta\in\underline{M}_\theta$. If $\underline{N}_\theta = \underline{J}^{M}_{\beta}\theta$, $\omega\beta\in M_\theta$, then $\underline{N}'_\theta = \underline{J}^{M'}_{\beta}\theta$, say; but then $\pi_{O\alpha}\sigma:\underline{M}'_O \to_{\Sigma_0} \underline{M}'$ is not cofinal, contradicting corollary 8.20.

Suppose $\underline{M}_\theta = \underline{J}^{N}_{\beta}\theta$, then we saw in the proof of lemma 10.7 that there is $A\in\Sigma_{n+1}(\underline{M})$ such that $A\cap\omega\rho^{n+1}_N \notin N$. But then $A\cap\sigma^{-1}(\kappa)\in\Sigma_{n+1}(\underline{M}_\theta)$ so $A\cap\sigma^{-1}(\kappa)\in N$. But $\sigma^{-1}(\kappa)\geq\omega\rho^{n+1}_N$. Contradiction! □

Fix θ for a while.

<u>Corollary 10.10</u>. $\rho^{n+1}_N = \rho^{n+1}_M$.

Proof: $\rho^{n+1}_N = \rho^{n+1}_{N_\theta} = \rho^{n+1}_{M_\theta} = \rho^{n+1}_M$. □

<u>Lemma 10.11</u>. $\sigma(p^{n+1}_M) = p^{n+1}_N$.

Proof: By the argument of lemma 10.7 $\sigma^{-1}(p_N^{n+1})\in P_{M^n}$ so $p_M^{n+1}\leq_*\sigma^{-1}(p_N^{n+1})$, i.e. $\sigma(p_M^{n+1})\leq_* p_N^{n+1}$. But by lemma 8.19

$$\pi_{0\theta}\sigma(p_M^{n+1})\geq_*\pi'_{0\theta}(p_M^{n+1}). \tag{10.4}$$

But $\pi'_{0\theta}(p_M^{n+1})=p_{M_\theta}^{n+1}$ by lemma 10.5 so

$$\pi_{0\theta}\sigma(p_M^{n+1})\geq_* p_{M_\theta}^{n+1}=\pi_{0\theta}(p_N^{n+1}). \tag{10.5}$$

Hence $\sigma(p_M^{n+1})\geq_* p_N^{n+1}$. □

Corollary 10.12. $A_M^{n+1}=A_N^{n+1}$.

Corollary 10.13. $\underline{M}^{n+1}=\underline{N}^{n+1}$.

Lemma 10.14. *$\underline{M}$ is n+1-sound.*

Proof: We have to show $\underline{M}^n\models V=h(\omega\rho_M^{n+1}\cup p_M^{n+1})$. Suppose $x\in M^n$; then $\sigma(x)\in X$, $\underline{N}^n\models\sigma(x)=h(i,\langle\vec{\gamma},p_N^{n+1}\rangle)$, say. So $\underline{M}^n\models x=h(i,\langle\vec{\gamma},p_M^{n+1}\rangle)$. $(\vec{\gamma}\in(\omega\rho_N^{n+1})^{<\omega}, i\in\omega)$. □

Lemma 10.15. $\pi_{0\theta}\tilde{\sigma}=\pi'_{0\theta}$.

Proof: $\underline{M}$ is n+1-sound and $\tilde{\sigma}(p_M^m)=p_N^m$ $(m\leq n+1)$. □

Lemma 10.16. *$\underline{N}=\underline{M}_\alpha$, for some α.*

Proof: Let $\kappa'=\sigma^{-1}(\kappa)$. Let $\kappa'_\alpha,\kappa_\alpha$ be the critical points of $\underline{M}_\alpha$, $\underline{N}_\alpha$ respectively; so $\kappa'=\kappa'_0$. Let α be least such that $\kappa'_\alpha\geq\kappa$. Suppose $\kappa'_\alpha>\kappa$. Suppose

$$\underline{M}_\alpha^n\models\kappa=\pi'_{0\alpha}(f)(\kappa'_{\alpha_1}\ldots\kappa'_{\alpha_n}) \tag{10.6}$$

where $f\in M^n$, $f:\kappa'^n\to\kappa'$ and $\alpha_1<\ldots<\alpha_n<\alpha$. Then

$$\underline{M}_\theta^n\models\kappa=\pi'_{0\theta}(f)(\kappa'_{\alpha_1}\ldots\kappa'_{\alpha_n}) \tag{10.7}$$

so by lemma 10.15

$$\underline{N}_\theta^n\models\kappa=\pi_{0\theta}\sigma(f)(\kappa'_{\alpha_1}\ldots\kappa'_{\alpha_n}). \tag{10.8}$$

But $\kappa'_{\alpha_1}\ldots\kappa'_{\alpha_n}<\kappa$ so

$$\underline{N}_\theta^n\models\kappa=\pi_{0\theta}(\sigma(f)(\kappa'_{\alpha_1}\ldots\kappa'_{\alpha_n})) \tag{10.9}$$

so $\kappa\in\mathrm{rng}(\pi_{0\theta})$. If $\kappa=\pi_{0\theta}(\delta)$ and $\delta<\kappa$ then $\kappa=\delta<\kappa$. But if $\kappa=\pi_{0\theta}(\delta)$ and $\delta\geq\kappa$ then $\kappa=\pi_{1\theta}\pi_{01}(\delta)\geq\pi_{01}(\delta)\geq\pi_{01}(\kappa)>\kappa$. Contradiction!

Hence $\kappa'_\alpha=\kappa$. Define $\tau:M_\alpha^n\to N^n$ by

$$\tau(h_{M_\alpha^n}(i,\langle\vec{\gamma},p_{M_\alpha}^{n+1}\rangle))=h_{N^n}(i,\langle\vec{\gamma},p_N^{n+1}\rangle) \tag{10.10}$$

where $i\in\omega$, $\vec{\gamma}\in\kappa$. Since

$$\underline{M}_\alpha^n \models \phi(h_{M_\alpha^n}(i,\langle\vec{\gamma},p_M^{n+1}\rangle)) \leftrightarrow \underline{M}_\theta^n \models \phi(h_{M_\theta^n}(i,\langle\vec{\gamma},p_M^{n+1}\rangle)) \leftrightarrow \underline{N}^n \models \phi(h_{N^n}(i,\langle\vec{\gamma},p_N^{n+1}\rangle)) \quad (10.11)$$

where ϕ is Σ_1, therefore τ is well-defined and $\tau:M_\alpha^n \to_{\Sigma_1} N^n$. But $\underline{N} \models V=h(\kappa \cup p_N^{n+1})$ so τ is surjective. Hence $\tau:\underline{M}_\alpha^n \cong \underline{N}^n$, so $\underline{M}_\alpha^n=\underline{N}^n$, so $\underline{M}_\alpha=\underline{N}$. □

Lemma 10.17. $\tilde{\sigma}=\pi'_{0\alpha}$.

Proof: $\pi'_{0\alpha}(p_M^m)=p_{M_\alpha}^m=p_N^m$ for $m\leq n+1$. □

Definition 10.18. *$\underline{M}$ is called the core of $\underline{N}$. $\underline{M}$ is said to be a core mouse if and only if it is the core of some $\underline{N}$.*

Note that if $\underline{M}$ is a core mouse and $\underline{N}$ is a mouse iterate of $\underline{M}$ then $\underline{M}$ is the core of $\underline{N}$.

Lemma 10.19. *Suppose $\underline{N}$ is a mouse with core $\underline{M}$. Let $\langle\langle \underline{M}_\alpha\rangle_{\alpha\in On},\langle\pi_{\alpha\beta}\rangle_{\alpha\leq\beta\in On}\rangle$ be the mouse iteration of $\underline{M}$ and suppose $\underline{N}=\underline{M}_\alpha$. Let $n=n(\underline{M})$. Let κ_α be the critical point of $\underline{M}_\alpha$. Then*

$$N^n=h_{N^n}(\omega\rho_N^{n+1}\cup\{\kappa_\gamma:\gamma<\alpha\}\cup p_N^{n+1}).$$

Proof: By lemma 7.1 $N^n=h_{N^n}(\pi_{0\alpha}"M^n\cup\{\kappa_\gamma:\gamma<\alpha\})$. But

$$\pi_{0\alpha}"M^n=\pi_{0\alpha}"h_{M^n}(\omega\rho_N^{n+1}\cup p_M^{n+1}) \quad (10.12)$$
$$=h_{N^n}(\omega\rho_N^{n+1}\cup p_N^{n+1}) \quad \square$$

Definition 10.20. *Suppose $\underline{N}$ is a mouse with core $\underline{M}$, and suppose the mouse iteration of $\underline{M}$ is $\langle\langle \underline{M}_\alpha\rangle_{\alpha\in On},\langle\pi_{\alpha\beta}\rangle_{\alpha\leq\beta\in On}\rangle$ and $\underline{N}=\underline{M}_\alpha$. Then C_N denotes $\{\kappa_\gamma:\gamma<\alpha\}$ where κ_γ is the critical point of $\underline{M}_\gamma$.*

Lemma 10.21. *C_N are Σ_1 indiscernibles for $\langle \underline{N}^n,p_N^{n+1},\gamma\rangle_{\gamma\in\omega\rho_N^{n+1}}$ where $n=n(\underline{N})$.*

Proof: Immediate by lemma 7.3 and the fact that $\mathrm{rng}(\pi_{0\alpha})\supseteq\omega\rho_N^{n+1}\cup p_N^{n+1}$. □

Corollary 10.22. *C_N are Σ_{n+1} indiscernibles for $\langle \underline{N},p_N^1\ldots p_N^{n+1},\gamma\rangle_{\gamma\in\omega\rho_N^{n+1}}$, $n=n(\underline{N})$.*

Lemma 10.23. *$\kappa\in C_N \leftrightarrow \kappa\notin h_{N^n}(\kappa\cup p_N^{n+1}) \quad \kappa>\omega\rho_N^{n+1} \quad (n=n(\underline{N}))$.*

Proof: Suppose $\underline{M}$ is the core of $\underline{N}$ with mouse iteration $\langle\langle \underline{M}_\alpha\rangle_{\alpha\in On},\langle\pi_{\alpha\beta}\rangle_{\alpha\leq\beta\in On}\rangle$, κ_γ the critical point of $\underline{M}_\gamma$. So $C_N=\{\kappa_\gamma:\gamma<\alpha\}$ if $\underline{N}=\underline{M}_\alpha$.

Suppose $\kappa=\kappa_\gamma$. If $\kappa_\gamma=h_{N^n}(i,\langle\vec{\delta},p_N^{n+1}\rangle)$ $(i\in\omega,\vec{\delta}\in\kappa_\gamma)$ then $\kappa_\gamma=$ $=\pi_{\gamma\alpha}(h_{M_\gamma^n}(i,\langle\vec{\delta},p_{M_\gamma}^{n+1}\rangle))$ so $\kappa_\gamma\in\mathrm{rng}(\pi_{\gamma\alpha})$, which is impossible; so $\kappa_\gamma\notin h_{N^n}(\kappa_\gamma\cup p_N^{n+1})$. Obviously $\kappa_\gamma>\omega\rho_N^{n+1}$.

Conversely suppose $\kappa\notin C_N$. Let γ be least such that $\kappa_\gamma>\kappa$. Suppose $\kappa>\omega\rho_N^{n+1}$. Then there are $\vec{\delta}<\kappa$ such that

$$\underline{M}_\gamma^n\models\kappa=h(i,\langle\vec{\delta},p_{M_\gamma}^{n+1}\rangle) \qquad (10.13)$$

so

$$\underline{N}^n\models\kappa=h(i,\langle\vec{\delta},p_N^{n+1}\rangle). \qquad (10.14)$$

Thus $\kappa\in h_{N^n}(\kappa\cup p_N^{n+1})$. □

Corollary 10.24. *$C_N\in\Pi_1(\underline{N}^n)$.*

This fact will be of importance in chapter 11.

The fact that the core of a mouse is itself a mouse may seem strange. We are accustomed to measurability being a large cardinal property; yet if $\underline{M}$ is a core mouse with κ critical, $n=n(\underline{M})$ and $\omega\rho_M^{n+1}<\kappa$ then $\underline{M}\models$"κ is inaccessible"; but there is a $\Sigma_{n+1}(\underline{M})$ map of $\omega\rho_M^{n+1}$ onto κ. This is not a contradiction, of course; although $\omega\rho_M^{n+1}=\kappa$ implies that $\underline{M}$ is a core mouse, the converse is not true. Indeed, we shall later on give an example of a core mouse with $\rho_M=1$.

This implies that $\underline{M}$ cannot be "continued", i.e. there is no premouse $\underline{N}$ with $\underline{M}=\underline{J}_\alpha^N$ for some $\omega\alpha\in N$. This does not only apply to core mice: by lemma 10.19 there must be a lot of κ_α for κ not to be collapsed definably. In fact, continuability is a very strong property: we shall see that if $\underline{M}=\underline{J}_\alpha^N$, $\omega\alpha\in N$, $\underline{N}$ a premouse with measure U, then $C_M\in U$. This topic is taken up again in the next chapter.

THE STRUCTURES N^m, $m>n$

To conclude the fine-structural analysis of mice we must look at the projectum below κ. Let $\underline{N}$ be a mouse with mouse iteration $\langle\langle\underline{N}_\alpha\rangle_{\alpha\in On},\langle\pi_{\alpha\beta}\rangle_{\alpha\leq\beta\in On}\rangle$; $n=n(\underline{N})$.

Lemma 10.25. *For $m\geq 0$ $\rho_N^{n+m+1}=\rho_{N^{n+1}}^m$.*

Proof: If m=0 this is trivial. $\rho_N^{n+m+2}=\bigcap\{\rho_{N^{n+m+1}}p: p\in PA_{n+m+1}^N\}$. Now $\rho_{N^{n+m+1}}p=\rho_{(N^n)^{m+1}}p'$ where p' are the last m+1 members of p. This follows from lemma 9.6. So $\rho_N^{n+m+2}=\bigcap\{\rho_{(N^n)^{m+1}}p': p'\in PA_{m+1}^{N^n}\}$. But $\rho_{N^{n+1}}^{m+1}=\bigcap\{\rho_{(N^{n+1})^m}p'': p''\in PA_m^{N^{n+1}}\}$. Hence it suffices to show $\rho_{(N^{n+1})^m}p''=\rho_{(N^n)^{m+1}}p'$, where p" are the last m members of p'. This would clearly be true if $A_{N^n}^p$ were rudimentary in $\underline{N}^{n+1}$ for all $p\in N^n$. But $A=A_{N^n}^p$ is $\Sigma_1(\underline{N}^n)$, and $\underline{N}^n$ is an iterate of $\underline{M}^n$, so A is $\Sigma_1(\underline{M}^n)$, and therefore $\Sigma_1(\underline{M}^n)$ with parameters from $p_M^{n+1} \cup \omega\rho_M^{n+1}$, and hence $\Sigma_1(\underline{N}^n)$ with parameters from $p_N^{n+1} \cup \omega\rho_M^{n+1}$; but then it is rudimentary in $\underline{N}^{n+1}$. □

Corollary 10.26. *Let $\underline{N}$ be a mouse with $n=n(\underline{N})$ and mouse iteration $\langle\langle \underline{N}_\alpha \rangle_{\alpha\in On}, \langle \pi_{\alpha\beta} \rangle_{\alpha\leq\beta\in On}\rangle$. Then for $m>n+1$*

(i) $\rho_N^m = \rho_{N_\alpha}^m$.

(ii) $p_N^m = p_{N_\alpha}^m$.

(iii) $A_N^m = A_{N_\alpha}^m$.

(iv) $\underline{N}^m = \underline{N}_\alpha^m$.

Proof: By corollary 10.13 $\underline{N}^{n+1} = \underline{N}_\alpha^{n+1}$. □

Lemma 10.27. *If $\underline{N}$ is a mouse and $n=n(\underline{N})$ then $\underline{N}^{n+1}$ is k-sound for all k, and $\rho_N^{n+k+1} = \rho_N^{n+k}$.*

Proof: By induction on k. If k=0 it is trivial. Suppose m=n+k+1 and suppose $\underline{N}^{n+1}$ is k-sound. Then

$$\rho_N^m = \rho_N^{n+k+1} = \rho_{N^{n+1}}^{k+1} \quad \text{(lemma 9.6)} \qquad (10.15)$$
$$= \rho_N^{n+k+2} \quad \text{(lemma 10.25)}.$$

It is sufficient to show $\underline{N}^m$ 1-sound. So let $X = h_{N^m}(\omega\rho_N^{m+1} \cup p_N^{m+1})$. Let $\sigma: \underline{M}' \cong \underline{N}^m | X$ where M' is transitive. So $\sigma: \underline{M}' \to_{\Sigma_1} \underline{N}^m$. Let $\underline{M}$ be such that $\underline{M}' = \underline{M}^{mp}$, $\underline{M}$ is p-sound, $\tilde{\sigma}: \underline{M} \to_{\Sigma_{m+1}} \underline{N}$, $\tilde{\sigma} \supseteq \sigma$ and $\tilde{\sigma}(p) = \langle p_N^1 \ldots p_N^m \rangle$. Then just as in lemma 10.7, $\underline{M}$ is a mouse. Furthermore $p = \langle p_M^1 \ldots p_M^m \rangle$ so $\underline{M}$ is m-sound, and in particular a core mouse. Let the mouse iterations of $\underline{M}, \underline{N}$ be $\langle\langle \underline{M}_\alpha \rangle_{\alpha\in On}, \langle \pi'_{\alpha\beta} \rangle_{\alpha\leq\beta\in On}\rangle, \langle\langle \underline{N}_\alpha \rangle_{\alpha\in On}, \langle \pi_{\alpha\beta} \rangle_{\alpha\leq\beta\in On}\rangle$ respectively. Just as in lemma 10.9 there is θ such that $\underline{M}_\theta = \underline{N}_\theta$. So $\underline{M} = \text{core}(\underline{M}_\theta) = \text{core}(\underline{N}_\theta) = \text{core}(\underline{N})$. Say $\underline{N} = \underline{M}_\alpha$. Then by corollary 10.26 $\underline{N}^m = \underline{M}'$. But $\underline{M}'$ is 1-sound by the argument of lemma 10.14 so $\underline{N}^m$ is 1-sound. □

Corollary 10.28. *If $\underline{N}$ is a core mouse then $\underline{N}$ is n-sound for all $n<\omega$.*

n-ITERABILITY

We conclude this chapter with some criteria for n-iterability.

Definition 10.29. *E_n is defined inductively on $N^{np_1 \ldots p_n}$ as follows*

$E_0(x,y) \leftrightarrow x \in y$

$E_{m+1}(x,y) \leftrightarrow x,y \in \omega \times (\omega\rho_N^{m+1}) \wedge \langle k, \langle x,y \rangle\rangle \in A^{m+1 p_1 \ldots p_{m+1}}$

where s(k) is the formula "$x = \langle i, \bar{x} \rangle \wedge y = \langle j, \bar{y} \rangle \wedge E_m(h(i, \langle \bar{x}, p_{m+1} \rangle), h(j, \langle \bar{y}, p_{m+1} \rangle))$".

Then E_n is rudimentary in N^{np}. Clearly

Lemma 10.30. *Suppose $\underline{N}$ is p-sound, $p \in PA_n^N$. Then $\underline{N}$ is well-founded if and only if $(E_n)_{N^{np}}$ is.*

Most of the results of chapter 8 can now bw reproduced with E_n in place of $\in$. Lemma 8.3 presents a problem though; the natural approach is to build the maps σ_α, as there, and take n-completions. However we do not know that $\sigma_\alpha : \underline{N}_\alpha \to_{\Sigma_0} \underline{M}_{f(\alpha)}$ cofinally.

We do have the following, a special case of which was used in lemma 10.7. The proof is the same as the proof of n-iterability in lemma 10.7.

Lemma 10.31. *Suppose $\sigma : M \to N$ is the n-completion of $\sigma | M^{np}$ to $\underline{M}$ and $\underline{M}$ is p-sound. If $\underline{N}$ is n-iterable then $\underline{M}$ is.*

More useful, though, is

Lemma 10.32. *Suppose $\underline{N}$ is an iterable premouse and for some $\omega\alpha \in N$, letting $\underline{N}' = J_\beta^N$ $\sigma : M \to N'$ is the n-completion of $\sigma | M^{np'}$ to $\underline{M}$ ($\underline{M}$ p'-sound). Then $\underline{M}$ is n-iterable provided $\underline{M}$ is n-suitable (that is, provided $\underline{M}$ is acceptable and the critical point of $\underline{M}$ is in $\omega\rho_M^n$).*

Proof: Let $\langle \langle \underline{N}_\alpha \rangle_{\alpha \in On}, \langle \pi_{\alpha\beta} \rangle_{\alpha \leq \beta \in On} \rangle$ be the iterated ultrapower of $\underline{N}$ and let $\langle \langle \underline{M}_\alpha \rangle_{\alpha \in On}, \langle \pi'_{\alpha\beta} \rangle_{\alpha \leq \beta \in On} \rangle$ be an n-iteration of $\underline{M}$. Let κ'_α be the critical point of $\underline{M}'_\alpha$ and let κ_α be the critical point of $\underline{N}_\alpha$. Let U'_α, U_α be the measures of $\underline{M}_\alpha, \underline{N}_\alpha$ respectively. We shall define $\underline{N}'_\alpha = J_{\beta_\alpha}^{N_\alpha}$ and maps $\sigma_\alpha : \underline{M}_\alpha^{np'_\alpha} \to_{\Sigma_0} \underline{N}_\alpha^{np_\alpha}$ such that

(i) $\sigma_\beta \pi'_{\alpha\beta} = \pi_{\alpha\beta}\sigma_\alpha \quad (\alpha \leq \beta)$

(ii) $\sigma_0 = \sigma$

where $p'_\alpha = \pi'_{0\alpha}(p')$, $p_\alpha = \pi_{0\alpha}\sigma(p')$.

β_α is to be $\pi_{0\alpha}(\beta)$ so $\underline{N}'_\alpha = \pi_{0\alpha}(\underline{N}')$. The maps are defined by induction. σ_0 is σ.

Suppose σ_α is defined. Then define $\sigma_{\alpha+1} : M_{\alpha+1}^{np'_{\alpha+1}} \to N_{\alpha+1}$ by

$\sigma_{\alpha+1}(\pi'_{\alpha\alpha+1}(f)(\kappa'_\alpha)) = \pi_{\alpha\alpha+1}(\sigma_\alpha(f))(\kappa_\alpha) \quad (f \in M_\alpha^{np'_\alpha}, f : \kappa'_\alpha \to M_\alpha^{np'_\alpha})$.

Claim 1 $\sigma_{\alpha+1} : M_{\alpha+1}^{np'_{\alpha+1}} \to H_\rho^{N'_{\alpha+1}}$, where $\rho = \omega\rho_{N'_{\alpha+1}}^n$.

Proof: Since $\pi_{\alpha\alpha+1} : N' \to_e N'_{\alpha+1}$, $N'_{\alpha+1}$ is acceptable. Given $f \in M_\alpha^{np'_\alpha}$, $\sigma_\alpha(f) \in N_\alpha'^{p'_\alpha}$ by induction hypothesis, so for all $\xi < \kappa_\alpha$ $\sigma_\alpha(f)(\xi) \in N'_\alpha$ and $\underline{N}'_\alpha \models \overline{TC(\sigma_\alpha(f)(\xi))} < \omega\rho_{N'_\alpha}^n$. But these statements are $\Sigma_0(\underline{N}_\alpha)$ so

$\pi_{\alpha\alpha+1}(\sigma_\alpha(f))(\kappa_\alpha)\in N'_{\alpha+1}$ and $\underline{N}'_{\alpha+1}\models \overline{TC(\pi_{\alpha\alpha+1}(\sigma_\alpha(f))(\kappa_\alpha))}<\rho$, which proves the claim. □(Claim 1)

Claim 2 $\sigma_{\alpha+1}:\underline{M}^{np'_{\alpha+1}}_{\alpha+1}\to_{\Sigma_0}\underline{N}'^{np_{\alpha+1}}_{\alpha+1}$.

Proof: Suppose ϕ is a Σ_0 formula; for simplicity, of one argument.

$$\begin{aligned}\underline{M}^{np'_{\alpha+1}}_{\alpha+1}\models\phi(\pi'_{\alpha\alpha+1}(f)(\kappa'_\alpha)) &\leftrightarrow \underline{M}^{np'_\alpha}_\alpha\models\{\xi:\phi(f(\xi))\}\in U'_\alpha \\ &\leftrightarrow \underline{N}^{np_\alpha}_\alpha\models\{\xi:\phi(\sigma_\alpha(f)(\xi))\}\in U_\alpha \\ &\leftrightarrow \underline{N}_\alpha\models\{\xi:\underline{N}^{np_\alpha}_\alpha\models\phi(\sigma_\alpha(f)(\xi))\}\in U_\alpha \\ &\leftrightarrow \underline{N}_{\alpha+1}\models\text{"}\pi_{\alpha\alpha+1}(\underline{N}^{np_\alpha}_\alpha)\models\phi(\pi_{\alpha\alpha+1}(\sigma_\alpha(f))(\kappa_\alpha))\text{"} \\ &\leftrightarrow \pi_{\alpha\alpha+1}(\underline{N}^{np_\alpha}_\alpha)\models\phi(\sigma_{\alpha+1}(\pi'_{\alpha\alpha+1}(f)(\kappa'_\alpha))).\end{aligned}\tag{10.16}$$

But $\pi_{\alpha\alpha+1}(\underline{N}'^{np_\alpha}_\alpha)=\underline{N}'^{np_{\alpha+1}}_{\alpha+1}$. □(Claim 2)

$\sigma_{\alpha+1}\pi'_{\alpha\alpha+1}=\pi_{\alpha\alpha+1}\sigma_\alpha$, so (i) holds.

In the limit case let $\sigma_\lambda(\pi'_{\alpha\lambda}(x))=\pi_{\alpha\lambda}\sigma_\alpha(x)$. Details are left to the reader. Since E_n is rudimentary in $\underline{N}^{np}$

$$\underline{M}^{np'_\alpha}_\alpha\models E_n(x,y)\leftrightarrow\underline{N}'^{np_\alpha}_\alpha\models E_n(\sigma_\alpha(x),\sigma_\alpha(y)).\tag{10.17}$$

But $\underline{N}'_\alpha$ is well-founded so $\underline{M}_\alpha$ is. □

Corollary 10.33. *If $\underline{N}$ is an iterable premouse, $\omega\beta\in N$, $\underline{M}=\underline{J}^N_\beta$, $\underline{M}$ acceptable, κ the critical point of $\underline{N}$ and $\omega\rho^n_M>\kappa$ then $\underline{M}$ is n-iterable. If in addition $\omega\rho^{n+1}_M\le\kappa$ and $\underline{M}$ is of the form $\underline{J}^U_\alpha$ then $\underline{M}$ is a mouse.*

Lemma 10.34. *If $\underline{N}$ is a transitive n-suitable premouse and its measure is countably complete then $\underline{N}$ is n-iterable.*

Proof: As lemma 8.9; apply the argument to E_n rather than $\in$. □

Definition 10.35. *E^+_n denotes the uniformly Σ_1 relation of lemma 8.11 such that whenever $\langle\langle\underline{N}_\alpha\rangle_{\alpha\in On},\langle\pi_{\alpha\beta}\rangle_{\alpha\le\beta\in On}\rangle$ is an iterated ultrapower of $\underline{N}$ then $\{\langle x,y\rangle:\underline{N}_{\omega_1}\models E_n(x,y)\}$ is well-founded if and only if $\{\langle x,y\rangle:\underline{N}\models E^+_n(x,y)\}$ is $(\omega_1\in N)$*

Lemma 10.36. *Suppose $\underline{N}$ is an n-suitable p-sound premouse, $p\in PA^N_n,\omega_1\in N^{np}$. Then $\underline{N}$ is n-iterable if and only if $(E^+_n)_{N^{np}}$ is well-founded.*

Definition 10.37. *E^m_n is defined by induction on $m\ge n$ by*

$E^n_n=E^+_n$

$E^{m+1}_n(x,y)\leftrightarrow x,y\in\omega\times(\omega\rho^{m+1}_N)^{<\omega}\wedge\langle k'\langle x,y\rangle\rangle\in A^{m+1p_1\ldots p_{m+1}}$ where $s(k')$ is the formula "$x=\langle i,\overline{x}\rangle\wedge y=\langle j,\overline{y}\rangle\wedge E^m_n(h(i,\langle\overline{x},p_{m+1}\rangle),h(j,\langle\overline{y},p_{m+1}\rangle))$".

Lemma 10.38. *Suppose $\underline{N}$ is $\langle p_1\ldots p_m\rangle$-sound and n-suitable, $m\ge n,\omega_1\in N^n$. Then*

$\underline{N}$ is n-iterable if and only if $(E^m_n)_N^{\vec{p}}$ is well -founded.

Exercises

1. Show that if $\underline{N}$ is a mouse then the following are equivalent:
   (i) for some $\underline{M}$, $\underline{N}=\text{core}(\underline{M})$.
   (ii) $\underline{N}=\text{core}(\underline{N})$
   (iii) $\underline{N}$ is $n(\underline{N})+1$-sound
   (iv) $C_N=\phi$.

2. Show that if $\underline{M}$ is a mouse with κ critical and mouse iteration $\langle\langle \underline{M}_\alpha \rangle_{\alpha\in On}, \langle \pi_{\alpha\beta} \rangle_{\alpha\leq\beta\in On}\rangle$ then $\Sigma_\omega(\underline{M})\cap P(\kappa)\subseteq\Sigma_\omega(\underline{M}_\alpha)\cap P(\kappa)$.

3. Suppose $\underline{M}^+=\underline{J}^U_{\alpha+1}$ is a premouse and $\underline{M}=\underline{J}^U_\alpha$ is a mouse with $n(\underline{J}^U_\alpha)=0$. Let

$\eta:\underline{M}^+\to_U\underline{N}$

$\eta':\underline{M}\to_U\underline{N}'$.

Show that in general $\underline{N}'\neq\eta(\underline{M})$. Do $\underline{N}'$, $\eta(\underline{M})$ have a common iterate?

4. Suppose $\underline{N}$ is a mouse. Show that if $A\in\Sigma_n(\underline{N})\cap P(\rho)$ and $A\notin N$ then $\omega\rho^n_N\leq\rho$.

11: ACCEPTABILITY REVISITED

This chapter will prove that iterable premice are acceptable. The proof will be inductive; the reader can check immediately that the limit case is trivial: that is, if $\underline{J}^U_\lambda$ is an iterable premouse and λ is a limit, and if $\underline{J}^U_\alpha$ is acceptable for all $\alpha<\lambda$ then $\underline{J}^U_\lambda$ is acceptable.

We shall be assuming that $\underline{N}^+=\underline{J}^U_{\alpha+1}$ is an iterable premouse with κ critical. We shall be able to assume that $\underline{N}=\underline{J}^U_\alpha$ is acceptable. This implies by corollary 10.33 that either $\omega\rho^n_N>\kappa$ for all n - and this case will present few difficulties - or $\underline{N}$ is a mouse. The two cases $\omega\rho^n_N=\kappa$ and $\omega\rho^n_N<\kappa$ are handled separately; there seems to be an overlap between them, but there is not.

THE CASE $\omega\rho^{n+1}_N<\kappa$

Let $\overline{\underline{N}}$=core($\underline{N}$); let n=n($\underline{N}$); and let $\langle\langle\overline{\underline{N}}_\alpha\rangle_{\alpha\in On},\langle\pi_{\alpha\beta}\rangle_{\alpha\leq\beta\in On}\rangle$ be the mouse iteration of $\overline{\underline{N}}$. Suppose $\underline{N}=\overline{\underline{N}}_\lambda$; let κ_α be the critical point of $\overline{\underline{N}}_\alpha$; so $C_N=\{\kappa_\gamma:\gamma<\lambda\}$ and $\kappa=\kappa_\lambda$. Let $\overline{\kappa}=\kappa_0$. Recall that $C_N\in\Pi_1(\underline{N}^n)$, so $C_N\in N^+$.

Lemma 11.1. *$C_N\in U$.*

Proof: By lemma 10.23 $C_N=\{\kappa'<\kappa:\kappa'>\omega\rho^{n+1}_N \wedge \kappa'\notin h_{N^n}(\kappa'\cup p^{n+1}_N)\}$. Suppose $\kappa\setminus C_N\in U$. Define $f:\kappa\to\kappa$ by

$$f(\gamma)=\text{the least }\gamma'<\gamma\text{ such that }\gamma\in h_{N^n}(\gamma'\cup p^{n+1}_N) \qquad (11.1)$$
$$=0 \text{ if there is none.}$$

Clearly $\{\xi:\xi\in h_{N^n}(f(\xi)\cup p^{n+1}_N)\}\in U$; and $f\in N^+$. By chapter 5 exercise 6 there is β such that $\{\xi:f(\xi)=\beta\}\in U$. Hence $Y=\{\xi:\xi\in h_{N^n}(\beta\cup p^{n+1}_N)\}\in U$. But then $Y\in N^+$, $\overline{\overline{Y}}=\overline{\overline{\beta}}$ in N^+ and Y is cofinal in κ. Contradiction! □

This makes precise our remark in chapter 10 about continuability. Let $C^m_N=\{\kappa_\gamma:\gamma<\lambda \wedge \gamma \text{ is m-good}\}$. Then $C^m_N\in N^+$.

Lemma 11.2. *For each $m\in\omega$ $C^m_N\in U$.*

Proof: By induction on m. If m=0 use lemma 11.1. Suppose $C^m_N\in U$; $C^{m+1}_N=\{\kappa'\in C^m_N: \kappa'=\sup(\kappa'\cap C^m_N)\}$. Let $f:\kappa\to\kappa$ be defined by $f(\xi)=\sup(\xi\cap C^m_N)$;

so $f\in N^+$. $f(\xi)\leq\xi$, for all ξ. Suppose $\{\xi:f(\xi)<\xi\}\in U$. Then by chapter 5 exercise 6 there is β such that $\{\xi:f(\xi)=\beta\}\in U$. So $\{\xi:\beta=\sup(\xi\cap C_N^m)\}\in U$; hence C_N^m is bounded in κ. But $C_N^m\in U$!

So $\{\xi:f(\xi)=\xi\}\in U$; that is, $C_N^{m+1}\in U$. □

Corollary 11.3. *λ is m-good for all $m<\omega$.*

So by corollary 7.20

Corollary 11.4. *C_N^m are Σ_{m+1} indiscernibles for $\langle \underline{N}^n,\pi_{0\lambda}(x)\rangle_{x\in N^n}$.*

Corollary 11.5. *C_N^m are Σ_{m+1} indiscernibles for $\langle \underline{N}^n,p_N^{n+1},\gamma\rangle_{\gamma\in\omega\rho_N^{n+1}}$.*

Corollary 11.6. *C_N^m are Σ_{n+m+1} indiscernibles for $\langle \underline{N},p_N^1\ldots p_N^{n+1},\gamma\rangle_{\gamma\in\omega\rho_N^{n+1}}$.*

Lemma 11.7. *If $X\in P(\kappa)\cap N^+$ and $X\in\Sigma_{n+m+1}(\underline{N})$ then*
$X\in U \Leftrightarrow \exists\gamma<\kappa(C_N^m\setminus\gamma\subseteq X)$

Proof: If $C_N^m\setminus\gamma\subseteq X$ then $X\in U$ by lemma 11.2.

Conversely suppose $C_N^m\setminus\gamma\not\subseteq X$ for all $\gamma<\kappa$. Suppose $X=\{\gamma<\kappa:\phi(\gamma,p)\}$ where ϕ is Σ_{n+m+1}. Then for some Σ_{n+1} term t $p=t^N(\kappa_{i_1}\ldots\kappa_{i_p},p_N^1\ldots$ $\ldots p_N^{n+1},\gamma_1\ldots\gamma_q)$ where $i<\omega$, $i_1<\ldots<i_p<\lambda$, $\gamma_1\ldots\gamma_q<\omega\rho_N^{n+1}$. Let $\gamma<\kappa$, $\gamma>\kappa_{i_p}$, or $\gamma>\omega\rho_N^{n+1}$ if p=0. Pick j such that $\kappa_j>\gamma$ and $\kappa_j\in C_N^m$. By lemma 7.12 we may assume that $\langle i_1\ldots i_p\rangle$ is m-full. So if $\kappa_\alpha\in C_N^m\setminus\gamma$ $\langle i_1\ldots i_p,\alpha\rangle$ is m-full. Also if $\kappa_\alpha,\kappa_{\alpha'}\in C_N^m\setminus\gamma$ then $ch_m(\alpha)=$ $=ch_m(\alpha')=m$. Hence by corollary 7.19 $\kappa_\alpha\in X \Leftrightarrow \kappa_{\alpha'}\in X$. Since $C_N^m\setminus\gamma\not\subseteq X$ it follows that $C_N^m\setminus\gamma\subseteq\kappa\setminus X$, i.e. $\kappa\setminus X\in U$. So $X\notin U$. □

Lemma 11.7 says that we can predict whether a definable set will be in U just by looking at $U\cap N$. From this we deduce

Lemma 11.8. *Let $M=rud_{U\cap N}(N)$. Then $\langle M,U\rangle$ is amenable.*

Proof: Suppose $X\in M$, $X\subseteq P(\kappa)$. By lemma 2.43 there is p such that $X\subseteq\Sigma_{n+p+1}(\underline{N})$. Then

$$X\cap U=\{Y\in X:\exists\gamma<\kappa(C^p\setminus\gamma\subseteq Y)\}. \tag{11.2}$$

But $C^p\in\Sigma_\omega(\underline{N})$, so $C^p\in M$. Hence $X\cap U\in M$. □

Lemma 11.9. *$N^+=rud_{U\cap N}(N)$.*

Proof: Every x in N^+ is definable from parameters in $N\cup\{N\}$ by a function rudimentary in $\underline{N}^+$. But $\langle M,\in,U\rangle\models R$ so the same function has a value in M, and this value must be x by absoluteness. So $x\in M$. □

Corollary 11.10. $P(\kappa)\cap N^+=P(\kappa)\cap\Sigma_\omega(\underline{N})$.

Proof: This follows from lemma 11.9 by lemma 2.41. □

This result is important in many places. It does require the assumption $\omega\rho_N^{n+1}<\kappa$; after all, the measure must be unpredictable somewhere. Now put this case aside for a bit.

THE CASE $\omega\rho_N^{n+1}=\kappa$

Keep the notation introduced for the previous case. This time we shall show that $H_\kappa^N=H_\kappa^{N^+}$, and therefore that $\omega\rho_N^m=\kappa$ for all $m>n(\underline{N})$. Thus in this case the previous case cannot arise.

A small change of convention is needed. Consider the case where $\underline{N}^+=\underline{J}_{\kappa+1}^U$, so $\underline{N}=\underline{J}_\kappa^U$. The reader may easily check that $\underline{J}_\kappa^U=\underline{J}_\kappa^\phi$, so by lemma 4.27 $\underline{J}_\kappa^U$ is strongly acceptable. Also $\Sigma_\omega(\underline{J}_\kappa^U)\subseteq\underline{J}_{\kappa+1}$; hence since $\underline{J}_{\kappa+1}\models\kappa$ is a cardinal, $\omega\rho_{J_\kappa}=\kappa$. We want to apply the argument of the present case in this situation; but strictly speaking $\underline{J}_\kappa^\phi$ is not a premouse. So from now on we demand only that $\underline{N}$ be acceptable.

Let $\langle\langle\underline{N}_\alpha^+\rangle_{\alpha\in On},\langle\pi_{\alpha\beta}\rangle_{\alpha\leq\beta\in On}\rangle$ be the iterated ultrapower of $\underline{N}^+$; each N^+ is of the form $J_{\beta_\alpha+1}^{U_\alpha}$ and we may let $\underline{N}_\alpha=J_{\beta_\alpha}^{U_\alpha}$. Observe that $\underline{N}_\alpha$ is not asserted to be an iterate of $\underline{N}$, or a mouse iterate of $\underline{N}$; indeed, $\rho_{N_\alpha}>\rho_N$ as $\pi_{0\alpha}(\rho_N)=\rho_{N_\alpha}$. Anyway, $\underline{N}$ is not even necessarily a premouse. Clearly $\pi_{\alpha\beta}(\underline{N}_\alpha)=\underline{N}_\beta$. Let κ_α be the critical point of $\underline{N}_\alpha^+$. Since $\omega\rho_N^{n+1}=\kappa$, $\omega\rho_{N_\alpha}^{n+1}=\kappa_\alpha$. Let $\underline{\hat{H}}_\alpha=\underline{N}_\alpha^{n+1}$. $T_\alpha=T_{N_\alpha}^{n+1}$; $\underline{\hat{H}}'_\alpha=\langle\hat{H}_\alpha,T_\alpha\rangle$.

Lemma 11.11. $\pi_{\alpha\beta}|\hat{H}_\alpha:\underline{\hat{H}}_\alpha\to_e\underline{\hat{H}}_\beta$.

Proof: $\pi_{\alpha\beta}(\underline{\hat{H}}_\alpha)=\underline{\hat{H}}_\beta$. So

$$\underline{\hat{H}}_\beta\models\phi(\pi_{\alpha\beta}(x_1)\ldots\pi_{\alpha\beta}(x_m)) \Leftrightarrow \underline{N}_\beta^+\models\text{"}\underline{\hat{H}}_\beta\models\phi(\pi_{\alpha\beta}(x_1)\ldots\pi_{\alpha\beta}(x_m))\text{"} \quad (11.3)$$
$$\Leftrightarrow \underline{N}^+\models\text{"}\underline{\hat{H}}_\alpha\models\phi(x_1\ldots x_m)\text{"}$$
$$\Leftrightarrow \underline{\hat{H}}_\alpha\models\phi(x_1\ldots x_m) \quad □$$

Corollary 11.12. $\underline{\hat{H}}_\alpha\prec\underline{\hat{H}}_\beta$.

Proof: $\pi_{\alpha\beta}|\hat{H}_\alpha=id|\hat{H}_\alpha$. For $\hat{H}_\alpha=\bigcup\{J_{\kappa_\alpha}^a:a\subseteq\gamma<\kappa_\alpha \wedge a\in N_\alpha\}$. □

Lemma 11.13. $\underline{\hat{H}}_\alpha\models ZFC$.

Proof: $\underline{\hat{H}}_\alpha\models R^+$, so we need only check separation, collection and power set. Note that $\hat{H}_\alpha\in\hat{H}_\beta$. For $\underline{\hat{H}}_\beta\models\forall\gamma J_\gamma^B\beta$exists, where $B_\beta=A_{N_\beta}^{n+1}$. But

$\pi_{\alpha\beta}(B_\alpha)=B_\beta$ so $B_\alpha=B_\beta\cap\omega\times(\kappa_\alpha)^{<\omega}$. Hence $\underline{\hat{H}}_\beta\models\underline{J}^{B_\beta}_{\kappa_\alpha}$ exists.

(a) Separation. Suppose ϕ is a formula and $x\in\hat{H}_\alpha$. Since $\underline{\hat{H}}_\beta\models R$ there is $y\in\hat{H}_\beta$ such that

$$\underline{\hat{H}}_\beta\models\text{"}t\in y \leftrightarrow t\in x \wedge \underline{\hat{H}}_\alpha\models\phi(t,p)\text{"} \qquad (p\in H_\alpha) \tag{11.4}$$

But $\underline{\hat{H}}_\alpha\models\phi(t,p) \leftrightarrow \underline{\hat{H}}_\beta\models\phi(t,p)$ so

$$\underline{\hat{H}}_\beta\models\exists y(t\in y \leftrightarrow t\in x \wedge \phi(t,p)) \tag{11.5}$$

so the same statement holds in $\underline{\hat{H}}_\alpha$.

(It follows immediately that $\omega\rho^m_{N_\alpha}=\kappa_\alpha$ for all $m>n$.)

(b) Collection. Let $x\in\hat{H}_\alpha$. Suppose $u\in x$, $\alpha<\beta$. If $\underline{\hat{H}}_\beta\models\exists v\phi(u,v,p)$ then $\underline{\hat{H}}_\alpha\models\exists v\phi(u,v,p)$. Say $\underline{\hat{H}}_\alpha\models\phi(u,v,p)$. Then $\underline{\hat{H}}_\beta\models v\in\hat{H}_\alpha \wedge \phi(u,v,p)$. Hence $\underline{\hat{H}}_\beta\models\exists v\phi(u,v,p) \rightarrow (\exists v\in\hat{H}_\alpha)\phi(u,v,p)$. So for $x\in\hat{H}_\alpha$

$$\underline{\hat{H}}_\beta\models\exists y\forall u\in x(\exists v\phi(u,v,p) \rightarrow (\exists v\in y)\phi(u,v,p)). \tag{11.6}$$

So the same statement holds in $\underline{\hat{H}}_\alpha$.

(c) Power set. Let $x\in\hat{H}_\alpha$. Let θ be a regular cardinal greater than $2^{\bar{\bar{x}}}$. Then $\kappa_\theta=\theta$ and $\overline{\overline{P(x)\cap\hat{H}_\theta}}<\theta$. So $P(x)\cap\hat{H}_\theta\subseteq\underline{J}^B_\gamma$, some $\gamma<\theta$. So

$$\underline{\hat{H}}_\theta\models\exists y\ y\supseteq P(x). \tag{11.7}$$

So the same statement holds in $\underline{\hat{H}}_\alpha$. □

In case $\underline{N}^+=\underline{J}^U_{\kappa+1}$, $\underline{\hat{H}}_\alpha\models V=L$. This result will be generalised in part four.

Lemma 11.14. $\hat{H}_\alpha=H^{\hat{H}_\beta}_{\kappa_\alpha}$.

Proof: $\hat{H}_\alpha\subseteq\hat{H}_\beta$, so $\hat{H}_\alpha\subseteq H^{\hat{H}_\beta}_{\kappa_\alpha}$. Suppose $x\in\hat{H}_\beta$ and $\underline{\hat{H}}_\beta\models\overline{\overline{TC(x)}}<\kappa_\alpha$. Since $\hat{H}_\alpha\prec\hat{H}_\beta$ $\underline{\hat{H}}_\beta\models\kappa_\alpha$ is a limit cardinal, so $x\in H^{\hat{H}_\beta}_\gamma$, some $\gamma<\kappa_\alpha$. Say $\underline{\hat{H}}_\alpha\models y=H_\gamma$; then $\underline{\hat{H}}_\beta\models y=H_\gamma$, so $x\in y$, so $x\in H_\alpha$. □

The strategy for the proof of the result we are after is as follows. Suppose we could construct all of $\underline{N}^+$ in some $\underline{\hat{H}}_\alpha$. Then $H^{N^+}_\kappa\subseteq H^{\hat{H}_\alpha}_\kappa=\hat{H}_0\subseteq N$, and the result would be proved. $\underline{\hat{H}}_\alpha\models ZFC$, so we can certainly construct $\underline{N}$ in $\underline{\hat{H}}_\alpha$ if $\alpha>0$. If $\langle\underline{\hat{H}}_\alpha,U\rangle$ were amenable we should be done. But it isn't; for if it were $U\in\hat{H}_\alpha$, so $U\in N^+_\alpha$, but see chapter 5 exercise 1.

On the other hand, suppose $\langle X_\gamma:\gamma<\kappa\rangle\in N^+$ with $X_\gamma\subseteq\kappa$ for all $\gamma<\kappa$. Since $X_\gamma\in U \leftrightarrow \kappa\in\pi_{01}(X_\gamma)$ therefore

$$\{\gamma:X_\gamma\in U\}=\{\gamma:\kappa\in\pi_{01}(X_\gamma)\} \tag{11.8}$$

But also $\langle\pi_{01}(X_\gamma):\gamma<\kappa\rangle=\pi_{01}(\langle X_\gamma:\gamma<\kappa\rangle)|\kappa\in N^+_\alpha$. So at least we have

$\{\gamma : X_\gamma \in U\} \in N^+$. As there is a definable map of κ onto N in N^+ working with κ-sequences is no hardship.

Lemma 11.15. *$\underline{N}_\alpha \in \hat{H}_\beta$ and is uniformly definable over $\hat{\underline{H}}_\beta$ from κ_α provided $\alpha<\beta$.*

Proof: $\hat{\underline{H}}'_\alpha \in \hat{H}_\beta$ and is uniformly definable from κ_α. As $\hat{\underline{H}}'_\alpha \models MC_{n+1}$ there is $\underline{M}$ with $M \subseteq \hat{H}_\alpha$ and $p \in PA^{M}_{n+1}$ such that $\underline{M}^p = \hat{\underline{H}}_\alpha$ and $\underline{M}$ is p-sound, iterating lemma 3.26 n+1 times. And $\underline{M}$ is unique up to isomorphism; so $\underline{M} \cong \underline{N}_\alpha$; hence $\underline{M}$ is well-founded. But then $\underline{N}_\alpha$ is uniquely definable from $\underline{M}$ as its transitive collapse. □

Lemma 11.16. *$\{\kappa_\alpha : \alpha<\beta\}$ are Σ_ω indiscernibles for $\hat{\underline{H}}_\beta$.*

Proof: They are Σ_1 indiscernibles for N^+_β. □

Let X^m_α be the smallest elementary substructure of $\hat{\underline{H}}_{\alpha+m}$ containing $\kappa_\alpha \cup \{\kappa_\alpha \ldots \kappa_{\alpha+m-1}\}$. Let $\pi^m_\alpha : \underline{M}^m_\alpha \cong \hat{\underline{H}}_{\alpha+m} | X^m_\alpha$ where $\underline{M}^m_\alpha$ is transitive; so $\hat{\pi}^m_\alpha : \underline{M}^m \rightarrow_e \hat{\underline{H}}_{\alpha+m}$. Let $K^m_\alpha = H^{M^m_\alpha}_{\kappa^+_\alpha}$, where κ^+_α denotes the successor of κ_α in $\hat{\underline{H}}_{\alpha+m}$.

Lemma 11.17. *$\pi^m_\alpha | K^m_\alpha = id | K^m_\alpha$.*

Proof: We have to show $X^m_\alpha \cap H^{\hat{H}_{\alpha+m}}_{\kappa^+_\alpha}$ transitive. Suppose $x \in X^m_\alpha$, $x \in H^{\hat{H}_{\alpha+m}}_{\kappa^+_\alpha}$. Then obviously $x \subseteq H^{\hat{H}_{\alpha+m}}_{\kappa^+_\alpha}$. Since $\hat{\underline{H}}_{\alpha+m} \models \overline{\overline{x}} \leq \kappa_\alpha$ there is $f \in \hat{H}_{\alpha+m}$ mapping κ_α onto x. Let f be least such in the canonical well-order of $\hat{\underline{H}}_{\alpha+m}$. Then if $y \in x$, y is definable from f and some $\gamma<\kappa_\alpha$; but f is definable from x. □

$\underline{M}^m_\alpha \models R$, for $\hat{\underline{H}}_{\alpha+m} \models R$. So $K^m_\alpha \models R$. $X^0_\alpha = \hat{H}_\alpha$ so $\underline{M}^0_\alpha = \hat{\underline{H}}_\alpha$. By lemma 11.17 $K^m_\alpha = X^m_\alpha \cap H^{\hat{H}_{\alpha+m}}_{\kappa^+_\alpha}$; and X^m_α, $\hat{H}_{\alpha+m}$ are in X^{m+1}_α. But $\hat{\underline{H}}_{\alpha+m+1} \models \overline{\overline{K}}^m_\alpha = \kappa_\alpha$ so $K^m_\alpha \in K^{m+1}_\alpha$. By lemma 11.15 $\underline{N}_\alpha \in X^1_\alpha$; and $\overline{\overline{N_\alpha}} = \kappa_\alpha$ in $\hat{\underline{H}}_{\alpha+1}$, since $\omega\rho^{n+1}_{N_\alpha} = \kappa_\alpha$. Thus $\underline{N}_\alpha \in K^1_\alpha$.

Let f^m_α be the least map of κ_α onto $P(\kappa_\alpha) \cap K^m_\alpha$ in the canonical well-order of $\hat{\underline{H}}_{\alpha+m+1}$. So f^m_α is definable over $\hat{\underline{H}}_{\alpha+m+1}$ in $\{\kappa_\alpha \ldots \kappa_{\alpha+m}\}$; and the definition is uniform for all α.

Lemma 11.18. *$\pi_{\alpha\beta}(f^m_\alpha(\gamma)) = f^m_\beta(\gamma)$ $(\gamma<\kappa_\alpha, \alpha \leq \beta)$*

Proof: Let $\sigma : \underline{N}^+_{\alpha+m+1} \rightarrow_{\Sigma_1} \underline{N}^+_{\beta+m+1}$ be the map of lemma 8.3 such that

(i) $\sigma\pi_{\alpha\alpha+m+1} = \pi_{\alpha\beta+m+1}$

(ii) $\sigma(\kappa_{\alpha+p}) = \kappa_{\beta+p}$ $(p \leq m)$

So by the uniform definability of f^m_α $f^m_\beta(\gamma) = \sigma(f^m_\alpha(\gamma))$ for $\gamma<\kappa_\alpha$. Thus

$$f^m_\beta(\gamma)=\sigma(f^m_\alpha(\gamma))=\sigma(\pi_{\alpha\alpha+m+1}(f^m_\alpha(\gamma))\cap\kappa_\alpha) \qquad (11.9)$$
$$=\pi_{\alpha\beta+m+1}(f^m_\alpha(\gamma))\cap\kappa_\beta$$
$$=\pi_{\alpha\beta}(f^m_\alpha(\gamma)). \qquad \square$$

Now $f^m_\alpha(\gamma)\in U_\alpha \leftrightarrow \kappa_\alpha\in\pi_{\alpha\alpha+1}(f^m_\alpha(\gamma)) \leftrightarrow \kappa_\alpha\in f^m_{\alpha+1}(\gamma)$.

Lemma 11.19. $U_\alpha\cap K^m_\alpha\in K^{m+2}_\alpha$.

Proof: $U_\alpha\cap K^m_\alpha=\{f^m_\alpha(\gamma):\gamma<\kappa_\alpha \wedge \kappa_\alpha\in f^m_{\alpha+1}(\gamma)\}$. But $f^m_{\alpha+1}$ is definable over $\underline{\hat{H}}_{\alpha+m+2}$ from $\kappa_{\alpha+1}\ldots\kappa_{\alpha+m+1}$. $\square$

Let $K_\alpha=\bigcup_{m<\omega} K^m_\alpha$. K_α is transitive and rudimentarily closed. By lemma 11.19 $\langle K_\alpha,U_\alpha\rangle$ is amenable. Thus $\langle K_\alpha,\in,U_\alpha\rangle\models R$. But $N_\alpha\cup\{N_\alpha\}\subseteq K_\alpha$, so $N^+_\alpha\subseteq K_\alpha$.

Lemma 11.20. $H^{N^+_\alpha}_{\kappa_\alpha}\subseteq\hat{H}_\alpha$.

Proof: $N^+_\alpha\subseteq K_\alpha\subseteq H^{H^+}_{\kappa_\alpha}\subseteq H$, where $H=\hat{H}_{\alpha+\omega}$. By lemma 11.14 $H^{N^+_\alpha}_{\kappa_\alpha}\subseteq\hat{H}_\alpha$. $\square$

Corollary 11.21. $H^{N^+}_\kappa=H^N_\kappa$.

THE CASE $\omega\rho^n_N>\kappa$ FOR ALL $n<\omega$

This is not very difficult.

Lemma 11.22. *Suppose $\omega\rho^n_N>\kappa$, all $n<\omega$. Then $N^+=\mathrm{rud}_{U\cap N}(\underline{N})$.*

Proof: Let $M=\mathrm{rud}_{U\cap N}(N)$. Then $\langle M,U\rangle$ is amenable. For given $X\in M$, if $X\subseteq\kappa$ then $X\in N$, since $P(\kappa)\cap M=P(\kappa)\cap\Sigma_\omega(\underline{N})$ by lemma 2.41. So given $Y\subseteq P(\kappa)$ $Y\in M$ we have $Y\cap U=Y\cap(U\cap N)$. But $U\cap N\in M$ so $Y\cap U\in M$.

Hence $\langle M,\in,U\rangle\models R$ so $N^+\subseteq M$. But clearly $M\subseteq N^+$. $\square$

Corollary 11.23. $P(\kappa)\cap N^+=P(\kappa)\cap\Sigma_\omega(\underline{N})=P(\kappa)\cap N$.

Proof: Like corollary 11.10. $\square$

So much for that.

ACCEPTABILITY

A lot of the difficulties that remain arise from the need to obtain a sequence of functions $\langle f_\xi:\delta\leq\xi<\lambda\rangle$ in $S_{\lambda+\omega}$ rather than individual functions. It is helpful to prove the simpler version first, though, as the strategy gets obscured by the details in the proof proper.

Lemma 11.24. *Suppose $\underline{N}^+$ is an iterable premouse and $\underline{N}$ is acceptable. Suppose $a \subseteq \delta$, $a \in N^+ \setminus N$. Suppose $u \in N^+$ and $u \subseteq P(\delta)$. Then $\underline{N}^+ \models \overline{\overline{u}} \leq \delta$.* $(\delta \in On_N)$

Proof: Case 1 For all $n < \omega$ $\omega\rho^n_{\underline{N}} \geq \kappa$.

Then by corollary 11.21 and corollary 11.23 $\delta \geq \kappa$. And there is n such that $\omega\rho^n_{\underline{N}} \leq \delta$. By lemma 9.7 there is a Σ_n $(\underline{N})$ map of a subset of δ onto N. If $u \in S^{N^+}_{\omega\lambda+k}$, $On_N = \omega\lambda$ then the result follows from lemma 2.42.

Case 2 For some $n < \omega$ $\omega\rho^n_N < \kappa$.

By corollary 11.21 letting $n = n(\underline{N})$, $\omega\rho^{n+1}_N < \kappa$. Hence by corollary 11.10 $a \in \Sigma_\omega(\underline{N})$. Let $\underline{M}$ be the core of $\underline{N}$ with mouse iteration $\langle\langle \underline{M}_\alpha \rangle_{\alpha \in On}, \langle \pi_{\alpha\beta} \rangle_{\alpha \leq \beta \in On}\rangle$, let κ_α be the critical point of $\underline{M}_\alpha$ and suppose $\underline{N} = \underline{M}_\lambda$.

We may assume $\delta < \kappa$. Let α be least such that $\delta \leq \kappa_\alpha$. Suppose $u \subseteq \Sigma_{n+p+1}(\underline{N})$ by lemma 2.43. Then $u \subseteq \Sigma_{p+1}(\underline{N}^n)$ so by corollary 7.18 $u \subseteq \Sigma_{p+1}(\underline{M}^n_\alpha)$. For some m $\omega\rho^m_{M_\alpha} \leq \delta$. We can get a subset of κ_α that codes all the $\Sigma_{p+1}(\underline{M}^n)$ subsets of δ; call it A. Then $A \in N^+$ by chapter 10 exercise 2. But then $\underline{N}^+ \models \overline{\overline{u}} \leq \kappa_\alpha$. Now if $\delta = \kappa_\alpha$ we are done. If $\delta < \kappa_\alpha$ then α is a successor or $\alpha = 0$. If $\alpha = \beta+1$ then there is an $\underline{N}$-definable map of $\kappa_\beta < \delta$ onto κ_α. If $\alpha = 0$ there is an $\underline{N}$-definable map of $\omega\rho^{m+1}_N$ onto κ_0. Either way $\underline{N}^+ \models \overline{\overline{u}} \leq \delta$. □

Now we must do it properly. Keep the terminology of the previous lemma.

Lemma 11.25. *$\underline{N}^+$ is strongly acceptable above κ (and hence the axiom of acceptability holds in $\underline{N}^+$ for all $\delta \geq \kappa$).*

Proof: Suppose $a \subseteq \delta$, $\kappa \leq \delta$, and $a \in N^+ \setminus N$. By corollaries 11.21 and 11.23 $\omega\rho^n_{\underline{N}} \leq \delta$ for some n so this follows from lemmas 9.7 and 9.8. □

So the proof is complete if $P(\gamma) \cap N = P(\gamma) \cap N^+$ for all $\gamma < \kappa$. Thus we may assume that $\underline{N}$ is a mouse and that $\omega\rho^{n(N)+1}_N < \kappa$. Let $n = n(\underline{N})$.

Let $\underline{M}$ be the core of $\underline{N}$ and let $\langle\langle \underline{M}_\alpha \rangle_{\alpha \in On}, \langle \pi_{\alpha\beta} \rangle_{\alpha \leq \beta \in On}\rangle$ be its mouse iteration. Suppose $\underline{N} = \underline{M}_\lambda$ and let κ_α be the critical point of $\underline{M}_\alpha$.

Suppose $a \subseteq \delta < \kappa$, $a \in N^+ \setminus N$. By corollary 11.10 $a \in \Sigma_\omega(\underline{N})$. It follows that for some m $\omega\rho^m_{\underline{N}} \leq \delta$. For otherwise, if $\delta < \omega\rho^m_N$ for all $m < \omega$, $a \in \Sigma_\omega(\underline{M})$ by corollary 7.18. But $a \in \underline{M}$ and $\underline{M}$ is k-sound for all k. If $a \in \Sigma_{k+1}(\underline{M})$ then $a \in \Sigma_1(\underline{M}^k) \setminus M^k$ so $\rho^{k+1}_M = \rho_M{}^k \leq \delta$. Contradiction! (cf exercise 10.4).

Lemma 11.26. *The axiom of acceptability holds in $\underline{N}^+$ for $\delta < \kappa$.*

Proof: By lemma 11.25 we need only produce a sequence $\langle f_\xi : \delta \le \xi < \kappa \rangle$. In fact we produce sequences $f_1 = \langle f^1_\xi : \omega\rho_N^{n+1} \le \xi < \kappa \rangle$, $f_{m+1} = \langle f^{m+1}_\xi : \omega\rho_N^{n+m+1} \le \xi < \omega\rho_N^{n+m} \rangle$ $(m \ge 1)$. Since for only finitely many m is $\rho^{n+m+1} < \rho^{n+m}$ this will suffice. Take $u \in N^+$. We may assume $u \subseteq P(\kappa) \cap N^+$. So by lemma 2.43 $u \subseteq \Sigma_{n+p+1}(\underline{N})$ say.

Case 1 The sequence f_1.

The relation $R(\kappa_\alpha, \hat{p}, i)$ is defined by

$$R(\kappa_\alpha, \hat{p}, i) \Leftrightarrow \hat{p} = \pi_{\alpha\lambda}(\bar{p}) \wedge \underline{M}^n_\alpha \models \phi_i(\bar{p}) \tag{11.10}$$

where ϕ_i is the Σ_{p+1} formula s(i). Since $\pi_{\alpha\lambda}{}''M^n_\alpha = h_N n(\kappa_\alpha \cup p_N^{n+1})$, letting $X_\alpha = h_N n(\kappa_\alpha \cup p_N^{n+1})$ $\underline{M}^n_\alpha \models \phi_i(\bar{p}) \Leftrightarrow \underline{N}^n | X \models \phi_i(\hat{p})$ $(\hat{p} = \pi_{\alpha\lambda}(\bar{p}))$; and the sequence $\langle \kappa_\alpha : \alpha < \lambda \rangle$ is definable over $\underline{N}$, therefore $R(\kappa_\alpha, \hat{p}, i)$ is definable over $\underline{N}$ (we make no claim about its complexity). Let $k(\xi)$ denote the least κ_α such that $\kappa_\alpha \ge \xi$. k is definable over $\underline{N}$. Let

$$g(\xi, \bar{\gamma}, i) = \{\gamma : R(k(\xi), \langle \gamma, \bar{\gamma}, p_N^{n+1} \rangle, i)\}. \tag{11.11}$$

So g is definable over $\underline{N}$. Since $k(\xi) \subseteq h_N n(\xi \cup p_N^{n+1})$, $\overline{\overline{k(\xi)}}^{N^+} \le \xi$, provided $\xi \ge \omega\rho_N^{n+1}$. Let $\langle t_\xi \rangle_{\xi<\kappa}$ be such that $t_\xi : \xi \to \omega \times (k(\xi))^{<\omega}$ onto, or $\xi < \omega\rho_N^{n+1}$; $\langle t_\xi \rangle_{\xi<\kappa} \in N^+$.

Let $f^1_\xi(\gamma) = g(\xi, \bar{\gamma}, i)$ if $t_\xi(\gamma) = \langle i, \bar{\gamma} \rangle$ and $g(\xi, \bar{\gamma}, i) \in u$

$= \xi$ otherwise.

We must show $f^1_\xi : \xi \to (u \cap P(\xi)) \cup \{\xi\}$ onto.

Suppose $a \in u \cap P(\xi)$, $\xi < \kappa$. Then $a \in \Sigma_{n+p+1}(\underline{N})$, so $a \in \Sigma_{p+1}(\underline{N}^n)$. By corollary 7.18 $a \in \Sigma_{p+1}(\underline{M}^n_{k(\xi)})$. Say $\bar{\gamma} \in (k(\xi))^{<\omega}$ such that

$$\gamma \in a \Leftrightarrow \underline{M}^n_{k(\xi)} \models \phi_i(\gamma, \bar{\gamma}, p^{n+1}_{M_{k(\xi)}}). \tag{11.12}$$

So

$$\gamma \in a \Leftrightarrow R(k(\xi), \langle \gamma, \bar{\gamma}, p_N^{n+1} \rangle, i). \tag{11.13}$$

Hence

$$\gamma \in a \Leftrightarrow \gamma \in g(\xi, \bar{\gamma}, i). \tag{11.14}$$

Suppose $\langle i, \bar{\gamma} \rangle = t_\xi(\hat{\gamma})$. Then $a = f^1_\xi(\hat{\gamma})$.

Case 2 The sequences f_{m+1} $(m \ge 1)$

Let $u' = u \cap P(\omega\rho_N^{n+m})$. So by corollary 7.18 $u' \subseteq \Sigma_{p+1}(\underline{M}^n)$. Hence $u' \subseteq \Sigma_{p+1-m}(\underline{M}^{n+m})$ (without loss of generality $m < p+1$). Let ϕ_i denote the Σ_{p+1-m} formula s(i). Let $A = \{\langle i, \gamma, \zeta \rangle : i < \omega \wedge \gamma \in (\omega\rho_M^{n+m+1})^{<\omega} \wedge \underline{M}^{n+m} \models \phi_i(\zeta, \gamma, p_M^{n+m+1})\}$. Then $A \in \Sigma_{p+1-m}(\underline{M}^n)$; hence $A \in \Sigma_{p+1}(\underline{N}^n | X_0)$, so $A \in \underline{N}^+$. Let $t \in N^+$, $t : \omega\rho_N^{n+m+1} \to \omega \times (\omega\rho_N^{n+m+1})^{<\omega}$ onto. Let

$$f_\xi^{m+1}(\gamma)=\{\zeta:\langle i,\bar{\gamma},\zeta\rangle\in A\} \text{ if this is in } u \quad (t(\gamma)=\langle i,\bar{\gamma}\rangle) \tag{11.15}$$
$$=\xi \text{ otherwise.}$$

This suffices. □

Corollary 11.27. *If $\underline{N}$ is an iterable premouse then $\underline{N}$ is acceptable.*
Proof: Induction based on lemma 4.27, lemma 11.25 and lemma 11.26. □

Corollary 11.28. *If $\underline{N}$ is an iterable premouse with κ critical and for some n $\omega\rho^n_{\underline{N}}\leq\kappa$ then $\underline{N}$ is a critical premouse.*

Corollary 11.29. *If $\underline{N}$ is an iterable premouse with κ critical and for some $\omega\alpha\in\underline{N}$, $\kappa<\omega\alpha$ $\underline{M}=\underline{J}^N_\alpha$ with $\omega\rho^n_{\underline{M}}\leq\kappa$ some n, then $\underline{M}$ is a mouse.*
Proof: By corollary 10.33 and corollary 11.27. □

Exercises

1. Is every core mouse strongly acceptable?

2. Show that $\langle C^m_N : m\in\omega\rangle\notin N^+$ in lemma 11.2.

3. Show that if $\underline{N}$ is an acceptable premouse with $\omega\rho_N<\kappa$ then $cf(\omega\rho_N)=cf(On_N)$. Deduce that if κ is regular and $\underline{N}$ is an acceptable premouse then $\rho_N=\kappa$. Show that if $\underline{M}=\underline{J}^U_{\kappa+1}$ and $\omega\rho_M<\kappa$ then $\omega\rho_{J^U_{\kappa+\alpha}}<\kappa$ for all $0<\alpha<\kappa$.

4. Suppose $V=L[U]$, U a normal measure on κ. Show that there is $\kappa'<\kappa^+$ such that if $a\subseteq\gamma<\kappa$ then $a\in J^U_{\kappa'}$. Is $J^U_{\kappa'}$ a mouse if κ' is chosen minimally?

5. Find α,A,B such that $\underline{J}^A_\alpha$ is strongly acceptable, $\underline{J}^B_\alpha$ is not, but $J^A_\alpha=J^B_\alpha$.

12: $0^{\#}$

At last we are in a position to construct a mouse. The existence of a mouse turns out to be equivalent to a well-known large cardinal axiom, that $0^{\#}$ exists. Each is equivalent to the existence of a non-trivial map $j:L\rightarrow_e L$.

We are going to prove

(i) if there is an iterable premouse then $0^{\#}$ exists;

(ii) if $0^{\#}$ exists there is a non-trivial $j:L\rightarrow_e L$;

(iii) if there is a non-trivial $j:L\rightarrow_e L$ then there is a mouse.

<u>Definition 12.1</u>. *A sharp is an iterable premouse $\underline{N}$ with κ critical and $On_N=\kappa+\omega$.*

<u>Lemma 12.2</u>. *If there is an iterable premouse then there is a sharp.*

Proof: Trivial. □

<u>Lemma 12.3</u>. *If $\underline{N}$ is a sharp with κ critical then $H^N_\kappa\subseteq L$.*

Proof: $\omega\rho_{J_\kappa}=\kappa$. So by corollary 11.21 $H^N_\kappa\subseteq J_\kappa\subseteq L$. □

Note that in addition $P(\kappa)\cap L\subseteq N$.

<u>Lemma 12.4</u>. *If $\underline{N}$ is a sharp then $\underline{N}$ is a mouse with $n(\underline{N})=0$.*

Proof: There is a $\Sigma_1(\underline{N})$ map of κ onto $\kappa+\omega$ so $\omega\rho_N\leq\kappa$. $\underline{N}$ is acceptable by corollary 11.27. □

<u>Lemma 12.5</u>. *If $\underline{N}$, $\underline{N}'$ are sharps then $\underline{N}$, $\underline{N}'$ have a common iterate.*

Proof: Let $\theta>\overline{\overline{N}},\overline{\overline{N}}'$; let $\langle\langle\underline{N}_\alpha\rangle_{\alpha\in On},\langle\pi_{\alpha\beta}\rangle_{\alpha\leq\beta\in On}\rangle,\langle\langle\underline{N}'_\alpha\rangle_{\alpha\in On},\langle\pi'_{\alpha\beta}\rangle_{\alpha\leq\beta\in On}\rangle$ be the iterated ultrapowers of $\underline{N},\underline{N}'$ respectively. So $\underline{N}_\theta,\underline{N}'_\theta$ are comparable. But $On_{N_\theta}=On_{N'_\theta}=\theta+\omega$. So $\underline{N}_\theta=\underline{N}'_\theta$. □

Let $\underline{M}$ be the core of some, and hence of all, sharps. This is fixed from now on - obviously $\underline{M}$ is the unique core sharp - and κ is its critical point.

<u>Lemma 12.6</u>. *$\rho_M=1$; $p_M=\phi$.*

Proof: Let $X=h_M(\kappa)$. Then $X\prec_{\Sigma_1}\underline{M}$. Say $\sigma:\underline{N}=\underline{M}|X$. So $\sigma:\underline{N}\rightarrow_{\Sigma_1}\underline{M}$. Let $\bar{\kappa}=$ $=\sigma^{-1}(\kappa)$. Then $\underline{N}$ is an iterable premouse at $\bar{\kappa}$; so $\underline{N}$ is a sharp.

Hence $\underline{N}$ is an iterate of $\underline{M}$. But $\bar{\kappa}\leq\kappa$ so $\underline{N}=\underline{M}$. But then there is a parameter free map of ω onto M; so $\rho_M=1$, $p_M=\phi$. □

Corollary 12.7. *If $\underline{N}$ is a sharp then $\rho_N=1$, $p_N=\phi$.*

$A_M\subseteq\omega\times(\omega)^{<\omega}$ could be coded as a real and would be a reasonable definition of $0^\#$. In fact $0^\#$ is a different real, but $0^\#\in L[A_M]$ and $A_M\in L[0^\#]$, as we shall see.

Let $\langle\langle\underline{M}_\alpha\rangle_{\alpha\in On},\langle\pi_{\alpha\beta}\rangle_{\alpha\leq\beta\in On}\rangle$ be the iterated ultrapower of $\underline{M}$; let κ_α be the critical point of $\underline{M}_\alpha$; let $C=\{\kappa_\alpha:\alpha\in On\}$. We now apply the apparatus of the previous chapter. Let $\underline{N}_\alpha=J^{M_\alpha}_{\kappa_\alpha}$, so $\underline{M}_\alpha=\underline{N}^+_\alpha$. Note that $\hat{H}_\alpha=N_\alpha$ in this case. Hence by corollary 11.12 $\underline{J}_{\kappa_\alpha}\prec\underline{J}_{\kappa_\beta}$, when $\alpha\leq\beta$, and $\underline{J}_{\kappa_\alpha}\models ZFC$ by lemma 11.13. By lemma 11.16 $\{\kappa_\alpha:\alpha<\beta\}$ are Σ_ω indiscernibles for $\underline{J}_{\kappa_\beta}$. The structures K^m_α are defined as before; $K^m_\alpha\subseteq K^{m+1}_\alpha$, $K^m_\alpha\prec\underline{J}_{\kappa^+_\alpha}$ and $K_\alpha=\bigcup_{m\in\omega}K^m_\alpha$. Also $M_\alpha\subseteq K_\alpha$ and $\langle K_\alpha,U_\alpha\rangle$ is amenable. It follows that $M_\alpha\subseteq J_{\kappa^+_\alpha}$; we shall soon see that it is not a member.

Lemma 12.8. *If $x\in J_{\kappa_\alpha}$ then x is definable in L from ordinals in $C\cap\kappa_{\alpha+\omega}$.*

Proof: $x\in M_\alpha$, so $\underline{M}_\alpha\models x=h(i,\langle\kappa_{\alpha_1}\ldots\kappa_{\alpha_p}\rangle)$ for some $\alpha_1<\ldots<\alpha_p<\alpha$. Let $x=h(i,y)\leftrightarrow\exists zH(z,x,i,y)$ be some definition of h with H Σ_0. Then $\langle K_\alpha,U_\alpha\rangle\models\exists zH(z,x,i,\langle\kappa_{\alpha_1}\ldots\kappa_{\alpha_p}\rangle)$. Suppose $z\in K^m_\alpha$, some $z\in M_\alpha$. We may assume $z\in S^{M_\alpha}_{\kappa_\alpha+n}\in K^m_\alpha$, $\langle\kappa_{\alpha_1}\ldots\kappa_{\alpha_p}\rangle,x\in K^m_\alpha$. But then x is the unique $x\in J_{\kappa_\alpha}$ such that $\exists z\in K^m_\alpha$ such that $z\in S^{M_\alpha}_{\kappa_\alpha+n}\wedge H(z,x,i,\langle\kappa_{\alpha_1}\ldots\kappa_{\alpha_p}\rangle)$. $S^{M_\alpha}_{\kappa_\alpha+n}$ is definable in $\langle K^m_\alpha,U_\alpha\cap K^m_\alpha\rangle$ and $\langle K^m_\alpha,U_\alpha\cap K^m_\alpha\rangle$ is definable from $\kappa_\alpha\ldots\kappa_{\alpha+m+1}$. □

Lemma 12.9. *(a) C contains all uncountable cardinals; and for all uncountable θ*

(b) $C\cap\theta$ is of order-type θ;

(c) $C\cap\theta$ is closed unbounded in θ provided θ is regular;

(d) $C\cap\theta$ are indiscernibles for $\underline{J}_\theta$;

(e) every $a\in J_\theta$ is J_θ-definable from $C\cap\theta$.

Proof: (a) If $\theta>\omega$ is a cardinal then $\theta=\kappa_\theta$ by lemma 8.15.

(b) $C\cap\theta=\{\kappa_\alpha:\alpha<\theta\}$.

(c) $C\cap\theta$ is closed by lemma 6.15 and unbounded by (b).

(d) By lemma 11.16.

(e) By lemma 12.8. □

<u>Lemma 12.10</u>. *C is the unique class satisfying (a)-(e) of lemma 12.9.*

Proof: Suppose C' were another. Let $C'=\{\kappa'_\alpha:\alpha\in On\}$. Define for regular θ $\sigma:J_\theta\to J_\theta$ by

$$\sigma(t^{L_\theta}(\kappa_{\alpha_1}\ldots\kappa_{\alpha_p}))=t^{L_\theta}(\kappa'_{\alpha_1}\ldots\kappa'_{\alpha_p}) \quad (12.1)$$

where t is any term, $\alpha_1<\ldots<\alpha_p<\theta$. $C\cap C'\cap\theta$ is closed unbounded in θ.

Hence given some $\beta_1<\ldots<\beta_p<\theta$ with each $\kappa_{\beta_k}\in C'$,

$$\begin{aligned}\underline{J}_\theta\models\phi(\kappa_{\alpha_1}\ldots\kappa_{\alpha_p}) &\Leftrightarrow \underline{J}_\theta\models\phi(\kappa_{\beta_1}\ldots\kappa_{\beta_p}) \quad (12.2)\\ &\Leftrightarrow \underline{J}_\theta\models\phi(\kappa'_{\beta'_1}\ldots\kappa'_{\beta'_p}) \quad \text{(say)}\\ &\Leftrightarrow \underline{J}_\theta\models\phi(\kappa'_{\alpha_1}\ldots\kappa'_{\alpha_p}).\end{aligned}$$

By this and property (e) of C, σ is well-defined. Also by (12.2) $\sigma:\underline{J}_\theta\to_e\underline{J}_\theta$; and by property (e) of C', σ maps onto J_θ. Hence σ is an isomorphism, so $\sigma=id|J_\theta$. So $C\cap\theta=C'\cap\theta$ for all regular uncountable θ. So C=C'. □

<u>Definition 12.11</u>. *Suppose there is a class C satisfying (a)-(e) of lemma 12.9. Let $\{\phi_n:n<\omega\}$ enumerate the formulae of L, where ϕ_n has m_n free variables. Let*

$$0^\#=\{n:\underline{J}_{\aleph_\omega}\models\phi_n(\aleph_1\ldots\aleph_{m_n})\}.$$

"$0^\#$ exists" means that a class satisfying (a)-(e) of lemma 12.9 exists.

<u>Lemma 12.12</u>. *If there is an iterable premouse then $0^\#$ exists.*

Note 1: $0^\#$ is read "O-sharp" or "zero-sharp".

Note 2: Suppose $0^\#$ exists: let M be an inner model with $0^\#\in M$. Is M a model of $0^\#$ exists? The terminology would be misleading if this were not so; and later on we shall show how to construct an iterable premouse from $0^\#$.

Note 3: $\underline{J}_{\aleph_\omega}\prec L$, of course, since L is a direct limit of $\langle\langle\underline{J}_\theta\rangle_{\theta\text{ regular}},\langle id|J_\theta\rangle\rangle$. But it would not be legitimate to replace $\underline{J}_{\aleph_\omega}$ by L in definition 12.11.

Note 4: Suppose $\underline{M}$ were a constructible sharp. Since some iterate of $\underline{M}$ contains $J_{\aleph_1}$, $P(\omega)\cap L\subseteq M$. But $\rho_M=1$ so there is $A\in P(\omega)\cap\Sigma_1(\underline{M})\setminus M$. Since $\underline{M}\in L$, $A\in L$; contradiction! Once we have proved the remark in note 2 we shall have shown $0^\#\notin L$. In fact $0^\#\in\Sigma_1(\underline{M})\setminus M$ for all sharps $\underline{M}$. Indeed, as we said, $A_{\underline{M}}\subseteq\omega\times(\omega)^{<\omega}$ would have been a plausible candidate for $0^\#$.

Our next aim must be to extract from $0^\#$ a non-trivial $j:L\to_e L$. This is rather easy. In fact any order-preserving map $\tau:On\to On$ yields a map of L to L; simply define $j(t^L(\kappa_{\alpha_1}\ldots\kappa_{\alpha_n}))=t^L(\kappa_{\tau(\alpha_1)}\ldots\kappa_{\tau(\alpha_n)})$, where t is any term and $C=\{\kappa_\alpha:\alpha\in On\}$.

Lemma 12.13. *If $0^{\#}$ exists then there is a non-trivial $j:L\to_e L$.*
Proof: Let $\tau(\alpha)=\alpha+1$ and define j as above. □

In the example in the proof the critical point of j is κ_0, which is countable. So we could have added the requirement that j have a countable critical point.

Third and most difficult is the construction of a sharp from $j:L\to_e L$ which is non-trivial.Since a sharp is a mouse this will complete our circle of inferences.

Suppose then that $j:L\to_e L$ has critical point κ. Let $U=$ $=\{X\in P(\kappa)\cap L:\kappa\in j(X)\}$. The proof of lemma 5.10 shows that U is a normal measure on κ in L.

Lemma 12.14. *$\langle J_{\kappa^+},U\rangle$ is amenable (where κ^+ denotes the successor of κ in L).*
Proof: Suppose $X\in J_{\kappa^+}$; assume $X\subseteq P(\kappa)$. Then $J_{\kappa^+}\models\overline{\overline{X}}\leq\kappa$. Suppose $f\in J_{\kappa^+}$, $f:\kappa\to X$ onto. Then $j(f)\in L$. But

$$f(\gamma)\in U \leftrightarrow \kappa\in j(f(\gamma)) \qquad (12.3)$$
$$\leftrightarrow \kappa\in j(f)(\gamma)$$

for $\gamma<\kappa$. So $X\cap U=\{f(\gamma):\kappa\in j(f)(\gamma) \wedge \gamma<\kappa\}\in L$. But $X\cap U\in H^L_{\kappa^+}$ so $X\cap U\in J_{\kappa^+}$. □

So $J^U_{\kappa+1}\subseteq J_{\kappa^+}$. Hence $U\cap J^U_{\kappa+1}$ is a normal measure on κ in $J^U_{\kappa+1}$. That is, $J^U_{\kappa+1}$ is a premouse. The difficulty arises in showing that it is iterable.

We are going to prove that L is iterable by U, that is, that the weak iterated ultrapower of L by U preserves well-foundedness. Since we shall need to generalise this result later on we shall work with an arbitrary model M and a normal measure U on κ in M such that

(i) $\underline{M}\models ZFC + V=L[D]$ (some class D);
(ii) $\langle\underline{N},U\rangle$ is amenable, where $\underline{N}=\underline{J}^M_{\kappa^+}$;
(iii) $N=H^M_{\kappa^+}$.

Definition 12.15. *$\underline{M}$ is κ-maximal provided that whenever $\pi:\underline{M}\to_e\underline{M}'$, $\underline{M}'$ is transitive and $\pi|\kappa=id|\kappa$ then $\underline{M}=\underline{M}'$.*

Note that L is 0-maximal. Suppose from now on that $\underline{M}$ is κ-maximal.

Let $\langle\langle\underline{M}_\alpha\rangle_{\alpha\in On},\langle\pi_{\alpha\beta}\rangle_{\alpha\leq\beta\in On}\rangle$ be the weak iterated ultrapower of $\underline{M}$. Let λ be least such that $\underline{M}_\lambda$ is not well-founded. Until just before lemma 12.20 we are working for a contradiction.

Lemma 12.16. *λ is a limit ordinal.*
Proof: $\lambda\neq 0$, of course. Suppose $\lambda=\mu+1$. Since $\underline{M}_\mu$ is transitive $\underline{M}_\mu=\underline{M}$.So

$$M=\{\pi_{0\mu}(f)(\kappa_{h_1}\ldots\kappa_{h_p}):f\in M \wedge h_1<\ldots<h_p<\mu\} \qquad (12.4)$$

where κ_h is the critical point of π_{hh+1}. Now $\mu \neq 0$. For we may define $\sigma: \underline{M}_1 \to \underline{M}$ by

$$\sigma(\pi_{01}(f)(\kappa)) = j(f)(\kappa) \tag{12.5}$$

and check that $\sigma: \underline{M}_1 \to_{\Sigma_1} \underline{M}$ is well-defined, so $\underline{M}_1$ is well-founded. Thus

$$M = \{\pi_{01}(g)(\kappa) : g \in M\}. \tag{12.6}$$

Now by (12.4) and (12.6)

$$M = \{\pi_{0\mu}\pi_{01}(f)(\kappa_{h_1} \ldots \kappa_{h_p}, \kappa_\mu) : h_1 < \ldots < h_p < \mu,\ f \in M\}. \tag{12.7}$$

Define $\sigma: \underline{M} \to \underline{M}_\lambda$ by

$$\sigma(\pi_{0\mu}\pi_{01}(f)(\kappa_{h_1} \ldots \kappa_{h_p}, \kappa_\mu)) = \pi_{0\lambda}(f)(\kappa_{h_1} \ldots \kappa_{h_p}, \kappa_\mu)) \tag{12.8}$$

Claim $\underline{M} \models \phi(\pi_{0\mu}\pi_{01}(f)(\kappa_{h_1} \ldots \kappa_{h_p}, \kappa_\mu)) \leftrightarrow \underline{M}_\lambda \models \phi(\pi_{0\lambda}(f)(\kappa_{h_1} \ldots \kappa_{h_p}, \kappa_\mu))$ where ϕ is any formula (with one free variable for simplicity).

Proof: $\underline{M}_\lambda \models \phi(\pi_{0\lambda}(f)(\kappa_{h_1} \ldots \kappa_{h_p}, \kappa_\mu)) \leftrightarrow \{\langle \xi_1 \ldots \xi_p, \zeta \rangle : \underline{M} \models \phi(f(\xi_1 \ldots \xi_p, \zeta))\} \in$
$\in U^{p+1}$ (lemma 7.7)

$\leftrightarrow \{\langle \xi_1 \ldots \xi_p \rangle : \{\zeta : \underline{M} \models \phi(f(\xi_1 \ldots \xi_p, \zeta)\} \in U\} \in U^p$

$\leftrightarrow \underline{M} \models \phi(\pi_{0\mu}\pi_{01}(f)(\kappa_{h_1} \ldots \kappa_{h_p}, \kappa_\mu))$. □(Claim)

So $\sigma: \underline{M} \to_e \underline{M}_\lambda$ is well-defined. But σ is surjective. So $\sigma: \underline{M} \cong \underline{M}_\lambda$. Hence $\underline{M}_\lambda$ is well-founded. Contradiction! □

Let $X = \{x \in M : 0 \leq j < \lambda \to \pi_{0j}(x) = x\}$. X is a proper class by chapter 8 exercise 2.

Lemma 12.17. *$\kappa \subseteq X$ and $X \prec \underline{M}$.*

Proof: $\kappa \subseteq X$ is clear. Suppose $x_1 \ldots x_n \in X$ and t is any term. Then for all $j < \lambda$, $\pi_{0j}(t^M(x_1 \ldots x_n)) = t^M(x_1 \ldots x_n)$, so $t^M(x_1 \ldots x_n) \in X$. □

Definition 12.18. *$\underline{M}$ is κ-minimal provided that whenever $\pi: \underline{M}' \to_e \underline{M}$, $\underline{M}'$ is transitive and $\pi|\kappa = id|\kappa$ then $\underline{M}' = \underline{M}$.*

Note that L is 0-minimal. Assume from now on that $\underline{M}$ is κ-minimal. Let X_i be the smallest elementary substructure of $\underline{M}$ containing $X \cup \{\kappa_j : j < i\}$ (for $i < \lambda$). Let $\pi_i' : \underline{M}_i' \cong \underline{M}|X_i$ where M_i' is transitive. Then $\pi_i' : \underline{M}_i' \to_e \underline{M}$ and $\pi_i'|\kappa = id|\kappa$ so $\underline{M}_i' = \underline{M}$. Let $\pi_{ij}' = \pi_j'^{-1}\pi_i'$.

Lemma 12.19. *$\pi_{ij}' = \pi_{ij}$ for all $i \leq j < \lambda$.*

Proof: Claim 1 For $a \subseteq \kappa$, $\pi_{0i}(a) = \pi_0'(a) \cap \kappa_i$.

Proof: $a = \pi_0'(a) \cap \kappa$ as $\kappa \subseteq X$. So $\pi_{0i}(a) = \pi_{0i}(\pi_0'(a) \cap \kappa) = \pi_0'(a) \cap \kappa_i$

(as $\pi'_0(a)\in X$). □(Claim 1).

<u>Claim 2</u> $\kappa_i \subseteq X_i$.

Proof: Suppose $\nu<\kappa_i$, $\nu=\pi_{0i}(f)(\kappa_{h_1}\ldots\kappa_{h_p})$ $(h_1<\ldots<h_p<i)$. Then we may assume $f:\kappa^p\to\kappa$, so by claim 1 $\pi_{0i}(f)=\pi'_0(f)|\kappa_i^p$. Thus $\nu=\pi'_0(f)(\kappa_{h_1}\ldots\kappa_{h_p})$ which is in X_i. □(Claim 2).

<u>Claim 3</u> $x\in X_i \Rightarrow \pi_{kj}(x)=x$, all k,j with $i\le k\le j<\lambda$.

Proof: For $\pi_{kj}(\kappa_h)=\kappa_h$ for $h<i$. □(Claim 3)

<u>Claim 4</u> $X_i\cap\kappa_j=\kappa_i$ for all $j\ge i$, $j<\lambda$.

Proof: Suppose $\gamma>\kappa_i, \gamma\in X_i$, $\gamma<\kappa_j$, some j. Let μ be greatest such that $\kappa_\mu\le\gamma$. Then $\mu\ge i$, $\gamma<\kappa_{\mu+1}$, $\mu+1<\lambda$. So

$$\pi_{\mu\mu+1}(\gamma)\ge\pi_{\mu\mu+1}(\kappa_\mu)=\kappa_{\mu+1}>\gamma.$$

But $\pi_{\mu\mu+1}(\gamma)=\gamma$ by claim 3. Contradiction! □(Claim 4)

<u>Claim 5</u> $\pi'_{ij}(\kappa_i)=\kappa_j$.

Proof: Suppose $\pi'_{ij}(\kappa_i)<\kappa_j$. Then $\pi'_{ij}(\kappa_i)=\pi'_i(\kappa_i)$; so $\pi'_i(\kappa_i)<\kappa_j$. So $\pi'_i(\kappa_i)\in X_i\cap\kappa_j$, so by claim 4 $\pi'_i(\kappa_i)<\kappa_i$. Contradiction!

So suppose $\pi'_{ij}(\kappa_i)>\kappa_j$. So $\pi_j'^{-1}\pi'_i(\kappa_i)>\kappa_j$, i.e. $\pi'_i(\kappa_i)>\pi'_j(\kappa_j)$. Note that $\pi'_i(\kappa_i)$ is the least member of X_i greater than κ_i. Suppose $\pi'_j(\kappa_j)=t(\kappa_{h_1}\ldots\kappa_{h_p},x)$ $(h_1<\ldots<h_p, x\in X)$. Then

$$\underline{M}\models(\exists\xi_1\ldots\xi_p<\kappa_j)(\kappa_j<t(\xi_1\ldots\xi_p,x)<\pi'_i(\kappa_i)). \tag{12.9}$$

Applying π_{ij}^{-1} to (12.9)

$$\underline{M}\models(\exists\xi_1\ldots\xi_p<\kappa_i)(\kappa_i<t(\xi_1\ldots\xi_p,x)<\pi'_i(\kappa_i)).$$

But by claim 2 $t(\xi_1\ldots\xi_p,x)\in X_i$. Contradiction! □(Claim 5)

<u>Claim 6</u> If $a\subseteq\kappa_i$ then $\pi_{ij}(a)=\pi'_{ij}(a)$.

Proof: $a=\pi'_i(a)\cap\kappa_i$. So

$$\begin{aligned}\pi_{ij}(a)&=\pi_{ij}(\pi'_i(a)\cap\kappa_i)=\pi'_i(a)\cap\kappa_j \\ &=\pi_j'^{-1}\pi'_i(a)\cap\kappa_j \\ &=\pi'_{ij}(a).\end{aligned} \tag{12.10}$$

□(Claim 6)

Hence $X\in U_i \leftrightarrow \kappa_i\in\pi'_{ij}(a)$, where U_i is the i-th normal measure. Now we may define inductively $\sigma_i:\underline{M}\to\underline{M}$ by

$$\sigma_0=\mathrm{id}|M$$

$$\sigma_{i+1}(\pi_{ii+1}(f)(\kappa_i))=\pi'_{ii+1}(f)(\kappa_i)$$

$$\sigma_\mu(\pi_{i\mu}(x))=\pi'_{i\mu}(x) \quad (\mu \text{ a limit}, \mu<\lambda).$$

Note that $\sigma_\beta \pi_{\alpha\beta} = \pi'_{\alpha\beta}\sigma_\alpha \quad \alpha \leq \beta < \lambda$.

We verify inductively that σ_i is well-defined, $\sigma_i : \underline{M} \to_e \underline{M}$ and σ_i is surjective. For i=0 this is clear.

For i=j+1 σ_i is well-defined and $\sigma_i : \underline{M} \to_e \underline{M}$ by claim 6. If $x \in M$ then $\pi'_i(x) = t^M(\kappa_{\alpha_1} \ldots \kappa_{\alpha_p}, \kappa_j, \hat{x})$, say $(\alpha_1 < \ldots < \alpha_p, \hat{x} \in X)$. So

$$x = \pi'^{-1}_i (t^M(\kappa_{\alpha_1} \ldots \kappa_{\alpha_p}, \kappa_j, \hat{x}))$$
$$= \pi'^{-1}_i \pi_j(f)(\kappa_j) \quad \text{where } f(\xi) = t^M(\kappa_{\alpha_1} \ldots \kappa_{\alpha_p}, \xi, \hat{x})$$
$$= \pi'_{ji}(f)(\kappa_j).$$

Thus σ_i is surjective, as σ_j is.

If i is a limit then clearly σ_i is well-defined and $\sigma_i : \underline{M} \to_e \underline{M}$. If $x \in M$ and $\pi'_i(x) = t^M(\kappa_{\alpha_1} \ldots \kappa_{\alpha_p}, \hat{x})$ $(\alpha_1 < \ldots < \alpha_p, \hat{x} \in X)$ then for some $j < i$ $\pi'_i(x) \in X_j$. So σ_i is surjective.

Hence each σ_i is the identity; so $\pi_{ij} = \pi'_{ij}$, $i \leq j < \lambda$. □

But now define $\pi : \underline{M}_\lambda \to \underline{M}$ by $\pi(\pi_{i\lambda}(x)) = \pi'_i(x)$. Then

$$\underline{M}_\lambda \models \phi(\pi_{i\lambda}(\vec{x})) \Leftrightarrow \underline{M} \models \phi(\vec{x})$$
$$\Leftrightarrow \underline{M} \models \phi(\pi'_i(\vec{x})).$$

And $\pi_{i\lambda}(x) = \pi_{j\lambda}(y) \Rightarrow \pi_{ij}(x) = y$ (assume $i \leq j$)

$$\Rightarrow \pi'_{ij}(x) = y$$
$$\Rightarrow \pi'_i(x) = \pi'_j(y).$$

So $\pi : \underline{M}_\lambda \to \underline{M}$ is well-defined and $\pi : \underline{M}_\lambda \to_e \underline{M}$. Thus $\underline{M}_\lambda$ is well-founded. Contradiction!

In summary

<u>Lemma 12.20</u>. *Suppose (i)* $\underline{M} \models ZFC + V = L[D]$

(ii) $\langle \underline{N}, U \rangle$ *is amenable, where* $\underline{N} = J^M_{\kappa^+}$

(iii) U *is a normal measure on* κ *in* $\underline{M}$

(iv) $N = H^M_{\kappa^+}$

(v) $\underline{M}$ *is* κ*-maximal and* κ*-minimal*

(vi) $j : \underline{M} \to_e \underline{M}$, $U = \{X : \kappa \in j(X)\}$.

Then letting $\langle \langle \underline{M}_\alpha \rangle_{\alpha \in On}, \langle \pi_{\alpha\beta} \rangle_{\alpha \leq \beta \in On} \rangle$ *be the weak iterated ultrapower of* $\underline{M}$ *by* U*, each* $\underline{M}_\alpha$ *is well-founded.*

This is a long-winded way of proving a standard result, but the techniques will be useful later on. Return now to our premouse $J^U_{\kappa+1}$.

<u>Lemma 12.21</u>. $J^U_{\kappa+1}$ *is iterable.*

Proof: $J^U_{\kappa+1} \subseteq L$. Let $\langle\langle \underline{M}_\alpha\rangle_{\alpha\in On}, \langle \pi_{\alpha\beta}\rangle_{\alpha\leq\beta\in On}\rangle$ be an iterated ultrapower of $\underline{J}^U_{\kappa+1}$ and let $\langle\langle L\rangle_{\alpha\in On}, \langle \pi'_{\alpha\beta}\rangle_{\alpha\leq\beta\in On}\rangle$ be the iterated ultrapower of L by U. We can get embeddings of M_α into L, σ_α, such that

(i) $\sigma_0 = id | J^U_{\kappa+1}$

(ii) $\sigma_\alpha : \underline{M}_\alpha \to_{\Sigma_0} \langle L, U_\alpha\rangle$

(iii) $\sigma_\beta \pi_{\alpha\beta} = \pi'_{\alpha\beta}\sigma_\alpha$.

the proof is like many others and the details are left to the reader. (Actually it is not hard to show that each σ_α is $id|M_\alpha$: see exercise 4) □

Hence $J^U_{\kappa+1}$ is a sharp.

$0^\#$ IS ABSOLUTE

The proof that if $0^\# \in M$ then $M \models$ "$0^\#$ is $0^\#$" can be done by an argument from descriptive set theory. Here we give a different proof that appeals to the absoluteness of well-foundedness.

Lemma 12.22. *If $\underline{M}$ is an inner model and $\underline{M} \models x=0^\#$ then $x=0^\#$.*

Proof: $\underline{M} \models 0^\#$ exists. Hence there is an iterable premouse in M. Hence there is an iterable premouse, so $0^\#$ exists. But then $x, 0^\#$ each satisfy definition 12.11 so $x=0^\#$. □

The other direction involves a bit of work.

Lemma 12.23. *If $x=0^\#$, $\underline{M}$ is an inner model and $x\in M$ then $\underline{M} \models x=0^\#$.*

Proof: Let $\overline{M}$ be the unique core sharp; let $\langle\langle \overline{\underline{M}}_\alpha\rangle_{\alpha\in On}, \langle \pi_{\alpha\beta}\rangle_{\alpha\leq\beta\in On}\rangle$ be its iterated ultrapower. Let κ_α be the critical point of $\overline{\underline{M}}_\alpha$. Let $\langle t_n : n<\omega\rangle$ enumerate terms of L, and suppose t_n is m_n-ary. Let $U' = \{t_n(\kappa_0 \ldots \kappa_{m_n-1}) : \kappa_0 \in t_n(\kappa_1 \ldots \kappa_{m_n}) \wedge t_n(\kappa_0 \ldots \kappa_{m_n-1}) \subseteq \kappa_0\}$. All t_n are interpreted in $J_{\aleph_1}$ and $\langle \kappa_\alpha : \alpha\in On\rangle$ enumerates C.

Claim 1 $U' \in M$.

Proof: Let s_n index the formula "$v_0 \in t_n(v_1 \ldots v_{m_n}) \wedge t_n(v_0 \ldots v_{m_n-1}) \subseteq v_0$"

Let $A = \{n : s_n \in 0^\#\}$. $A\in M$ and $t_n(\kappa_0 \ldots \kappa_{m_n-1}) \in U' \leftrightarrow n\in A$.

Say $R(p,q) \leftrightarrow t_p(\kappa_0 \ldots \kappa_{m_p-1}) < t_q(\kappa_0 \ldots \kappa_{m_q-1}) < \kappa_0$. Then $R\in M$.

Let $Q = \{p : (\forall q<p)(t_q(\kappa_0 \ldots \kappa_{m_q-1}) \neq t_p(\kappa_0 \ldots \kappa_{m_p-1}))\}$. Then $Q\in M$. So $R' = R\cap Q^2 \in M$. R' is a well-order of Q so $\underline{M} \models$ "R' is a well-order of Q".

Let $f\in M$, $f : \langle Q, R'\rangle \cong \langle \overline{\kappa}, \in\rangle$. Clearly $\overline{\kappa} \leq \kappa_0$. Given $\xi < \kappa_0$, suppose $\xi = t_n(\kappa_{\alpha_1} \ldots \kappa_{\alpha_{m_n}})$. Then $\xi = t_n(\kappa_0 \ldots \kappa_{m_n-1})$. So $\overline{\kappa} = \kappa_0$.

It follows that $f(p)=t_p(\kappa_0\ldots\kappa_{m_p-1})$. Let $X_n=\{f(p):t_p(\kappa_0\ldots$ $\ldots\kappa_{m_p-1})\in t_n(\kappa_0\ldots\kappa_{m_n-1})\}$. Then $\langle X_n:n<\omega\rangle\in M$. But $U'=\{X_n:n\in A\}$.

□(Claim 1)

Claim 2 $U'=U$.

Proof: If $t_n(\kappa_0\ldots\kappa_{m_n-1})\in U$ then $\kappa_0\in t_n(\kappa_1\ldots\kappa_{m_n})$ as in lemma 11.19. So $t_n(\kappa_0\ldots\kappa_{m_n-1})\in U'$. Similarly for $\kappa_0\setminus t_n(\kappa_0\ldots\kappa_{m_n-1})$. So it suffices to show that if $X\in L$, $X\subseteq\kappa_0$ then $X=t_n(\kappa_0\ldots\kappa_{m_n-1})$, some n. But suppose $X=t_n(\kappa_{\alpha_1}\ldots\kappa_{\alpha_{m_n}})$. Then for $\xi<\kappa_0$,

$$\xi\in t_n(\kappa_{\alpha_1}\ldots\kappa_{\alpha_{m_n}}) \leftrightarrow \xi\in t_n(\kappa_0\ldots\kappa_{m_n-1}).$$ □(Claim 2)

hence $U\in M$. So $N\in M$. Hence $\underline{M}\models 0^{\#}$ exists and $\underline{M}\models x=0^{\#}$. □

So if $0^{\#}$ exists then $L[0^{\#}]\models 0^{\#}$ exists. In chapter 15 we shall see how to generalise these considerations.

Exercises

1. Replace L by $L[0^{\#}]$ throughout the chapter. What equivalents for the existence of non-trivial $j:L[0^{\#}]\to_e L[0^{\#}]$ are there?

2. Show that if $0^{\#}$ exists then $P(\omega)\cap L$ is countable.

3. Show that if there is a measurable cardinal then there is a premouse in L.

4. Suppose $0^{\#}$ exists. Show that, letting U be such that $J^U_{\kappa+1}$ is a sharp, each weak iterate of L by U is well-founded (hence L) and that the critical points of the weak iteration are the canonical indiscernibles for L.

13: THE MODEL L[U]

Suppose U is a normal measure on κ; then $U\cap L[U]$ is a normal measure on κ in L[U]. So L[U] is a premouse. By lemma 8.10 $L[U]\models$"each iterate of V by U is well-founded"; so each iterate really is well-founded; so L[U] is an iterable premouse. Fix U,κ such that $U\cap L[U]$ is a normal measure on κ in L[U]. We need not assume that U is a normal measure in V.

Lemma 13.1. *$L[U]\models GCH$*
Proof: L[U] is acceptable by corollary 11.27. By corollary 3.4 for each ν either $\overline{\overline{P(\nu)}}=\nu^+$ or for all $u\subseteq P(\nu)$ $\overline{\overline{u}}\leq\nu$. But if the latter held then since the power set axiom holds in L[U] , $\overline{\overline{P(\nu)}}\leq\nu$, which cannot be. So for all ν $\overline{\overline{P(\nu)}}=\nu^+$ in L[U]. □

We know that $L[U]\neq L$. In fact

Lemma 13.2. *$L[U]\neq L[A]$ for all $A\subseteq\kappa$.*
Proof: Let $j:L[U]\to_U M$. Then $L[A]=L[j(A)\cap\kappa]\subseteq M$ so $M=L[U]$. But see chapter 5 exercise 1. □

Lemma 13.3. *κ is the largest measurable cardinal in L[U].*
Proof: A slight generalisation of the previous lemma. If $\lambda>\kappa$ and $V\in L[U]$, V is a normal measure on λ in L[U] let $j:L[U]\to_V M$. Then $M=L[j(U)]=L[U]$; but $V\notin M$ by chapter 5 exercise 1. □

Lemma 13.4. *κ is the only measurable cardinal in L[U].*
Proof: Suppose $\lambda<\kappa$ and V is a normal measure on λ in L[U]. Let $j:L[U]\to_V L[U']$; then U' will be normal on $j(\kappa)$ in L[U']so L[U'] is an iterable premouse. Let Ω,Ω' be comparable iterates of L[U], L[U'] respectively: then $\Omega=\Omega'$, since each is a proper class. Now on the one hand $V\in L[U]$ so $V\in\Omega$ as $\kappa>2^\lambda$(in L[U]). On the other hand $V\notin L[U']$ so $V\notin\Omega'$. Contradiction! □

Lemma 13.5. *$U\cap L[U]$ is the only normal measure on κ in L[U].*
Proof: Suppose V were another. Let $j:L[U]\to_V L[U']$; L[U']is an iterable premouse so L[U], L[U'] have a common iterate, Ω, say. Suppose $j:L[U]\to_e\Omega$, $j':L[U']\to_e\Omega$ are the iteration maps. Let

$$C=\{\xi: j(\xi)=j'(\xi)=\bar{j}(\xi)=\xi \wedge \xi>\kappa\}. \qquad (13.1)$$

By chapter 8 exercise 2 C is a proper class. Let X be the smallest elementary substructure of L[U] that contains $\kappa \cup C$. Suppose $\sigma: L[U'']=L[U]|X$. Then $\sigma|\kappa\cup\{\kappa\}=id|\kappa\cup\{\kappa\}$, and if $X\subseteq\kappa$ then $\sigma(X)=X$, so $U''=U\cap L[U'']$. It follows that $L[U'']=L[U]$. In $L[U'']$ every subset of κ is definable from parameters in $\kappa\cup\sigma^{-1}$"C; so in L[U] every subset of κ is definable from parameters in $\kappa\cup C$. Suppose $x\subseteq\kappa$, $x=$ $=t^{L[U]}(\xi_1\ldots\xi_n,\zeta_1\ldots\zeta_m)$ $(\xi_1<\ldots<\xi_n<\kappa,\zeta_1<\ldots<\zeta_m\in C)$. Then $j(x)=$ $=t^Q(\xi_1\ldots\xi_n,\zeta_1\ldots\zeta_m)$ and $j'\bar{j}(x)=t^Q(\xi_1\ldots\xi_n,\zeta_1\ldots\zeta_m)$ so $j(x)=j'\bar{j}(x)$.

$$\begin{aligned} x\in U &\Leftrightarrow \kappa\in j(x) \qquad (13.2)\\ &\Leftrightarrow \kappa\in j'\bar{j}(x)\\ &\Leftrightarrow \kappa\in\bar{j}(x) \qquad (\kappa<\bar{j}(\kappa) \text{ so } j'(\kappa)=\kappa)\\ &\Leftrightarrow x\in V. \end{aligned}$$

Thus $U\cap L[U]=V$. □

L[U] is not a mouse but in many ways it behaves like one. In particular we can obtain a core L[U] of which all models L[U] are iterates. We need a preliminary lemma.

Lemma 13.6. *If L[U], L[V] are such that U,V are normal measures on κ in L[U], L[V] respectively, then L[U]=L[V].*

Proof: Let Q be a common iterate of L[U], L[V]. Let j,j' be the iteration maps. Let $C=\{\xi: j(\xi)=j'(\xi)=\xi\}$. C is a proper class and just as in the previous lemma every x in L[U] which is a subset of κ is definable from parameters in $\kappa\cup C$; similarly for L[V].

Consider $x\in L[U]$, $x\subseteq\kappa$. Suppose $x=t^{L[U]}(\xi_1\ldots\xi_n,\zeta_1\ldots\zeta_m)$ $(\xi_1<\ldots<\xi_n<\kappa,\zeta_1<\ldots<\zeta_m\in C)$. Then $j(x)=t^Q(\xi_1\ldots\xi_n,\zeta_1\ldots\zeta_m)$. If $y=t^{L[V]}(\xi_1\ldots\xi_n,\zeta_1\ldots\zeta_m)$ then $j'(y)=t^Q(\xi_1\ldots\xi_n,\zeta_1\ldots\zeta_m)$ so $j(x)=$ $=j'(y)$. Suppose $x\in U$. Then $y\in V$. But for $\gamma<\kappa$

$$\begin{aligned} \gamma\in x &\Leftrightarrow \gamma\in j(x) \qquad (13.3)\\ &\Leftrightarrow \gamma\in j'(y)\\ &\Leftrightarrow \gamma\in y \end{aligned}$$

so x=y. Hence $x\in V$. So $U\subseteq V$. Similarly $V\subseteq U$ so U=V, L[U]=L[V]. □

Of course, lemma 13.5 could have been derived directly from the proof of this lemma.

Now there may (so far as we know) be lots of different normal measures on κ; but for any two, U,V we have shown that $U\cap L[U]=$ $V\cap L[V]$. Cutting down to L[U] loses the sets where U and V differ. In part six we shall return to the problem of inner models with different measures at a point.

Lemma 13.7. *Suppose U, V are normal measures on κ,λ in L[U], L[V] respectively. If $\kappa<\lambda$ then L[V] is an iterate of L[U].*

Proof: Let $\langle\langle L[U_\alpha]\rangle_{\alpha\in On},\langle\pi_{\alpha\beta}\rangle_{\alpha\leq\beta\in On}\rangle$ be the iterated ultrapower of L[U]; let $\kappa_\alpha=\pi_{0\alpha}(\kappa)$. Let α be least such that $\kappa_\alpha\geq\lambda$, and suppose (to get a contradiction) that $\kappa_\alpha>\lambda$.

Let $Q=L[U_\theta]$ be a common iterate of L[U], L[V]. α must be a successor ordinal; say $\alpha=\beta+1$. Suppose $\lambda=\pi_{\beta\alpha}(f)(\kappa_\beta)$. If $f=$ $=t^{L[U_\beta]}(\xi_1\ldots\xi_n,\zeta_1\ldots\zeta_m)$ where $\xi_1<\ldots<\xi_n<\kappa_\beta$, $\zeta_1\ldots\zeta_m\in C=\{\xi:\pi_{0\theta}(\xi)=$ $=\pi'(\xi)=\xi\}$ and π' is the iteration map from L[V] to Q, then

$$\pi_{\beta\alpha}(f)=t^{L[U_\alpha]}(\xi_1\ldots\xi_n,\zeta_1\ldots\zeta_m) \tag{13.4}$$

so there is a term t' such that

$$\lambda=t'^{L[U_\alpha]}(\xi_1\ldots\xi_n,\kappa_\beta,\zeta_1\ldots\zeta_m) \tag{13.5}$$

so $\lambda=t'^{Q}(\xi_1\ldots\xi_n,\kappa_\beta,\zeta_1\ldots\zeta_m)=\pi'(t'^{L[V]}(\xi_1\ldots\xi_n,\kappa_\beta,\zeta_1\ldots\zeta_m))$ so $\lambda\in rng(\pi')$. Contradiction! So $\kappa_\alpha=\lambda$. The result therefore follows by lemma 13.6. □

Compare the proof of lemmas 13.7 and 13.6 with that of lemma 10.16. The large class of fixed points C enables us to carry through the argument.

Definition 13.8. *(i) If U is a normal measure on κ in L[U] then L[U] is called a ρ-model.*

(ii) Let L[U] be the ρ-model with minimal critical point. L[U] is called the core ρ-model.

(iii) The critical point of the core ρ-model is called the critical measure ordinal.

$0^\dagger$

$0^\dagger$ is intended to code embeddings of ρ-models just as $0^\#$ coded those of L. It may also give insight into why ρ-models behave like mice.

The theory of double mice will not be developed here in any detail. One reason is that it is similar in most respects to the theory of single mice and it would be tedious to repeat all the details. Another is that we really do not need any of this material later on (other than the main definition 13.17).

Definition 13.9. *Suppose $\underline{M}=\langle M,\in_M,U,V\rangle$. $\underline{M}$ is a double premouse provided $\underline{M}\models R^+$*

U,V are normal measures on κ,λ respectively in M and $\kappa<\lambda$.

Definition 13.10. *Let $\langle h_i : i\in On\rangle$ be such that $h_i\in 2$, all $i\in On$. Let $\underline{M}$ be a double premouse. Then $\langle\langle \underline{M}_\alpha\rangle_{\alpha\in On},\langle \pi_{\alpha\beta}\rangle_{\alpha\leq\beta\in On}\rangle$ is called an iterated ultrapower of $\underline{M}$ with index $\langle h_i : i\in On\rangle$ provided*

(i) $\langle\langle \underline{M}_\alpha\rangle_{\alpha\in On},\langle \pi_{\alpha\beta}\rangle_{\alpha\leq\beta\in On}\rangle$ is a commutative system;

(ii) $\underline{M}_0=\underline{M}$; $\underline{M}=\langle M_\alpha,\in_{M_\alpha},U_\alpha,V_\alpha\rangle$;

(iii) $\pi_{\alpha\alpha+1}:\underline{M}_\alpha\to_{U_\alpha}\underline{M}_{\alpha+1}$ if $h_\alpha=0$;

(iv) $\pi_{\alpha\alpha+1}:\underline{M}_\alpha\to_{V_\alpha}\underline{M}_{\alpha+1}$ if $h_\alpha=1$;

(v) $\langle \underline{M}_\lambda,\langle \pi_{\alpha\lambda}\rangle_{\alpha\leq\lambda}\rangle$ is a direct limit of $\langle\langle \underline{M}_\alpha\rangle_{\alpha<\lambda},\langle \pi_{\alpha\beta}\rangle_{\alpha\leq\beta<\lambda}\rangle$ when λ is a limit ordinal.

It is easy to see that for each $\langle h_i : i\in On\rangle$ with $h_i\in 2$, all $i\in On$, there is an iterated ultrapower of $\underline{M}$ with index $\langle h_i : i\in On\rangle$.

Definition 13.11. *A double premouse $\underline{M}$ is iterable if and only if for all $h:On\to 2$ there is an iterated ultrapower of $\underline{M}$, $\langle\langle \underline{M}_\alpha\rangle_{\alpha\in On},\langle \pi_{\alpha\beta}\rangle_{\alpha\leq\beta\in On}\rangle$ with index h such that for all α $\underline{M}_\alpha$ is transitive.*

Definition 13.12. *A double premouse is a dagger provided that $\underline{M}$ is iterable and, letting $\underline{M}=\langle M,\in_M,U,V\rangle$, V is a normal measure on the largest limit ordinal of $\underline{M}$ in $\underline{M}$.*

If there is a dagger then there is an inner model with a measurable cardinal; indeed, every uncountable cardinal is the critical point of some ρ-model.

Lemma 13.13. *Suppose $\underline{M}=J_\alpha^{U,V}$ is a dagger. Then U is a normal measure in $L[U]$.*
Proof: Suppose U,V are normal measures on κ,λ respectively in $\underline{M}$. If U is not a normal measure in $L[U]$ then there is a least β such that U is not a normal measure in J_β^U. Let $h:On\to 2$ be defined by $h(\alpha)=1$, all α. Let $\langle\langle \underline{M}_\alpha\rangle_{\alpha\in On},\langle \pi_{\alpha\beta}\rangle_{\alpha\leq\beta\in On}\rangle$ be the iterated ultrapower of $\underline{M}$ with index h. Say $\underline{M}_\alpha=\langle M_\alpha,\in_{M_\alpha},U,V_\alpha\rangle$. Let $\lambda_\alpha=\pi_{0\alpha}(\lambda)$. Pick α with $\lambda_\alpha>\beta$. Then $J_\beta^U=J_\beta^M$ so U is a normal measure on κ in J_β^U. □

Now a literal repetition of the proof of corollary 11.21 shows that if $\underline{M}$ is a dagger and $\underline{M}\models\lambda$ is the largest limit ordinal then $H_\lambda^M=J_\lambda^M$; and so $\underline{M}$ is acceptable.

Suppose $\underline{M}$, $\underline{M}'$ are daggers. Let θ,θ' be regular cardinals such that $\theta'>\theta>\overline{\overline{M}},\overline{\overline{M}}'$. Let $h:On\to 2$ with $h(\alpha)=0$ $(\alpha<\theta)$, $h(\alpha)=1$ $(\alpha\geq\theta)$. Let $\langle\langle \underline{M}_\alpha\rangle_{\alpha\in On},\langle \pi_{\alpha\beta}\rangle_{\alpha\leq\beta\in On}\rangle$, $\langle\langle \underline{M}'_\alpha\rangle_{\alpha\in On},\langle \pi'_{\alpha\beta}\rangle_{\alpha\leq\beta\in On}\rangle$ be the iterated ultrapowers of $\underline{M},\underline{M}'$ respectively with index h. Then it is easily seen that $\underline{M}_{\theta'}=\underline{M}'_{\theta'}$. (They are comparable just as in lemma 8.18; but $On_{M_{\theta'}}=\theta'+\omega=On_{M'_{\theta'}}$). So any two daggers have a common iterate.

Suppose $\underline{M}$ is a dagger. Let $X=h_M(\phi)$. Then if $\sigma:\underline{M}'\cong\underline{M}|X$ and $\underline{M}'$ is transitive, $\underline{M}'$ will be a dagger. Clearly $\rho_{M'}=1, p_{M'}=\phi$. It is easily seen that $A_{M'}$ is $\Sigma_1(\underline{M})$ with no parameters and $A_{M'}$ $\notin M$. So for any dagger $\underline{M}$, $\rho_M=1$ and $p_M=\phi$. In fact the proof of lemma 10.16 adapts to show that $\underline{M}$ is an iterate of $\underline{M}'$ and, indeed, an iterate with the special property that there is an index h for the iteration that is monotone. Actually this is not a special property at all: every iterate has such an index. This is left to the reader.

Now let $\underline{M}'$ be fixed as the core dagger. Let $\langle\langle \underline{M}^h_\alpha\rangle_{\alpha\in On},\langle\pi^h_{\alpha\beta}\rangle_{\alpha\leq\beta\in On}\rangle$ denote its iterated ultrapower with index h. Let θ, θ' be uncountable cardinals. Let $h(\alpha)=0$ $(\alpha<\theta)$ and $h(\alpha)=1$ $(\alpha\geq\theta)$. Let $\underline{M}=\underline{M}^h_{\theta'}$. (We should have to improve this terminology if we wanted to study double mice in any depth!) Suppose $\underline{M}'=\langle M',\in,U',V'\rangle$ where U',V' are normal measures on κ,λ respectively in $\underline{M}'$. Let $\kappa_\alpha=\pi_{0\alpha}(\kappa)$ $\lambda_\alpha=\pi_{0\alpha}(\lambda)$ $(\alpha<\theta')$. Let $C=\{\kappa_\alpha:\alpha<\theta\}$, $D=\{\lambda_\alpha:\theta\leq\alpha<\theta'\}$. Just as in lemma 12.9

Lemma 13.14. *(a) $C\cup\{\theta\}\cup D$ contains all uncountable cardinals less than θ'.*

(b) C has order type θ; D has order type θ'.

(c) C is closed unbounded in θ if θ is regular; D is closed unbounded in θ' if θ' is regular.

(d) If ϕ is any formula, $\alpha_1<\ldots<\alpha_n,\alpha'_1<\ldots<\alpha'_n$ are in C, $\beta_1<\ldots<\beta_m,\beta'_1<\ldots<\beta'_m$ are in D then

$$\underline{J}^U_{\theta'}\models\phi(\kappa_{\alpha_1}\ldots\kappa_{\alpha_n},\lambda_{\beta_1}\ldots\lambda_{\beta_m})\leftrightarrow\phi(\kappa_{\alpha'_1}\ldots\kappa_{\alpha'_n},\lambda_{\beta'_1}\ldots\lambda_{\beta'_m}).$$

(e) Every a $J^U_{\theta'}$ is $\underline{J}^U_{\theta'}$-definable from $C\cup D$ (where $\underline{M}=\langle M,\in,U,V\rangle$). Furthermore C,D are the unique classes with this property.

Now let κ'_α be the critical point of the α-th ρ-model. Let $C'=\{\kappa'_\alpha:\alpha<\theta\}$. By lemma 13.13 $C\subseteq C'$.

Lemma 13.15. *Let ϕ be any formula of L_U and suppose $\alpha_1<\ldots<\alpha_n,\alpha'_1<\ldots<\alpha'_n<\theta$. Then for all $\lambda_1\ldots\lambda_m\in D$*

$$\underline{J}^U_{\theta'}\models\phi(\kappa'_{\alpha_1}\ldots\kappa'_{\alpha_n},\lambda_1\ldots\lambda_m)\leftrightarrow\phi(\kappa'_{\alpha'_1}\ldots\kappa'_{\alpha'_n},\lambda_1\ldots\lambda_m).$$

Proof: Suppose $\lambda\in D$. Let $\langle\langle L[U'_\alpha]\rangle_{\alpha\in On},\langle\pi'_{\alpha\beta}\rangle_{\alpha\leq\beta\in On}\rangle$ be the iterated ultrapower of the core ρ-model. If λ is a limit cardinal with $cf(\lambda)>$ $>\theta$ then $\pi'_{\alpha\beta}(\theta)=\theta$ for all $\alpha\leq\beta<\theta$. But U is a normal measure on θ in

$L[U]$ so $U'=U$. Picking $\lambda_1\ldots\lambda_m \in D$ cardinals of cofinality $>\theta$,

$$\underline{J}^{U}_{\theta'} \models \phi(\kappa'_{\alpha_1}\ldots\kappa'_{\alpha_n}, \lambda_1\ldots\lambda_m) \Leftrightarrow \{\langle \xi_1\ldots\xi_n\rangle : \underline{J}^{U'}_{\theta}\models\phi(\xi_1\ldots\xi_n, \lambda_1\ldots\lambda_m)\}\in U_0'^{n} \Leftrightarrow \underline{J}^{U}_{\theta'}\models\phi(\kappa'_{\alpha_1}\ldots\kappa'_{\alpha_n}, \lambda_1\ldots\lambda_m). \qquad (13.6)$$

But then this is true for <u>all</u> $\lambda_1\ldots\lambda_m \in D$. □

<u>Corollary 13.16</u>. *$C'=C$.*

It follows that if we let h be the constant zero function and let $\kappa_\alpha=\pi^h_{0\alpha}(\kappa)$ then $\langle\kappa_\alpha:\alpha\in On\rangle$ enumerates the critical points of all ρ-models. In particular, κ is the critical measure ordinal.

<u>Definition 13.17</u>. *Let θ,θ' be uncountable cardinals, $\theta<\theta'$. Let C,D, and U be as in lemma 13.14. Let $\langle\kappa_\alpha:\alpha<\theta\rangle$,$\langle\lambda_\beta:\beta<\theta'\rangle$ enumerate C,D respectively. Let $\{\phi_n:n<\omega\}$ enumerate the formulae of L_U with m(n) free variables of even index and p(n) of odd index. Then (arranging the even before the odd)*

$$0^{\dagger}=\{n:\underline{J}^{U}_{\theta'}\models\phi_n(\kappa_0\ldots\kappa_{m(n)-1},\lambda_0\ldots\lambda_{p(n)-1})\}.$$

$0^\dagger$ is plainly independent of our choice of θ'. We may take the statement "$0^\dagger$ exists" to assert the existence of a dagger.

<u>Lemma 13.18</u>. *If $0^\dagger$ exists then every ρ-model may be non-trivially elementarily embedded in itself.*

Proof: Every $f:On\to On$ which is strictly increasing generates a map $\pi:L[U]\to L[U]$ such that $\pi|\kappa=id|\kappa$, $\pi(\lambda_\alpha)=\lambda_{f(\alpha)}$ where $h:On\to 2$ is the function $h(\alpha)=0$ $(\alpha<\theta)$, $h(\alpha)=1$ $(\alpha\geq\theta)$, $\underline{M}^h_\theta=\langle M^h_\theta,\in,U,V_\theta\rangle$, say, and $\lambda_\alpha=\pi^h_{0\theta+\alpha}(\lambda)$. □

Finally we shall obtain a dagger from some $j:L[U]\to_e L[U]$ that is non-trivial, where $L[U]$ is a ρ-model with κ critical and $j|\kappa=id|\kappa$. Let $\langle\langle L[U_\alpha]\rangle_{\alpha\in On},\langle\pi_{\alpha\beta}\rangle_{\alpha\leq\beta\in On}\rangle$ be the iterated ultrapower of $L[U]$. Let $V=\{X\in L[U]:\lambda\in j(X)\wedge X\subseteq\lambda\}$ where λ is the critical point of j. Then V is a normal measure on λ in $J^{U,V}_{\lambda+1}$. By lemma 12.20 if we let $\langle\langle\underline{M}_\alpha\rangle_{\alpha\in On},\langle\pi_{\alpha\beta}\rangle_{\alpha\leq\beta\in On}\rangle$ be an iterated ultrapower of $J^{U,V}_{\lambda+1}$ by V then each $\underline{M}_\alpha$ is well-founded. It follows just as for $0^\#$ that $J^{U,V}_{\lambda+1}\cap P(\lambda)\subseteq L[U]$, so that $J^{U,V}_{\lambda+1}$ is a double premouse. The reader is left to check (exercise 1) that iterability will follow once we know that for all monotone h and all iterated ultrapowers of $J^{U,V}_{\lambda+1}$ $\langle\langle\underline{M}^h_\alpha\rangle_{\alpha\in On},\langle\pi^h_{\alpha\beta}\rangle_{\alpha\leq\beta\in On}\rangle$ $\underline{M}^h_\alpha$ is well-founded (all α).

Observe that there are maps $j_\alpha : L[U_\alpha] \to_e L[U_\alpha]$ such that

(i) $j_\alpha | \kappa_\alpha+1 = id | \kappa_\alpha+1$ (κ_α the critical point of $L[U_\alpha]$)

(ii) $j_\beta \pi_{\alpha\beta} = \pi_{\alpha\beta} j_\alpha$ (all $\alpha \leq \beta$)

(iii) $j_0 = j$.

Let $V_\alpha = \{X \in L[U_\alpha] : X \subseteq \lambda_\alpha \wedge \lambda_\alpha \in j_\alpha(X)\}$, where λ_α is the critical point of j_α. (In fact $\lambda_\alpha = \pi_{0\alpha}(\lambda)$. If $\gamma < \pi_{0\alpha}(\lambda)$ then $\gamma < \pi_{0\alpha}(\bar{\gamma})$ for some $\bar{\gamma} < \lambda$ since λ must be regular in $L[U]$. Say $\gamma = \pi_{0\alpha}(f)(\kappa_{\alpha_1} \ldots \kappa_{\alpha_n})$ where $\alpha_1 < \ldots < \alpha_n < \alpha$ and $f : \kappa^n \to \bar{\gamma}$. Then $j_\alpha(\gamma) = j_\alpha(\pi_{0\alpha}(f)(\kappa_{\alpha_1} \ldots \kappa_{\alpha_n})) = \pi_{0\alpha} j(f)(\kappa_{\alpha_1} \ldots \kappa_{\alpha_n}) = \pi_{0\alpha}(f)(\kappa_{\alpha_1} \ldots \kappa_{\alpha_n})$. But $j_\alpha(\pi_{0\alpha}(\lambda)) = \pi_{0\alpha}(j(\lambda)) > \pi_{0\alpha}(\lambda)$). Then if $\langle \langle \underline{M}^\alpha_\beta \rangle_{\beta \in On}, \langle \pi^\alpha_{\beta\gamma} \rangle_{\beta \leq \gamma \in On} \rangle$ is the weak iterated ultrapower of $L[U_\alpha]$ by V_α each $\underline{M}^\alpha_\beta$ is well-founded.

Given $h : On \to 2$ which is monotone let θ be least such that $h(\theta) = 1$ (if there is one). If $\alpha \leq \theta$ we shall embed $\underline{M}^h_\alpha$ into $L[U_\alpha]$; if $\alpha > \theta$ we shall embed $\underline{M}^h_\alpha$ into $\underline{M}^\theta_{\alpha-\theta}$. Suppose $\underline{M}^h_\alpha = \langle M^h_\alpha, \in_{M^h_\alpha}, U^h_\alpha, V^h_\alpha \rangle$. For $\alpha \leq \theta$ let $\sigma_\alpha = id | J^{U,V}_{\lambda+1}$ if $\alpha = 0$; let σ_α be defined by

$$\sigma_\alpha(\pi^h_{\beta\alpha}(f)(\hat{\kappa}_\beta)) = \pi_{\beta\alpha}(\sigma_\beta(f))(\kappa_\beta) \tag{13.7}$$

where $\hat{\kappa}_\alpha = \pi^h_{0\alpha}(\kappa)$, if $\alpha = \beta+1$. If α is a limit $\sigma_\alpha(\pi^h_{\beta\alpha}(x)) = \pi_{\beta\alpha}(\sigma_\beta(x))$. Then $\sigma_\theta : \langle M^h_\theta, \in_{M^h_\theta}, U^h_\theta \rangle \to_{\Sigma_0} L[U_\theta]$ and each $\underline{M}^h_\alpha$ is well-founded ($\alpha < \theta$). (Actually σ_θ is $id | M^h_\theta$.)

Now define $\sigma_{\theta+\alpha} : M^h_{\theta+\alpha} \to M^\theta_\alpha$ by

$$\sigma_{\theta+\alpha+1}(\pi^h_{\theta+\alpha\,\theta+\alpha+1}(f)(\hat{\lambda}_\alpha)) = \pi^\theta_{\alpha\alpha+1}(\sigma_{\theta+\alpha}(f))(\pi^\theta_\alpha(\lambda_\theta)) \tag{13.8}$$

(where $\hat{\lambda}_\alpha = \pi^h_{\theta+\alpha}(\lambda)$)

$$\sigma_{\theta+\alpha}(\pi^h_{\theta+\beta\,\theta+\alpha}(x)) = \pi^\theta_{\beta\alpha}(\sigma_{\theta+\beta}(x)) \quad (\alpha \text{ a limit}).$$

It is necessary to check that

$$\sigma_{\theta+\alpha} : \underline{M}^h_{\theta+\alpha} \to_{\Sigma_0} \langle J^M_\delta, V^\alpha_\theta \rangle \tag{13.9}$$

where $\delta = \pi^\theta_\alpha(\lambda_\theta)^+$ and V^α_θ is the α-th iteration of V_θ. (13.9) follows from

$$\sigma_\theta : \underline{M}^h_\theta \to_{\Sigma_0} \langle J^{U_\theta}_{\lambda^+_\theta}, V_\theta \rangle \tag{13.10}$$

in the usual way. So we must show $\sigma_\theta(x \cap V^h_\theta) = \sigma_\theta(x) \cap V_\theta$. This is proved by induction on θ. If $\theta = 0$ it is immediate. If $\theta = \theta'+1$ then

$$\begin{aligned} \pi^h_{\theta'\theta}(f)(\hat{\kappa}_{\theta'}) \in V^h_\theta &\leftrightarrow \{\xi : f(\xi) \in V^h_{\theta'}\} \in U^h_\theta \\ &\leftrightarrow \{\xi : \sigma_{\theta'}(f)(\xi) \in V_{\theta'}\} \in U_\theta \\ &\leftrightarrow \{\xi : \lambda_{\theta'} \in j_{\theta'}(\sigma_{\theta'}(f)(\xi))\} \in U_\theta \\ &\leftrightarrow \pi_{\theta'\theta}(\lambda_{\theta'}) \in \pi_{\theta'\theta}(j_{\theta'}(\sigma_{\theta'}(f)))(\kappa_{\theta'}) \end{aligned} \tag{13.11}$$

$$\leftrightarrow \lambda_\theta \in j_\theta(\pi_{\theta',\theta}(\sigma_{\theta'}(f))(\kappa_{\theta'}))$$
$$\leftrightarrow \lambda_\theta \in j_\theta(\sigma_\theta(\pi^h_{\theta',\theta}(f)(\kappa_{\theta'})))$$
$$\leftrightarrow \sigma_\theta(\pi^h_{\theta',\theta}(f)(\kappa_{\theta'}))\in V_\theta.$$

Suppose θ is a limit. Then

$$\pi^h_{\theta',\theta}(x)\in V^h_\theta \leftrightarrow x\in V^h_{\theta'} \qquad (13.12)$$
$$\leftrightarrow \sigma_{\theta'}(x)\in V_{\theta'}$$
$$\leftrightarrow \lambda_{\theta'}\in j_{\theta'}(\sigma_{\theta'}(x))$$
$$\leftrightarrow \lambda_\theta \in \pi_{\theta',\theta}(j_{\theta'}(\sigma_{\theta'}(x)))$$
$$\leftrightarrow \lambda_\theta \in j_\theta(\pi_{\theta',\theta}(\sigma_{\theta'}(x)))$$
$$\leftrightarrow \pi_{\theta',\theta}(\sigma_{\theta'}(x))\in V_\theta.$$

The proof is complete. We have shown

Lemma 13.19. *If there is a non-trivial elementary embedding of a ρ-model into itself whose critical point is greater than that of the ρ-model then there is a dagger.*

We could prove analogues of lemma 12.22 and lemma 12.23; hence there is a minimal model $L[0^\dagger]$ in which $0^\dagger$ exists. Hence every ρ-model is contained in $L[0^\dagger]$.

SOME DESCRIPTIVE SET THEORY

The rest of the chapter is not needed for any subsequent results in this book.

Definition 13.20. *A mouse $\underline{M}$ with $\rho_M=1$ is called a real mouse.*

For example, a sharp is a real mouse. Note that an iterable premouse $\underline{M}$ with $\rho_M=1$ must be a real mouse. Fix some recursive function $\tau:\omega \leftrightarrow (\omega\times(\omega)^{<\omega})$. Let A^τ denote $\{n:\tau(n)\in A\}$, when $A\subseteq(\omega\times(\omega)^{<\omega})$.

Lemma 13.21. *There is a Π^1_2 formula ϕ such that*

$\phi(a) \leftrightarrow$ there is a real mouse $\underline{M}$ with $a=(A^p_M)^\tau$ (some $p\in PA^1_M$)

Proof: By lemma 3.26 there is a Π_2 sentence σ such that

$$\langle J_1, A\rangle\models\sigma \leftrightarrow \text{there is an acceptable premouse } \underline{M} \text{ with } \rho_M=1 \qquad (13.13)$$
$$\text{and } T^p_M\cap(\omega\times(\omega)^{<\omega})=A.$$

σ can be written as an arithmetic condition ψ on A^τ. The complexity arises from the iterability requirement.

If $R\subseteq\omega^2$ is a well-order of a subset of ω then $\underline{M}_R$ denotes the (or rather an) α-th iterate of $\underline{M}$ when α is the order-type of R. $\phi(a)$ is to be the formula

$$\psi(a) \wedge \forall R(R \text{ well-orders a subset of } \omega \Rightarrow \Rightarrow \underline{M}_R \text{ is well-founded}) \tag{13.14}$$

where $\underline{M}$ is some p-sound premouse with $\rho_M=1$ and $(A_M^p)^\tau=a$.
Now "R well-orders a subset of ω" is

$$(\forall b\subseteq\omega)(\exists n\in b)(\forall m\in b)(-Rmn) \tag{13.15}$$

together with the arithmetic requirement that R be a linear order. By lemma 8.6 we need only consider well-orders R of ω, and the lemma follows from the following claim.

Claim There is a Π_1^1 formula χ such that

$$\chi(a,R) \Leftrightarrow \text{there is a p-sound premouse } \underline{M} \text{ with } \rho_M=1, (T_M^p\cap(\omega\times(\omega)^{<\omega}))^\tau=a \text{ and } \underline{M}_R \text{ is well-founded.}$$

Proof: The proof is very like that of lemma 8.11. Let $\hat{R}$ be the field of R. Given $v\in[\hat{R}]^n$ and $m\leq n$ define a function f^{uv} by

$$f^{uv}(\gamma_1\ldots\gamma_n)=f(\gamma_{i_1}\ldots\gamma_{i_m}) \tag{13.16}$$

where $v=\{k_1\ldots k_n\}, k_1R\ldots Rk_n$ and $u=\{k_{i_1}\ldots k_{i_m}\}$ $(i_1<\ldots<i_m)$.
There is a formula ψ^* such that

$$\underline{M}\models\psi^*(f,g,n) \Leftrightarrow \underline{M}_R\models\pi_{OR}(f)(\kappa_{\alpha_1}\ldots\kappa_{\alpha_n})\in\pi_{OR}(g)(\kappa_{\alpha_1}\ldots\kappa_{\alpha_n}) \tag{13.17}$$

where π_{OR} is the iteration map of $\underline{M}$ to $\underline{M}_R$ and κ_α is the critical point of the α-th iterate of $\underline{M}$.

Let $Z_q=\{\langle i,n_1\ldots n_p\rangle : h_M(i,\langle n_1\ldots n_p\rangle)$ is a function of $\kappa\to\kappa^q$ where $\kappa=\kappa_0.\}$

Let $T_R'=\{\langle x,u\rangle\in J_1 : x\in Z_q \wedge u\in[\hat{R}]^q \text{ (some q)}\}$. Define a relation on T_R' by

$$S(\langle x,u\rangle,\langle y,v\rangle) \Leftrightarrow \underline{M}\models\psi^*((h_M(x))^{uu\cup v},(h_M(y))^{vu\cup v},\overline{\overline{u\cup v}}). \tag{13.18}$$

Then since $\psi^*((h_M(x))^{uu\cup v},(h_M(y))^{vu\cup v},\overline{\overline{u\cup v}})$ is a Σ_1 property of x,y, u,v,$R\cap(u\cup v)^2$ S is arithmetic in R,a. But $\chi(a,R) \Leftrightarrow$ S is well-founded so $\chi(a,R)$ is Π_1^1. □(Claim)

□

Lemma 13.22. *$O^\#$ is Δ_3^1.*

Proof: There is a recursive function t such that for any sharp $\underline{M}$ $n\in O^\# \Leftrightarrow t(n)\in(A_M^\phi)^\tau$. And "a is $(T_M^\phi\cap(\omega\times(\omega)^{<\omega}))^\tau$ for some sharp $\underline{M}$" is Π_2^1, for it is $\phi(a) \wedge i\in a$, where s(i) is the sentence asserting that the critical point is the largest ordinal. Call this Π_2^1 formula $\psi(a)$. Then

$$n\in O^\# \Leftrightarrow \exists a(\psi(a) \wedge t(n)\in a) \quad (\Sigma_3^1) \tag{13.19}$$
$$\Leftrightarrow \forall a(\psi(a) \Rightarrow t(n)\in a) \quad (\Pi_3^1)$$

□

Note 1: This is the best possible result, for if $O^\#$ were Π_2^1 or Σ_2^1 then it would be in L by the Schönfield absoluteness theorem.

Note 2: $x=0^{\#}$ is Π^1_2, although this is not clear from our approach. We should have done better to define $0^{\#}$ as the unique a satisfying the formula ψ of lemma 13.22.

Lemma 13.23. *Suppose $a\in L[U]$, where U is a ρ-model, and $a\subseteq\omega$. Then there is a real mouse $\underline{M}$ with $a\in M$.*

Proof: $a\in J^U_{\kappa^+}$ where κ is the critical point of L[U] and κ^+ is calculated in L[U]. If $a\in J^U_\kappa$ then $a\in J^U_{\kappa+1}$ which is a real mouse. Otherwise let β be least such that $a\in J^U_{\beta+1}=M$, say. $\underline{M}$ is an iterable premouse, hence acceptable. By chapter 3 exercise 6 if $\rho_M>1$ then $\beta+\omega\in M$; contradiction! So $\rho_M=1$. □

This would be true even if we insisted $p_M=\phi$.

Let $<_M$ be the canonical well-order of the real mouse $\underline{M}$. If $\underline{N}$ is another real mouse, let $<_N$ be its canonical well-order; let $\underline{M}'$, $\underline{N}'$ be comparable iterates with canonical well-orders $<_{M'}, <_{N'}$. Then for $a,b\in P(\omega)\cap M\cap N$

$$a<_M b \Leftrightarrow a<_{M'} b \Leftrightarrow a<_{N'} b \Leftrightarrow a<_N b.$$

Lemma 13.24. *In L[U] there is a Δ^1_3 well-order of the reals.*

Proof: Say $a<^*b$ if there is a real mouse $\underline{M}$ in which $a<_M b$. We have just seen that this defines a linear order. In fact it is a well-order, for given $u\subseteq P(\omega)$, $u\in L[U]$, let $\gamma+1$ be least$>\kappa$ such that $J^U_{\gamma+1}\cap u\neq\phi$. Then $P(\omega)\cap J^U_\gamma\neq P(\omega)\cap J^U_{\gamma+1}$, so $\underline{J}^U_{\gamma+1}$ is a real mouse. Take $a\in u$ least in its canonical well-order. Then a will be $<^*$ minimal in u. (We are using the result of Chapter 8 exercise 3.)

Now suppose $\underline{M}$ is a real mouse and let $\hat{a}=(T^p_M\cap(\omega\times(\omega)^{<\omega}))^\tau$. Then

$$a<_M b \Leftrightarrow (\exists i,x,j,y)(a=h(i,x) \wedge b=h(j,y) \wedge h(i,x)Wh(j,y)) \qquad (13.20)$$

which is arithmetic in $\hat{a}$. So

$$a<^*b \Leftrightarrow \exists\underline{M}(\underline{M}\text{ is a real mouse} \wedge a<_M b) \qquad (\Sigma^1_3) \qquad (13.21)$$

$$\Leftrightarrow \forall\underline{M}(\underline{M}\text{ is a real mouse} \Rightarrow a<_M b) \qquad (\Pi^1_3). \quad □$$

This is known to be the best possible.

Exercises

1. Suppose $\underline{M}$ is a double premouse. Show that if $\underline{N}$ is an iterate of $\underline{M}$ then there is $h:On\to 2$ monotone such that $\underline{N}=\underline{M}^h_\alpha$ for some α.

2. Show that Consis(ZFC+there is a measurable cardinal) ⇒ Consis(ZFC +" the critical measure ordinal is uncountable".

3. Show that $0^\dagger$ is Δ^1_3. Use the method of this chapter to show that if V=L then there is a Δ^1_2 well-order of the reals.

PART FOUR: THE CORE MODEL

This short but essential part introduces the core model. Once mice are to hand the definition (14.3) is very simple. Chapter 14 gives the analysis of the Ω_κ structure of K and proves that K is a model for GCH and is O-maximal.

Chapter 15 generalises the work of chapter 12 and shows on the one hand that all manner of # iterations can be coded by mice but on the other that before we get very far the situation degenerates into chaos. Mice are, we think, a better way to describe sharps than reals and we single out some "sharplike" mice $M^\#$. At this stage $M^\#$ is only defined if $M \models V=K$; other cases require the covering lemma and are dealt with in part five.

Chapter 16 introduces a new technique to show that elementary non-trivial embeddings of models M for V=K into themselves imply the existence of $M^\#$ unless K has its "maximal" form when there is a ρ-model. The ρ-model is a sort of $K^\#$, and we have seen in chapter 13 that it behaves like a real mouse (it is assumed that the reader is becoming inured to the absurdity of the terminology).

Chapter 17 covers as many applications as can be covered without the covering lemma. The covering lemma and the singular cardinal hypothesis are the most important applications of K and we reach these in part five. Chapter 17 is not needed for the proof of the covering lemma and could be omitted.

We may now define K, the core model. Perhaps the simplest definition would be to make K the union of all mice; we should then have to prove $K \models ZFC$, and this is dull work, so instead we adopt a definition that gives $K \models ZFC$ immediately but leaves us to prove that K is the union of all mice. This has a couple of advantages: firstly we get a model even if there are no mice. Secondly, we get a structure with levels, indeed, a model of R^{+}. Although it is not a model of RA we can still get GCH.

Lemma 14.1. *If $\underline{M}, \underline{N}$ are core mice with mouse iterations $\langle\langle \underline{M}_{\alpha} \rangle_{\alpha \in On}, \langle \pi'_{\alpha\beta} \rangle_{\alpha \leq \beta \in On}\rangle$, $\langle\langle \underline{N}_{\alpha} \rangle_{\alpha \in On}, \langle \pi_{\alpha\beta} \rangle_{\alpha \leq \beta \in On}\rangle$ respectively and if $\underline{N}_{\omega}$, $\underline{M}_{\omega}$ have the same critical point κ, then $\underline{N} = \underline{M}$.*

Proof: Let $\theta > \bar{\bar{M}}, \bar{\bar{N}}$ be regular. If $M_{\theta} \in N_{\theta}$ then $C_{M_{\omega}} = C_{M_{\theta}} \cap \kappa \in \Pi_1(\underline{M}_{\theta}^{n(M)})$ so $C_{M_{\omega}} \in N_{\omega}$. But then $\underline{N}_{\omega} \models cf(\kappa) = \omega$; but $\underline{N}_{\omega} \models \kappa$ inaccessible. Similarly $N_{\theta} \notin M_{\theta}$. So $\underline{M}_{\theta} = \underline{N}_{\theta}$. So $\underline{M} = \underline{N}$. □

Definition 14.2. *$D = \{\langle \xi, \kappa \rangle$: $\underline{N}$ is a mouse with critical point κ, $otp(C_N) = \omega \wedge \wedge \xi \in C_N\}$.*

By lemma 14.1 $\underline{N}$ is unique if it exists.

Definition 14.3. *$\underline{K} = L[D]$*

So $\underline{K} \models ZFC$.

Lemma 14.4. *If $\underline{N}$ is a mouse then $\underline{N} \in K$.*

Proof: We may assume $otp(C_N) = \omega$, since $\underline{N} \in K$ if and only if some mouse iterate of core$(\underline{N}) \in K$. Suppose $\underline{N} = \underline{J}_{\alpha}^{U}$. It suffices to show $U \in K$. Suppose κ is the critical point of $\underline{N}$. Then for all $x \in P(\kappa) \cap N$

$$x \in U \Leftrightarrow C_N \setminus \gamma \subseteq x, \text{ some } \gamma < \kappa. \quad (14.1)$$

So there is $U' \in K$ such that $U' \cap N = U$. It follows that $U \in K$. □

Definition 14.5. *$\underline{K}_{\alpha} = \underline{J}_{\alpha}^{D}$.*

<u>Lemma 14.6</u>. *$K=L \Leftrightarrow O^\#$ does not exist.*
Proof: If $O^\#$ exists then by lemma 14.4 $O^\# \in K$. But $O^\# \notin L$. So $K=L$ implies that $O^\#$ does not exist. On the other hand if $O^\#$ does not exist then there are no mice by lemma 12.12. So $D=\phi$; so $K=L$. □

Suppose until further notice (corollary 14.14) that $O^\#$ exists.

<u>Definition 14.7</u>. *$Q_\kappa = \bigcup\{N : \underline{N}$ is a mouse with critical point κ and $\omega\rho_N^{n(N)+1} < \kappa\}$.*
$F_\kappa = \{X \subseteq \kappa$: X contains a closed unbounded set$\}$.

Q_κ is only interesting if κ is regular $>\omega$.

<u>Lemma 14.8</u>. *If κ is regular then $\underline{Q}_\kappa = \langle Q_\kappa, \in, F_\kappa \cap Q_\kappa \rangle$ is a premouse.*
Proof: Since κ is regular every mouse $\underline{N}$ with critical point κ and $\omega\rho_N^{n(N)+1} < \kappa$ is the κth mouse iterate of its core. By lemma 8.16 each such mouse is of the form $\underline{J}_\alpha^F \kappa$ for some $\alpha > \kappa$. So letting $\theta = On \cap Q_\kappa$ $\underline{Q}_\kappa = \underline{S}_\theta^F \kappa$ and F_κ is a normal measure on κ in Q_κ. □

Call this θ θ_κ. Since $F_\kappa \cap Q_\kappa$ is countably complete $\underline{Q}_\kappa$ is iterable.

<u>Lemma 14.9</u>. $\omega\rho_{Q_\kappa} \leq \kappa$.
Proof: Let $\theta_\kappa = \omega\bar{\theta}$ and suppose $\bar{\theta}$ is a limit ordinal, otherwise the result is trivial. It suffices to show that $Q_\kappa \models On = h(\kappa)$. So let $X = h_{Q_\kappa}(\kappa)$. Suppose $\kappa < \omega\gamma < \theta_\kappa$ and for some n $\omega\rho_M^{n+1} < \kappa$ where $\underline{M} = \underline{J}_\gamma^F \kappa$. Then $\underline{M}$ is a mouse. Let $\bar{\kappa}$ be the ωth member of C_M. By lemma 14.1 γ is the unique $\bar{\gamma}$ in Q_κ such that

(i) $\omega\bar{\gamma} > \kappa$

(ii) for some n $\omega\rho^n_{J_{\bar{\gamma}}^F \kappa} < \kappa$

(iii) $\bar{\kappa}$ is the ωth member of $C_{J_{\bar{\gamma}}^F \kappa}$

Hence $\gamma \in X$.

Suppose $\kappa < \delta < \theta$. There is $\omega\gamma > \delta$ with $\omega\rho_M^{n(M)+1} < \kappa$, $\underline{M} = \underline{J}_\gamma^F \kappa$, $\omega\gamma < \theta$. So κ is definable over $\underline{M}$ from parameters in $\kappa \cup \{p_M^1 \ldots p_M^{n+1}\}$; but each p_M^k is definable over $\underline{M}$ and $\underline{M}$ is Σ_1-definable over Q_κ from γ. So $\delta \in X$. □

<u>Lemma 14.10</u>. $\omega\rho_{Q_\kappa} = \kappa$.
Proof: Suppose $\omega\rho_{Q_\kappa} < \kappa$. Let $C = C_{Q_\kappa}$. Since $C = \bigcap\{X \in F_\kappa : X \in h_{Q_\kappa}(\omega\rho_{Q_\kappa} \cup p_{Q_\kappa})\}$

C must contain a closed unbounded set. Let $Q^+=\mathrm{rud}_{F_\kappa\cap Q_\kappa}(Q_\kappa)$. Let $C^0=C$, $C^{m+1}=$ the limit points of C^m. So each C^m contains a closed unbounded set. Now given $X\in Q^+$, $X\subseteq\kappa$ there is m such that $X\in\Sigma_{m+1}(\underline{Q}_\kappa)$. So either $(\exists\gamma<\kappa)(C^m\setminus\gamma\subseteq X)$ or $(\exists\gamma<\kappa)(C^m\setminus\gamma\subseteq\kappa\setminus X)$. It follows that F_κ is a normal measure on κ in Q^+. But $\langle Q^+,F_\kappa\rangle$ is amenable. For given $X\in Q^+$, $X\subseteq P(\kappa)$ then $X\subseteq\Sigma_{m+1}(\underline{Q}_\kappa)$ for some m. So

$$X\cap F_\kappa=\{Y\in X:\exists\gamma<\kappa(C^m\setminus\gamma\subseteq Y)\}. \tag{14.2}$$

So $X\cap F_\kappa\in Q^+$. Hence $Q^+=\mathrm{rud}_{F_\kappa}(Q_\kappa)$. Thus $\underline{Q}^+=\langle Q^+,\in,F_\kappa\rangle$ is a premouse. Since F_κ is countably complete $\underline{Q}^+$ is iterable. So $\underline{Q}^+$ is acceptable. Now $P(\omega\rho_{Q_\kappa})\cap Q_\kappa\neq P(\omega\rho_{Q_\kappa})\cap Q^+$; if $\rho_{Q_\kappa}<\rho_{Q^+}$ it follows by chapter 3 exercise 6 that $\theta+\omega\rho_{Q_\kappa}\in Q^+$. Contradiction! So $\omega\rho_{Q^+}\leq\omega\rho_{Q_\kappa}<\kappa$. Hence $\underline{Q}^+$ is a mouse and $Q^+\subseteq Q$. Contradiction! □

It follows that $\bar\theta$ is a limit, so by the proof of lemma 14.9

<u>Lemma 14.11</u>. $p_{Q_\kappa}=\emptyset$.

<u>Lemma 14.12</u>. $K_\kappa\subseteq Q_\kappa$.
Proof: $\underline{K}_\kappa=\underline{J}^D_\kappa$. It suffices to show $K_\kappa\subseteq H^Q_\kappa\kappa$; and so it suffices to show that for all $\gamma<\kappa$ $D\cap\gamma^2\in Q_\kappa$. Suppose $\underline{M}$ is a mouse with critical point $\kappa'<\gamma$ such that otp $C_M=\omega$. Let $\underline{M}'$ be its κth mouse iterate and let $\xi_{\kappa'}=On\cap M'$. Let $\xi=\sup\limits_{\kappa'<\gamma}\xi_{\kappa'}$. Clearly $\xi\leq\theta$.
<u>Claim</u> $\xi\neq\theta$
Proof: Suppose it were. $\{\xi_{\kappa'}:\kappa'<\gamma\}$ is $\Sigma_1(\underline{Q}_\kappa)$; and for all $\kappa'<\gamma$ there is a map in $J^{F_\kappa}_{\xi_{\kappa'}+1}$ of γ onto $P(\gamma)\cap J^{F_\kappa}_{\xi_{\kappa'}}$; so we obtain a $\Sigma_1(\underline{Q}_\kappa)$ map of γ onto $P(\gamma)\cap Q_\kappa$. But then $\omega\rho_{Q_\kappa}\leq\gamma<\kappa$. Contradiction! □(Claim)
So $\xi<\theta$. But $D\cap\gamma^2$ is $\Sigma_1(\underline{J}^{F_\kappa}_\xi)$. So $D\cap\gamma^2\in Q_\kappa$. □

<u>Lemma 14.13</u>. $K_\kappa=H^Q_\kappa\kappa$.
Proof: $K_\kappa\subseteq H^Q_\kappa\kappa$ is proved. Since $H^Q_\kappa\kappa=\bigcup\{J^a_\kappa: a\subseteq\gamma<\kappa \wedge a\in Q_\kappa\}$ it suffices to show that $a\subseteq\gamma<\kappa$, $a\in Q_\kappa \rightarrow a\in K_\kappa$. But suppose ξ is least with $\xi\geq\kappa$ and $a\in J^{F_\kappa}_{\xi+1}$. Then $\underline{J}^{F_\kappa}_{\xi+1}$ is a mouse with $\omega\rho_{J^{F_\kappa}_{\xi+1}}\leq\gamma$. Let $\underline{M}$ be its core with mouse iteration $\langle\langle\underline{M}_\alpha\rangle_{\alpha\in On},\langle\pi_{\alpha\beta}\rangle_{\alpha\leq\beta\in On}\rangle$. Let κ_α be the critical point of $\underline{M}_\alpha$. Since $\kappa_\kappa=\kappa$ there is $\alpha<\kappa$ such that $\gamma<\kappa_\alpha$. So $a\in M_\alpha$; it suffices, then, to prove that $\underline{M}_\alpha\in K_\kappa$. Certainly $\underline{M}_\omega\in K_\kappa$; so $\underline{M}\in K_\kappa$. But the theorem about existence of mouse iterates holds in K_κ (it was proved in ZF^-) and so $\underline{M}_\alpha\in K_\kappa$. □

Corollary 14.14. *$K=\bigcup\{M: \underline{M}$ is a mouse$\}$*

Corollary 14.14 is false if $0^\#$ does not exist: from now on the assumption that $0^\#$ does not exist is not needed.

Lemma 14.15. *If β is an infinite cardinal in K then $P(\beta)\subseteq K_{\beta^+}$ (in K).*
Proof: (In K) By lemma 14.13

$$K_{\beta^+}=H^{Q_{\beta^+}}_{\beta^+}.$$

So it suffices to show that if $a\subseteq\beta$ then $a\in Q_{\beta^+}$. Suppose $a\in Q_\kappa$, κ regular and greater than β^+. Let γ be least such that a is in $J^{\Omega_\kappa}_\gamma$. Then as in the proof of lemma 14.12 $\underline{M}=\underline{J}^{\Omega_\kappa}_\gamma$ is a mouse and $\omega\rho_M\leq\beta$. Let $\underline{M}'$ be the β^+-th iteration of the core of $\underline{M}$. Then $a\in M'$ and $\omega\rho_{M'}\leq\beta$ so $M'\subseteq Q_{\beta^+}$. If $0^\#$ does not exist the result is known. □

Corollary 14.16. *$K\models GCH$*

Corollary 14.17. *If $0^\#$ exists and β is an uncountable cardinal in K then $K_\beta=\bigcup\{N:\underline{N}\in K_\beta \wedge \underline{N}$ is a mouse$\}$.*

K is not absolute. For example if $V=L[0^\#]$ then $K=V$ but $K^L=L$. In fact

Lemma 14.18. *If M is an inner model then $K^M=K\cap M$.*
Proof: $\underline{N}$ is a mouse if and only if $M\models\underline{N}$ is a mouse. □

In general K is not κ-minimal for any κ. But K is 0-maximal.

Lemma 14.19. *Suppose $j:K\to_e M$ where M is transitive. Then $M=K$.*
Proof: Since $K\models\forall x\exists\underline{N}(\underline{N}$ a mouse and $x\in N)$ therefore

$$M\models\forall x\exists\underline{N}(\underline{N} \text{ a mouse and } x\in N). \tag{14.3}$$

So $M\subseteq K$.

Let κ be regular and let $\underline{Q}'_\kappa=j(\underline{Q}_\kappa)$. $\underline{Q}'_\kappa$ is a mouse with critical point $j(\kappa)$ and $\omega\rho_{Q'_\kappa}=j(\kappa)$. Let $\underline{N},\underline{N}'$ be comparable iterates of $\underline{Q}_\kappa,\underline{Q}'_\kappa$. If $N'\in N$ then, letting π' be the iteration map from $\underline{Q}'_\kappa$ to $\underline{N}'$, $\pi'j:\underline{Q}_\kappa\to_{\Sigma_0}\underline{N}$ non-cofinally, contradicting corollary 8.20. So $N\subseteq N'$. It follows that $P(\kappa)\cap Q_\kappa\subseteq Q'_\kappa$.

Now for any γ $\gamma^2\cap D\in Q_\kappa\subseteq Q'_\kappa$, where $\kappa>\gamma$ is regular. So $\gamma^2\cap D\in M$; so $K\subseteq M$. □

Exercises

1. Show that $cf(\theta_\kappa)=\kappa$ when κ is regular.

2. Given that $0^{\#}$ exists then by exercise 1 $\theta_{\kappa} \geq \kappa + \kappa$. Could they be equal?

3. Find core models K,K' where $Q_{\omega_1}^{K} = \Omega_{\omega_1}^{K'}$ but $K \neq K'$.

4. κ, κ' are regular and $\kappa < \kappa'$. $\underline{J}_{\psi}^{F}\kappa'$ is the κ'-th iterate of $\underline{\Omega}_{\kappa}$. $\kappa' < \gamma < \psi$ and $\omega\rho_{J_{\gamma}^{F}\kappa'} < \kappa'$. Is it necessarily the case that $\omega\rho_{J_{\gamma}^{F}\kappa'} < \kappa$?

5. In question 4 what is the relation between $J_{\psi}^{F}\kappa'$ and $\pi(Q_{\kappa})$ where π is the iteration map of Q_{κ}^{+} to its κ'-th iterate.

6. Suppose κ^{+} is a successor cardinal in K. Then $K_{\kappa^{+}} \not\models ZFC$ but $K_{\kappa^{+}} = H_{\kappa^{+}} = H_{\kappa^{+}}^{Q_{\kappa}^{+}}$. Why does this not contradict lemma 11.1 ?

15: EXAMPLES OF K

We may think of the Q_κ of the previous chapter as follows: start constructing at regular κ from F_κ and continue for as long as possible while new bounded subsets of κ are constucted and $J^{F}_{\gamma}\kappa$ is a mouse. θ_κ is a measure of how long this process goes on. It can be used as a measure of the large cardinal properties of the universe: for example, if $0^{\#}$ exists but $0^{\#\#}$ does not then $\theta_\kappa=\kappa+\kappa$; if all the # sequence exist then $\theta_\kappa\geq\kappa^2$; and so on.

This can be made precise.

Definition 15.1. *Suppose $\underline{M}$ is a mouse with critical point κ. $\underline{M}$ is called sharplike provided $H^M_\kappa \in M$.*

The reason for the term sharplike is that every sharp has this property. In fact all sharplike mice relate to some model of V=K in the way that a sharp relates to L.

Note that if $\underline{M}$ is a mouse with critical point κ and $On_M=\kappa+\omega.2$ then $\underline{M}$ is not sharplike.

Lemma 15.2. *If $J^U_{\lambda+1}$ is a mouse with critical point κ then $J^U_{\lambda+1}$ is sharplike if and only if $\omega\rho^n_{J^U_\lambda}\geq\kappa$ for all n.*

Proof: If $\omega\rho^n_{J^U_\lambda}\geq\kappa$ for all n then $H^{J^U_\lambda}_\kappa=H^{J^U_{\lambda+1}}_\kappa$ so $H^{J^U_{\lambda+1}}_\kappa\in J^U_{\lambda+1}$. If $\omega\rho^n_{J^U_\lambda}<\kappa$ for some n then $P(\omega\rho^n_{J^U_\lambda})\cap J^U_{\lambda+1}\notin J^U_{\lambda+1}$ so $H^{J^U_{\lambda+1}}_\kappa\notin J^U_{\lambda+1}$. □

Suppose $\underline{M}$ is a sharplike mouse with κ critical. Let $\langle\langle \underline{M}_\alpha\rangle_{\alpha\in On},\langle\pi_{\alpha\beta}\rangle_{\alpha\leq\beta\in On}\rangle$ be its mouse iteration. Let $\hat{H}_\alpha=H^{M_\alpha}_{\kappa_\alpha}$ where κ_α is the critical point of $\underline{M}_\alpha$. Then $\pi_{\alpha\beta}(\hat{H}_\alpha)=\hat{H}_\beta$ and $\pi_{\alpha\beta}|\hat{H}_\alpha=id|\hat{H}_\alpha$ so $\hat{H}_\alpha\prec\hat{H}_\beta$.

Lemma 15.3. *$\hat{H}_\alpha=H^{M_\beta}_{\kappa_\alpha}$ $(\alpha\leq\beta)$*

Proof: $\hat{H}_\alpha=\bigcup\{S^a_\gamma\colon a\subseteq\gamma<\kappa_\alpha,\ a\in M_\alpha\}$

$=\bigcup\{S^a_\gamma\colon a\subseteq\gamma<\kappa_\alpha,\ a\in M_\beta\}$

$=H^{M_\beta}_{\kappa_\alpha}$ by lemma 3.17.

(S^a_γ exists in $\underline{M}_\alpha$ since $a\in J^M_\delta$, some $\omega\delta\in M_\alpha$ with $\gamma<\omega\rho_{J^M_\delta}$; similarly for $\underline{M}_\beta$). □

Corollary 15.4. $\hat{H}_\alpha \in \hat{H}_\beta$.
Proof: $H^{M_\beta}_{\kappa_\alpha}\in\Sigma_\omega(\hat{H}_\beta)\subseteq M_\beta$; and $\overline{\overline{(H^{M_\beta}_{\kappa_\alpha})}}^{M_\beta}=\overline{\overline{(\bigcup\{S^a_\gamma:\ a\subseteq\gamma<\kappa_\alpha\})}}^{M_\beta}\leq\kappa_\beta$. □

It follows as in lemma 11.13 that $\hat{H}_\alpha\models$ZFC.

Lemma 15.5 $\hat{H}_\alpha\models V=K$
Proof: Let $H=H_0$. We must show that if $a\subseteq\gamma<\kappa$, $a\in M$ then $a\in K^H$. Suppose $a\in J^M_{\beta+1}\setminus J^M_\beta$. We may assume $\beta>\kappa$. So $\underline{J}^M_{\beta+1}$ is a mouse; call it $\underline{N}$. Since $H^N_\kappa\notin N$, $\beta+1\in M$. And $\omega\rho_{\underline{N}}\leq\gamma$ so $A_N\in H$. Thus core($\underline{N}$)$\in H$, and $H\models$ZFC so there is a mouse iterate of $\underline{N}'$ in H containing a. ($\underline{N}'$=core($\underline{N}$)). □

Definition 15.6. *Let* $K_M=\bigcup_{\alpha\in On}\hat{H}_\alpha$. *Let* D_M *have the same definition in* K_M *as D in K. Let* $\underline{K}_M=\langle K_M,\in,D_M\rangle$.

Since $\langle\hat{H}_\alpha:\alpha\in On\rangle$ is an elementary chain $\hat{H}_\alpha\prec K_M$ for all α; so $K_M\models$ZFC+V=K. It follows that $\underline{K}_M\models V=L[D_M]$. Before carrying the analysis further we introduce some more notation.

Definition 15.7. *Suppose* $\underline{M}$, $\underline{N}$ *are core mice.* $\underline{M}<\underline{N}$ *means that* $\underline{M},\underline{N}$ *have comparable mouse iterates* $\underline{M}',\underline{N}'$ *with* $M'\in N'$.

Lemma 15.8. *Suppose* $\underline{M},\underline{N}$ *are core mice and* $\underline{M}'=\underline{M}^{n(M)}$, $\underline{N}'=\underline{N}^{n(N)}$. $\underline{M}<\underline{N}$ *if and only if* $\underline{M}',\underline{N}'$ *have comparable iterates* $\underline{M}'',\underline{N}''$ *with* $M''\in N''$.
Proof: Suppose $\underline{M}<\underline{N}$. Let $\langle\langle\underline{M}_\alpha\rangle_{\alpha\in On},\langle\pi'_{\alpha\beta}\rangle_{\alpha\leq\beta\in On}\rangle,\langle\langle\underline{N}_\alpha\rangle_{\alpha\in On},\langle\pi_{\alpha\beta}\rangle_{\alpha\leq\beta\in On}\rangle$ be the mouse iterations of $\underline{M},\underline{N}$ respectively. Suppose $\underline{M}_\alpha,\underline{N}_\beta$ are comparable and $M_\alpha\in N_\beta$. Let $\underline{M}'_\alpha=\underline{M}^{n(M)}_\alpha$, $\underline{N}'_\beta=\underline{N}^{n(N)}_\beta$. Suppose κ is the common critical point of $\underline{M}_\alpha,\underline{N}_\beta$. $\underline{M}'_\alpha,\underline{N}'_\beta$ are comparable: and $P(\kappa)\cap M_\alpha\subseteq P(\kappa)\cap N_\beta$: also $P(\kappa)\cap M_\alpha\neq P(\kappa)\cap N_\beta$, as $\underline{M}_\alpha$ is a mouse. But $P(\kappa)\cap N_\beta=P(\kappa)\cap N'_\beta$ so $M_\alpha\in N'_\beta$. Thus $M'_\alpha\in N'_\beta$.

Conversely if $\underline{M}'_\alpha,\underline{N}'_\beta$ are comparable then $\underline{M}_\alpha$, $\underline{N}_\beta$ are comparable as $P(\kappa)\cap M_\alpha\subseteq P(\kappa)\cap M'_\alpha$, $P(\kappa)\cap N_\beta\subseteq P(\kappa)\cap N'_\beta$. Suppose $M'_\alpha\in N'_\beta$. Then $N_\beta\notin M_\alpha$ (otherwise $N'_\beta\in M'_\alpha$) and $\underline{N}_\beta\neq\underline{M}_\alpha$ (otherwise $\underline{M}'_\alpha=\underline{N}'_\beta$). So $M_\alpha\in N_\beta$. □

Lemma 15.9. *< is a strict linear order on the class of core mice.*
Proof: < is irreflexive. Suppose $\underline{M}_1<\underline{M}_2$ and $\underline{M}_2<\underline{M}_3$. Suppose $\underline{Q}_1$, $\underline{Q}_2$ are comparable iterates of $\underline{M}_1,\underline{M}_2$ with $Q_1\in Q_2$; similarly $\underline{Q}'_2,\underline{Q}'_3$. Let θ be regular, $\theta>\overline{\overline{Q}}_1,\overline{\overline{Q}}_2,\overline{\overline{Q}}_3$. Let $\underline{N}_1,\underline{N}_2,\underline{N}_3$ be the θth mouse iterates of $\underline{M}_1,\underline{M}_2,\underline{M}_3$. $\underline{N}_1$ and $\underline{N}_2$ are comparable. Suppose $N_2\subseteq N_1$. Let $\pi:Q_2\to N_2$ be

the iteration map. Then $\pi|\Omega_1:Q_1 \to_e \pi(\Omega_1) \in N_2 \subseteq N_1$. So $\pi|\Omega_1:Q_1 \to_{\Sigma_0} N_1$ non-cofinally. Hence $N_1 \in N_2$. Similarly $N_2 \in N_3$. So $N_1 \in N_3$.
$<$ is clearly connected. □

In the next lemma it is important to note that a well-order of a proper class is not taken to require that the initial segments be sets.

<u>Lemma 15.10</u>. *$<$ is a well-order*
Proof: Let A be a set of core mice. Let θ be regular such that $\theta > \overline{\overline{M}}$ for all $\underline{M} \in A$. Let $\underline{M}' = J^F_{\alpha_M}\theta$ denote the θth mouse iterate of $\underline{M}$. Pick $\underline{M}$ with α_M minimal and $\underline{M}$ will be minimal in A. □

<u>Definition 15.11</u>. *Suppose $\underline{M}$ is a core mouse. deg($\underline{M}$) denotes $\{\underline{N}:\underline{N}$ is a core mouse, $\underline{N} \in L[\underline{M}]$ and $\underline{M} \in L[\underline{N}]$.$\}$*

<u>Lemma 15.12</u>. *If $\underline{N} < \underline{M}$ then $\underline{N} \in L[\underline{M}]$.*
Proof: $A_N^{n(N)+1} \in M$. □

Hence deg($\underline{M}$) is an interval of $<$. The relevance of this to K_M will be seen from the next lemma.

<u>Lemma 15.13</u>. *core($\underline{M}$) is the $<$-minimal core mouse not in K_M.*
Proof: Suppose $\underline{M} \in K_M$. $K_M = \bigcup_{i<\infty} \hat{H}_i$; $A_M^n \in K_M$ $(n=n(\underline{M})+1)$ so $A_M^n \in \hat{H}_i$ say. But $A_M^n \in M_i \Rightarrow A_M^n \in M$; contradiction!

Suppose $\underline{N} < \underline{M}$. Let $\underline{N}'$, $\underline{M}'$ be comparable mouse iterates of $\underline{N}$, $\underline{M}$ with $N' \in M'$. So $A_{N'}^{n(N)+1} \in M'$. But $P(\omega\rho_N^{n(N)+1}) \cap M' \subseteq H^{M'}_{\omega\rho_N^{n(N)+1}}$ so $A_{N'}^{n(N)+1} \in K_M$. So $\underline{N} \in K_M$. □

This is a dual process: suppose $\underline{M} \models V=K$; then provided $M \neq K$ there is a sharplike mouse $\underline{N}$ such that $M = K_N$.

<u>Definition 15.14</u>. *Suppose $\underline{M} \models V=K, M \neq K$. $\underline{M}^{\#}$ denotes the $<$-least core mouse not in $\underline{M}$ ($\underline{M}$ an inner model).*

Note 1: $0^{\#}$ and $L^{\#}$ are interconstructible. $L^{\#}$ is the core sharp.
Note 2: If $\underline{M}$ is sharplike then $K_M^{\#} = \underline{M}$.
Fix $\underline{M}$ such that $\underline{M}$ is an inner model, $\underline{M} \models V=K$ but $M \neq K$.

<u>Lemma 15.15</u>. *$M^{\#}$ is sharplike.*

Proof: Suppose $a\in M^{\#}$, where $M^{\#}$ has critical point κ and $a\subseteq\gamma<\kappa$. Then there is a mouse $\underline{N}$ with $\text{core}(\underline{N})<\underline{M}$ and $a\in\Sigma_{\omega}(\underline{N})$. But $\underline{N}\in M$ so $a\in M$. Hence $\bigcup_{\gamma<\kappa} P(\gamma)\cap M^{\#}\subseteq M$. Also $P_M(\kappa)\subseteq M^{\#}$. For if $A\subseteq\kappa, A\in M$, suppose $A\in\underline{N}$ where $\underline{N}$ is a mouse and $\underline{N}\in M$. Let $\underline{M}'$, $\underline{N}'$ be comparable mouse iterates of $\underline{M}^{\#}, \underline{N}$. If $M'\in N'$ then by lemma 15.12 $\underline{M}^{\#}\in M$; so $A\in M'$, hence $A\in M^{\#}$.

It follows that κ is a cardinal in M. $M\models$GCH; so let f map κ onto $\bigcup_{\gamma<\kappa} P_M(\gamma)$, $f\in M$. Let $A=\{\langle\xi,\zeta\rangle:\xi\in f(\zeta)\}$; so $A\in M^{\#}$. Hence for $\gamma<\kappa$ $P_{M^{\#}}(\gamma)\in M^{\#}$. If $On_{M^{\#}}=\omega\alpha+\omega$ then for all $\gamma<\kappa$ $P_{M^{\#}}(\gamma)\subseteq J_{\alpha}^{M^{\#}}$ so $H_{\kappa}^{M^{\#}}=H_{\kappa}^{J_{\alpha}^{M^{\#}}}$. If $On_{M^{\#}}=\omega\lambda$, λ a limit, then there is α such that $A\in J_{\alpha}^{M^{\#}}$, so $H_{\kappa}^{M^{\#}}=H_{\kappa}^{J_{\alpha}^{M^{\#}}}$. □

Lemma 15.16. *$K_{M^{\#}}=M$.*

Proof: ($\supseteq$) Let $\langle\langle M_{\alpha}^{\#}\rangle_{\alpha\in On},\langle\pi_{\alpha\beta}\rangle_{\alpha\leq\beta\in On}\rangle$ be the mouse iteration of $M^{\#}$. Suppose κ_{α} is the critical point of $M_{\alpha}^{\#}$. Let $a\subseteq\gamma<\kappa_{\alpha}$, $a\in M$. Then $a\in M_{\alpha}^{\#}$ by the argument of lemma 15.15. So $a\in H_{\kappa_{\alpha}}^{M_{\alpha}^{\#}}$, so $a\in K_{M}^{\#}$. Thus for all γ $P(\gamma)\cap M\subseteq K_{M^{\#}}$. But then for all mice $\underline{N}\in M$ $A_{\underline{N}}\in K_{M^{\#}}$ so $M\subseteq K_{M^{\#}}$.

($\subseteq$) Suppose $a\in H_{\kappa_{\alpha}}^{M_{\alpha}^{\#}}$, $a\subseteq\gamma<\kappa_{\alpha}$. As before it suffices to show $a\in M$. But this was proved in lemma 15.15. □

It follows that there is a one-one correspondence between sharplike mice and models of V=K other than K itself. And $\underline{M}<\underline{M}' \Leftrightarrow K_{M}\subseteq K_{M'} \wedge$ $\wedge\ K_{M}\neq K_{M'}$. In fact $K_{M}=\bigcup\{M':\underline{M}' \text{ is a mouse and core}(\underline{M}')<\underline{M}\}\cup L$.

Lemma 15.17. *If $\underline{M}$ is sharplike then $\underline{M}$ is minimal in deg($\underline{M}$).*

Proof: Otherwise suppose $\underline{M}'<\underline{M}$, $\underline{M}'\in\text{deg}(\underline{M})$. Then $\underline{M}'\in K_{M}$; so $L[\underline{M}']\subseteq K_{M}$, so $\underline{M}\in K_{M}$. Contradiction! □

So each deg($\underline{M}$) contains at most one sharplike mouse. Some deg($\underline{M}$) contain no sharplike mice, however. The constructibility degree of $\{\langle n,m\rangle:n\in O^{\#m}\}$ is one of them.

Lemma 15.18. *deg($\underline{M}$) contains no <-maximal element.*

Proof: Suppose $\underline{N}\in\text{deg}(\underline{M})$ were one. Let θ be regular, $\theta>\overline{\overline{N}}$, and let $\underline{N}'$ be the θth mouse iterate of $\underline{N}$. In $L[\underline{M}]$ $\underline{N}'\in Q_{\theta}$. Contradiction! □

In fact deg($\underline{M}$) is a proper class; hence the caveat about well-ordered classes before lemma 15.10.

EXAMPLES

$O^{\#}$ and $L^{\#}$ are interchangeable: $L^{\#}$ is the core sharp. Which sharplike mouse comes next; that is, what is $L[O^{\#}]^{\#}$? By chapter 3 exercise 6   if $\lambda>\kappa$ and $P(\omega)\cap J^U_{\kappa+1}\neq P(\omega)\cap J^U_{\kappa+2}$ but $\omega\rho_{J^U_\lambda}=\kappa$ then $\omega\lambda$ must be closed under $\kappa+\mu$ for $\mu<\kappa$. The next candidate for sharphood after $J^U_{\kappa+1}$ must be $J^U_{\kappa+\kappa+1}$. (Throughout this discussion assume for the sake of definiteness that κ is some regular cardinal and U is F_κ). So the question is: does $\omega\rho_{J^U_{\kappa+\kappa}}=\kappa$? Since there is a simple map of κ onto $\kappa+\kappa$ $\omega\rho_{J^U_{\kappa+\kappa}}\leq\kappa$. If $\omega\rho_{J^U_{\kappa+\kappa}}=\gamma<\kappa$ and $J^U_{\kappa+\kappa+1}$ is an iterable premouse, argue as follows: suppose

$$x\in A_{J^U_{\kappa+\kappa}} \Leftrightarrow \underline{J}^U_{\kappa+\kappa}\models\exists t\phi(t,x,p) \qquad (\text{some } p\in[\kappa+\kappa]^{<\omega},\ \phi\ \Sigma_0).$$

Let t_x be least in the canonical well-order of $\underline{J}^U_{\kappa+\kappa}$ such that $\phi(t,x,p)$, if there is one. Let $\tau:\gamma\leftrightarrow\omega\times(\gamma)^{<\omega}$. Define $f:\gamma\to\kappa+\kappa$ by

$$f(\delta)=\text{the least } \mu \text{ such that } t_{\tau(\delta)}\in S^U_\mu \text{ (if one exists).}$$

If rng(f) were bounded in $\kappa+\kappa$ then $A_{J^U_{\kappa+\kappa}}$ would be in $J^U_{\kappa+\kappa}$; so $f:\gamma\to\kappa+\kappa$ cofinally. But $f\in J^U_{\kappa+\kappa+1}$ and $\underline{J}^U_{\kappa+\kappa+1}$ is a premouse so $\underline{J}^U_{\kappa+\kappa+1}\models cf(\kappa)=\gamma$. Contradiction! So $\omega\rho_{J^U_{\kappa+\kappa}}=\kappa$. Since $\underline{J}^U_{\kappa+\kappa+1}$ is an iterable premouse it follows that it is sharplike.

What is $K_{J^U_{\kappa+\kappa+1}}$? Certainly it contains $0^\#$. Suppose it also contained $\underline{N}$ (a mouse) with $\underline{N}\notin L[0^\#]$; let $\underline{N}$ be <-minimal. Then $\underline{N}$ is sharplike. But then for some κ', some mouse iterate $\underline{N}'=\underline{J}^{U'}_\beta$ of $\underline{N}$ with κ' critical, $\beta<\kappa'+\kappa'+1$. So $\beta=\kappa'+1$, but then $\underline{N}$ is a sharp so $\underline{N}\in L[0^\#]$. Contradiction! So $K_{J^U_{\kappa+\kappa+1}}$ is $L[0^\#]$. In fact in general if $\underline{M}$ is sharplike and a "successor" in the sense that for some mouse $\underline{M}'$ $\underline{N}<\underline{M}\Rightarrow\underline{N}\in L[\underline{M}']$, we deduce that $K_M=L[M']$.

$J^U_{\kappa+\kappa+1}$ is $L[0^\#]^\#$. Now apply the results of chapter 12 to see that $\rho_{J^U_{\kappa+\kappa+1}}=1$ and $p_{J^U_{\kappa+\kappa+1}}=\phi$. We may, then, extract a unique class C of generating indiscernibles for $L[0^\#]$ and a set $0^{\#\#}$. Furthermore since $L[0^\#]$ is 0-minimal and 0-maximal we get the equivalence of the the following:

(i) $0^{\#\#}$ exists

(ii) there is an iterable premouse $J^U_{\kappa+\kappa+1}$ with κ critical

(iii) there is a non-trivial $j:L[0^\#]\to_e L[0^\#]$.

$L[0^{\#\#}]=L[0^\#]^\#$. There is an obvious generalisation to $0^{\#n}$ for all $n\in\omega$.

Consider $\underline{J}^U_{\kappa\omega}$, then, and suppose it is an iterable premouse. Each $\underline{J}^U_{\kappa n+1}$ is an iterable premouse and is interconstructible with $0^{\#n}$. Furthermore $\underline{J}^U_{\kappa\omega}$ is a premouse whenever $\underline{J}^U_{\kappa n+1}$ is a premouse for all n. It is not sharplike ($\kappa\omega$ is a limit of points with projectum 1) but it (or rather its core) must be minimal in its degree. If we let $0^{\#\omega}=\{\langle n,m\rangle : n\in 0^{\#m}\}$ we see that $L[\underline{J}^U_{\kappa\omega}]$ is $L[0^{\#\omega}]$.

How far does all this go? Mice of the form $\underline{J}^U_{\kappa\lambda+1}$ where $\lambda<\kappa$ is a successor ordinal are sharplike and correspond with the model of V=K with λ-1 iterates of #, giving the # of that model. Transfinite iteration of # will soon lose the property that $0^{\#}\subseteq\omega$, of course. Consider $\underline{M}=\underline{J}^U_{\kappa(\omega_1+1)+1}$ for example: $0^{\#\omega_1}$ is $\{\langle\alpha,\beta\rangle : \alpha\in 0^{\#\beta}, \beta<\omega_1\}$. We can no longer prove $\rho_M=1$. For to do that we let $X=h_M(\phi)$, $\sigma:\underline{N}\cong\underline{M}|X$ with $\underline{N}$ transitive and iterated $\underline{N}$,$\underline{M}$ to comparable $\underline{\Omega}$, $\underline{\Omega}'$. These will only be equal if the iteration maps preserve ω_1; so we must insist that $\omega_1\subseteq X$. So we set $X=h(\omega_1)$ and prove $\rho_M=\omega_1$. This is not surprising. We get the equivalence for successor λ of:

(i) $0^{\#\lambda}$ exists

(ii) there is an iterable premouse of the form $J^U_{\kappa\lambda+1}$ with κ critical

(iii) there is a non-trivial $j:L[0^{\#\lambda-1}]\to_e L[0^{\#\lambda-1}]$ with $j|\lambda=id|\lambda$ provided $\lambda<\kappa$. ((iii) is necessary for minimality and maximality). For limit $\lambda<\kappa$ $0^{\#\lambda}$ exists is equivalent to the existence of a premouse of the form $\underline{J}^U_{\kappa\lambda}$ with κ critical.

The reader will see that this simple situation breaks down if $\lambda\geq\kappa$. Certainly $J^U_{\kappa^2}$ gives all the sharps up to κ, and is not sharplike. Consider, then, $J^U_{\kappa^2+1}$. $\omega\rho_{J^U_{\kappa^2}}=\kappa$, so $J^U_{\kappa^2+1}$ is sharplike. Where did all the $0^{\#\alpha}$ for $\alpha>\kappa$ go? They can easily be retrieved by iterating to $\kappa'>\kappa$; they simply will not occur in Ω_κ, for $J^U_{\kappa^2+1}$ is a different creature altogether. Call it $\underline{M}$ and consider K_M. Iterate $\underline{M}$ to some $\kappa'>\kappa$ and it will be of the form $J^{U'}_{\kappa'^2+1}$ (whereas $J^U_{\kappa^2}$ only goes to $J^{U'}_{\kappa'\kappa}$) and so exceeds all the $0^{\#\alpha}$, $\alpha<\kappa'$. Iterate far enough and it eventually exceeds any sharp. If $H=H^M_\kappa$ then $H\models$"for all α $0^{\#\alpha}$ exists". On the other hand if $\underline{N}<\underline{M}$ then $\underline{N}\in L[0^{\#\alpha}]$ for some α. K_M is the closure of L under all sharps - sometimes called $L^{\#}$ but $L^{\#\infty}$ in our notation. It is not of the form $L[\underline{M}]$ for any mouse $\underline{M}$. It recovers properties of L lost along the way - it is 0-minimal for example - and $J^U_{\kappa^2+1}$ is $(L^{\#\infty})^{\#}$. This corresponds to a subset of ω, and off we go again, up to $J^U_{\kappa^{2}2+1}$ - and presumably $J^U_{\kappa^3}$ corresponds with some still grander model.

In all the models defined so far the length has been nicely definable from the critical point. So in all our sharplike mice the projectum has fallen straight to the critical point: $n(\underline{N})=0$. This cannot continue all the way up to κ^+. There is a murky middle ground where coding by # sequences fails. For suppose a sharplike mouse $\underline{N}$ does not have $n(\underline{N})=0$; or even worse, that several levels pass between H^N_κ and a mouse (to be precise, $H^N_\kappa=H^{J^N_\beta}_\kappa$,

$\beta \in N$ but $\omega\beta$ is not the largest limit ordinal in $\underline{N}$). Certainly we shall get indiscernibles for K_N; and in $\underline{N}$ they will be generating indiscernibles; but in K_N they will not. The argument that turned Σ_1 in a sharp into elementary in L relied on Σ_1 heavily.

It may be that the equivalence with non-trivial elementary embeddings breaks down as well. This, after all, relied heavily on minimality and maximality. Maximality is no problem: if $\underline{M} \models V=K$ then either $M=K$, which is 0-maximal, or $M \neq K$, so $M^{\#}$ exists. Minimality cannot be so simply disposed of. In chapter 16 we shall take a sledgehammer to the problem and show that if $\underline{M} \models V=K$ then $M^{\#}$ exists if and only if there is $j:\underline{M} \to_e \underline{M}$ non-trivially, unless there is an inner model with a measurable cardinal (which is sharplike all but in name). The case where there is a ρ-model is of considerable interest: it lies at the opposite end of the core model spectrum from $0^{\#}$.

ρ-MODELS AND K

Let $L[U]$ be a ρ-model with critical point κ. Let $\langle\langle L[U_\alpha]\rangle_{\alpha \in On}, \langle \pi_{\alpha\beta}\rangle_{\alpha<\beta \in On}\rangle$ be its iterated ultrapower. Let κ_α be the critical point of $L[U_\alpha]$.

Lemma 15.19. $K = \bigcup_{\alpha \in On} H_{\kappa_\alpha}^{L[U_\alpha]}$.

Proof: Suppose that x is in $H_{\kappa_\alpha}^{L[U_\alpha]}$. There is $B \subseteq \kappa_\alpha, B \in L[U_\alpha]$ such that $H_{\kappa_\alpha}^{L[U_\alpha]} = J_{\kappa_\alpha}^B$ so we may assume $x \subseteq \gamma < \kappa_\alpha$. If $x \in J_{\kappa_\alpha}$ then $x \in K$. Otherwise let β be least such that $x \in J_{\beta+1}^U$. Then $\underline{J}_{\beta+1}^U$ is a mouse so $x \in K$.

Conversely suppose $\underline{M}$ is a mouse. Let $\underline{M}' = \underline{M}^{n(\underline{M})}$. Let $L[U_\alpha]$, $\underline{M}''$ be comparable iterates of $L[U_\alpha], \underline{M}'$ with $\kappa_\alpha > \overline{\overline{M}}$. Obviously $M'' \in L[U_\alpha]$. So the mouse iterate of $\underline{M}$ at κ_α, $\overline{M}$ is in $L[U_\alpha]$. So $\underline{M} \in L[U_\alpha]$; and so $\underline{M} \in H_{\kappa_\alpha}^{L[U_\alpha]}$. □

Lemma 15.20. $K = \bigcap_{\alpha \in On} L[U_\alpha]$.

Proof: If $x \subseteq \gamma < \kappa_\alpha$, $x \in L[U_\alpha]$ then $x \in L[U_\beta]$ $(\beta < \alpha)$ as $L[U_\alpha] \subseteq L[U_\beta]$. But if $\beta \geq \alpha$ then $\pi_{\alpha\beta}(x) = x$ so $x \in L[U_\beta]$.

Conversely if $x \in \bigcap_{\alpha \in On} L[U_\alpha]$ then for some α $\overline{\overline{TC(x)}} < \kappa_\alpha$. So $x \in H_{\kappa_\alpha}^{L[U_\alpha]}$. □

Suppose $\underline{M}$ is an inner model and $\underline{M} \models$ there is a ρ-model. Call this $L[U]$ so $\underline{M} \models K = \bigcap_{\alpha \in On} L[U_\alpha]$. But $L[U]$ really is a ρ-model, so $K = \bigcap_{\alpha \in On} L[U_\alpha]$. Hence $K^M = K$. Thus the existence of a ρ-model means that K is at its largest. If $\underline{M} \models V=K$ too then $L[U] \subseteq K$, so $L[U] = \bigcap_{\alpha \in On} L[U_\alpha]$; but $U \notin L[U_1]$, so this is impossible. Hence $K\models$"there is no ρ-model" even if there is one.

If there is a ρ-model then K is not κ-minimal for any κ. For take $\kappa<\kappa_\alpha$ and let $X=h_N(\kappa_\alpha\cup p)$ where $\underline{N}=\underline{J}^{U_\alpha}_{\kappa_\alpha^+}$ and p is such that $H^{L[U_\alpha]}_{\kappa_\alpha}$ is $\Sigma_1(\underline{N})$ in $\kappa\cup p$. Let $\sigma:\underline{M}\cong\underline{N}|X$ with M transitive; so $\sigma:\underline{M}\to_{\Sigma_1}\underline{N}$. $\underline{M}\models V=h(\kappa_\alpha\cup\sigma^{-1}(p))$ and $\underline{M}=\underline{J}^{U_\alpha}_\beta$ for some $\beta<\kappa_\alpha^+$ so $\underline{M}$ is a mouse; letting $\overline{H}=\sigma^{-1}(H^{L[U_\alpha]}_{\kappa_\alpha})$ $\underline{M}\models\forall x(x\subseteq\gamma<\kappa_\alpha\Rightarrow x\in\overline{H})$, so $\underline{M}$ is sharplike. If we let $\langle\langle\underline{M}_\alpha\rangle_{\alpha\in On},\langle\pi'_{\alpha\beta}\rangle_{\alpha\leq\beta\in On}\rangle$ be its iterated ultrapower and let maps $\sigma_\alpha:\underline{M}_\alpha\to_{\Sigma_1}L[U_\alpha]$ be defined such that $\sigma_0=\sigma$, $\sigma_\beta\pi'_{\alpha\beta}=\pi_{\alpha\beta}\sigma_\alpha$ then $\sigma_\alpha:H^{M_\alpha}_{\kappa'_\alpha}\to_e H^{L[U_\alpha]}_{\kappa_\alpha}$ where κ'_α is the critical point of M'_α. Given $x\in H^{M_\alpha}_{\kappa'_\alpha}$ let $\sigma(x)=\sigma_\alpha(x)$. Then $\sigma:K_M\to_e K$ but $K_M\neq K$. On the other hand $\sigma|\kappa=id|\kappa$. The existence of a ρ-model is by no means necessary for this result. The existence of a mouse $\underline{M}$ with $n(\underline{M})>0$ suffices.

It follows that there is no sentence that characterises the largest core model internally.

<u>Exercises</u>

1. Given the existence of a ρ-model show that there is a sharplike mouse $\underline{M}$ with $On_M=\omega\lambda$, λ a limit.

2. Show that each $\deg(\underline{M})$ is a proper class.

3. Is $\deg(\underline{M})\cap\{\underline{N}:\underline{N}<\underline{M}\}$ always a set?

4. Suppose there is a ρ-model. Show that there is $j:K\to_e K$ which is non-trivial. Suppose $K^M\neq K$; show that there is $j:K^M\to_e K^M$ which is non-trivial.

(Hint: Let $\underline{N}=(K^M)^\#$ and let $W=\{f\in N:\exists m\in\omega\ f:\kappa^m\to H^N_\kappa\}$ where κ is the critical point of $\underline{N}$. For $f\in W$ let $\tilde{f}=\bigcup_{i<\infty}\pi_{0i}(f)$, where $\underline{N}$ has mouse iteration $\langle\langle\underline{N}_\alpha\rangle_{\alpha\in On},\langle\pi_{\alpha\beta}\rangle_{\alpha\leq\beta\in On}\rangle$ and κ_α is the critical point of $\underline{N}_\alpha$. Show that:

(i) $\forall x\in K^M\exists f\in W\exists i_0<\ldots<i_{n-1}\ x=\tilde{f}(\kappa_{i_0}\ldots\kappa_{i_{n-1}})$;

(ii) if $f_1\ldots f_p:\kappa^m\to H^N$, $i_0<\ldots<i_{m-1}$, $j_0<\ldots<j_{m-1}$ and ϕ is m-ary

$$K^M\models\phi(\tilde{f}_1(\kappa_{i_0}\ldots\kappa_{i_{m-1}})\ldots\tilde{f}_p(\kappa_{i_0}\ldots\kappa_{i_{m-1}}))\Leftrightarrow$$
$$\Leftrightarrow\phi(\tilde{f}_1(\kappa_{j_0}\ldots\kappa_{j_{m-1}})\ldots\tilde{f}_p(\kappa_{j_0}\ldots\kappa_{j_{m-1}})).)$$

16: EMBEDDINGS OF K

This chapter contains two important results: the indiscernibles lemma and the embeddings lemma. The indiscernibles lemma will be further used in chapter 17. The embeddings lemma is one of the main tools in the proof of the covering lemma.

Definition 16.1. *Suppose* $\underline{A}$ *is a transitive model of* R^+ *and* $\kappa=On_{\underline{A}}$. *Let* $\underline{A}_{\omega\beta}$ *denote* $\underline{J}^A_\beta$. *I is a good set of indiscernibles for* $\underline{A}$ *if and only if for all* $\gamma\in I$

(a) $\underline{A}_\gamma \prec \underline{A}$

(b) $I\setminus\gamma$ *is a set of indiscernibles for* $\langle \underline{A},\xi\rangle_{\xi<\gamma}$.

The indiscernibles lemma states that if $\underline{A}=\langle K_\kappa, D\cap\kappa^2, A_1\ldots A_N\rangle$, I is a good set of indiscernibles for $\underline{A}$ (this requires that $\underline{A}\models R^+$) and cf(otp(I))>ω then there is I'$\in$K such that I' is a good set of indiscernibles for $\underline{A}$ and I$\subseteq$I'. In a sense this continues the work of the previous chapter: all # are contained in K. It also follows that although κ, the critical point of some ρ-model, cannot be measurable in K, it will be Ramsey. Model 2 in appendix 1 shows that cf(otp(I))>ω is necessary.

Fix a structure $\underline{A}$, then, where $\underline{A}=\langle K_\kappa, D\cap\kappa^2, A_1\ldots A_N\rangle$ and suppose I is a good set of indiscernibles for $\underline{A}$. Suppose cf(otp(I))>ω.

Lemma 16.2. *If* $\gamma\in I$ *then* γ *is strongly inaccessible in* $\underline{A}$.

Proof: Suppose γ were singular in $\underline{A}$. If cf(γ)=δ, $\delta<\gamma$ then for all $\gamma'\in I\setminus\gamma$ cf(γ')=δ. Let $\gamma'\in I,\gamma'>\gamma$. Suppose f is the least cofinal map of δ to γ in the canonical order of $\underline{A}$; f' defined similarly for γ'. Let $\delta'<\delta$ be least such that f'(δ')>γ. Say δ''=f(δ'), so $\delta''<\gamma$. This is a definable property of $\gamma,\delta,\delta',\delta''$ so f'(δ')=$\delta''<\gamma$; contradiction!

Suppose $\gamma\in I$, $\delta<\gamma$ but there is a map of $u\subseteq P(\delta)$ onto γ. Take $\gamma<\gamma'<\gamma''\in I$. Let f be the least map of some $u\subseteq P(\delta)$ onto γ'' in the canonical order of $\underline{A}$. Say γ=f(a),a$\in$u, γ'=f(a'), a'$\in$u. For $\delta'<\delta$ $\delta'\in a$ is a definable property of $\delta,\delta',\gamma,\gamma''$. So $\delta'\in a \Leftrightarrow \delta'\in a'$, i.e. a=a'. Contradiction! □

Corollary 16.3. $\underline{A}\models$*ZFC*

From now on β^+ denotes β^+ calculated in $\underline{A}$. If $\gamma\in I$ let $\underline{\bar{A}}^n_\gamma = \underline{A}_{\gamma^+}|X$ where X is the set of x in $\underline{A}_{\gamma^+}$ definable in $\underline{A}$ from $\gamma\cup\{\gamma,\gamma_1\ldots\gamma_n\}$ ($\langle\gamma,\gamma_1\ldots\gamma_n\rangle\in I^{n+1}$). So $\underline{\bar{A}}^n_\gamma \prec \underline{A}_{\gamma^+}$. But $\underline{\bar{A}}^n_\gamma$ is transitive (if $x\in\bar{A}^n_\gamma$ there is a definable map of γ onto x) so $\underline{\bar{A}}^n_\gamma = \underline{A}_{\delta^n_\gamma}$, some $\delta^n_\gamma < \gamma^+$.

Lemma 16.4. $\delta^n_\gamma < \delta^{n+1}_\gamma$.

Proof: δ^n_γ is $\underline{A}$-definable from $\gamma\cup\{\gamma,\gamma_1\ldots\gamma_{n+1}\}$, since

$$\bar{A}^n_\gamma = \{x\in\underline{A}_{\gamma^+}: x \text{ is } \underline{A}_{\gamma_{n+1}}\text{-definable from } \gamma\cup\{\gamma,\gamma_1\ldots\gamma_n\}\}. \tag{16.1}$$

□

Also $\underline{\bar{A}}^n_\gamma \prec \underline{\bar{A}}^{n+1}_\gamma$. Let $\underline{\bar{A}}_\gamma = \underline{A}|\bigcup_{n\in\omega}\bar{A}^n_\gamma$, so $\underline{\bar{A}}_\gamma \prec \underline{A}_{\gamma^+}$. Let δ_γ denote $\sup_{n\in\omega}\delta^n_\gamma = \mathrm{On}\cap\bar{A}_\gamma$.

Define maps $\pi^n_{\gamma\gamma'}: \underline{\bar{A}}^n_\gamma \to_e \underline{\bar{A}}^n_{\gamma'}$, for $\gamma,\gamma'\in I$, $\gamma\leq\gamma'$ by

$$\pi^n_{\gamma\gamma'}(t^A(\vec{\nu},\gamma,\gamma_1\ldots\gamma_n)) = t^A(\vec{\nu},\gamma',\gamma_1\ldots\gamma_n) \tag{16.2}$$

where t is a term of $L_{D,\bar{A}}$, $\vec{\nu}<\gamma\leq\gamma'<\gamma_1<\ldots<\gamma_n$, $\langle\gamma_1\ldots\gamma_n\rangle\in I^n$. The choice of $\gamma_1\ldots\gamma_n$ is irrelevant. Letting $\pi_{\gamma\gamma'} = \bigcup_{n\in\omega}\pi^n_{\gamma\gamma'}$, $\pi_{\gamma\gamma'}: \underline{\bar{A}}_\gamma \to_e \underline{\bar{A}}_{\gamma'}$; also $\pi_{\gamma\gamma'}|\gamma = \mathrm{id}|\gamma$ and $\pi_{\gamma\gamma'}(\gamma)=\gamma'$. Define U_γ by $X\in U_\gamma \Leftrightarrow \gamma\in\pi_{\gamma\gamma'}(X)$ where $X\in\bar{A}_\gamma\cap P(\gamma)$; the choice of $\gamma'>\gamma, \gamma'\in I$ is irrelevant.

Lemma 16.5. *$\langle\bar{A}_\gamma, U_\gamma\rangle$ is amenable.*

Proof: Let $U^n_\gamma = \bar{A}^n_\gamma\cap U_\gamma$. It is sufficient to show $U^n_\gamma\in\bar{A}_\gamma$ for all $n\in\omega$. But $\pi^n_{\gamma\gamma'}$ is $\underline{A}$-definable from $\gamma,\gamma',\gamma_1\ldots\gamma_{n+1}$, as is $\underline{\bar{A}}^n_\gamma$ so $U^n_\gamma\in\bar{A}^{n+2}_\gamma$. □

It also follows that $\pi_{\gamma\gamma'}(U^n_\gamma) = U^n_{\gamma'}$, so $\pi_{\gamma\gamma'}: \langle\underline{\bar{A}}_\gamma, U_\gamma\rangle \to_{\Sigma_0} \langle\underline{\bar{A}}_{\gamma'}, U_{\gamma'}\rangle$. In fact $\pi_{\gamma\gamma'}$ is cofinal so $\pi_{\gamma\gamma'}: \langle\underline{\bar{A}}_\gamma, U_\gamma\rangle \to_{\Sigma_1} \langle\underline{\bar{A}}_{\gamma'}, U_{\gamma'}\rangle$.

Now let $\gamma^* = \sup I$ and $I^* = I\cup\{\gamma^*\}$. Let $\langle\langle\underline{\bar{A}}_{\gamma^*}, U_{\gamma^*}\rangle, \langle\pi_{\gamma\gamma^*}\rangle_{\gamma<\gamma^*}\rangle$ be the direct limit of $\langle\langle\underline{\bar{A}}_\gamma, U_\gamma\rangle_{\gamma\in I}, \langle\pi_{\gamma\gamma'}\rangle_{\gamma\leq\gamma'\in I}\rangle$. (The direct limit because it is well-founded: any $a\subseteq\bar{A}_{\gamma^*}$ that is countable is contained in $\mathrm{rng}(\pi_{\gamma\gamma^*})$, some $\gamma\in I$, since $\mathrm{cf}(\mathrm{otp}(I))>\omega$). Let $\underline{\bar{A}}^n_{\gamma^*} = \pi_{\gamma\gamma^*}(\underline{\bar{A}}^n_\gamma)$ and $U^n_{\gamma^*} = \pi_{\gamma\gamma^*}(U^n_\gamma)$, any $\gamma\in I$ (choice irrelevant).

Lemma 16.6. *U_{γ^*} is countably complete.*

Proof: Suppose $X\subseteq\gamma^*$, $X\in\bar{A}_{\gamma^*}$. Suppose $X = \pi_{\gamma\gamma^*}(\hat{X})$, $\gamma\in I$. Then $X\in U_{\gamma^*} \Leftrightarrow \hat{X}\in U_\gamma \Leftrightarrow \gamma\in\pi_{\gamma\gamma^*}(\hat{X})$. So $X\in U_{\gamma^*} \Leftrightarrow I\setminus\gamma\subseteq X$; this proves the result as $\mathrm{cf}(\mathrm{otp}(I))>\omega$. □

For $\gamma\in I^*$ let $M_\gamma=\cup\{\underline{J}^{U_\gamma}_{\nu+1}: \underline{J}^{U_\gamma}_{\nu}\in\bar{A}_\gamma\}$. Then $\underline{M}_\gamma=\langle M_\gamma,\in,U_\gamma\rangle=\underline{J}^{U_\gamma}_{\beta_\gamma}$ say.

<u>Lemma 16.7</u>. *$M_\gamma\subseteq\bar{A}$ but $\underline{M}_\gamma\notin\bar{A}_\gamma$.*

Proof: Suppose $J^{U_\gamma}_{\nu}\in\bar{A}_\gamma$. Then since $\langle\bar{\underline{A}}_\gamma,U_\gamma\rangle\models R$ $J^{U_\gamma}_{\nu+1}\subseteq\bar{A}_\gamma$. But if $\underline{M}_\gamma\in\bar{A}_\gamma$ then $J^{U_\gamma}_{\beta_\gamma+1}\subseteq M_\gamma$; contradiction! □

$\underline{M}_{\gamma^*}$ is iterable as U_{γ^*} is countably complete; but $\pi_{\gamma\gamma^*}|M_\gamma:\underline{M}_\gamma\to_{\Sigma_0}\underline{M}_{\gamma^*}$ so each $\underline{M}_\gamma$ is iterable.

<u>Lemma 16.8</u>. *If $\gamma\in I^*$ and $n\in\omega$ then $P(\gamma)\cap M_\gamma\not\subseteq\bar{A}^n_\gamma$.*

Proof: We may assume $\gamma\in I$. Suppose $P(\gamma)\cap M_\gamma\subseteq\bar{A}^n_\gamma$. So $\underline{M}_\gamma=\underline{J}^{U^n_\gamma}_{\beta_\gamma}$. $\underline{A}\models$ZFC+V=K so $\underline{A}\models$there are no ρ-models. Thus $\underline{J}^{U^n_\gamma}_{\kappa}\models U^n_\gamma$ is not a normal measure ($\kappa=\gamma^+$). So for some τ with $\beta_\gamma\le\tau<\gamma^+$ $P(\gamma)\cap J^{U^n_\gamma}_{\tau+1}\not\subseteq\bar{A}^n_\gamma$; take τ minimal. $\bar{\underline{A}}_\gamma\prec\underline{A}_{\gamma^+}$ so $\tau\in\bar{A}_\gamma$. Hence $\beta_\gamma\in\bar{A}_\gamma$ so $M_\gamma=J^{U^n_\gamma}_{\beta_\gamma}\in\bar{A}_\gamma$; contradiction! □

<u>Lemma 16.9</u>. *$P(\gamma)\cap M_\gamma=P(\gamma)\cap\bar{A}_\gamma$ for $\gamma\in I^*$.*

Proof: $P(\gamma)\cap M_\gamma\subseteq P(\gamma)\cap\bar{A}_\gamma$ obviously. Assume $\gamma\in I$. So suppose $a\subseteq\gamma$, $a\in\bar{A}_\gamma$ but $a\notin M_\gamma$. So $a\notin L$; so $\underline{A}\models$ZFC+V=K+V$\neq$L. Hence $\underline{A}_{\gamma^+}\models$there is a mouse $\underline{N}$ with $\varkappa>\gamma$ critical and $a\in N$. And $\bar{\underline{A}}_\gamma\prec\underline{A}_{\gamma^+}$ so the same holds in $\bar{\underline{A}}_\gamma$.

Suppose $\bar{\underline{A}}_\gamma\models\underline{N}$ is a mouse with $\varkappa$ critical. Then $\underline{A}_{\gamma^+}\models\underline{N}$ is a mouse so $\underline{A}\models\underline{N}$ is a mouse. But $\omega_1\in A$ (as cf(otp(I)) is uncountable) so $\underline{N}$ is a mouse. Iterate $\underline{M}_\gamma$, $\underline{N}$ to $\underline{M}'$, $\underline{N}'$ respectively with θ critical where θ is regular and greater than $\overline{\overline{M_\gamma}},\overline{\overline{N}}$. If $N'\subseteq M'$ then $a\in P(\gamma)\cap N=P(\gamma)\cap N'\subseteq P(\gamma)\cap M'\subseteq P(\gamma)\cap M_\gamma$; but $a\notin M_\gamma$. So $M'\in N'$. Suppose $\underline{N}\in\bar{A}^n_\gamma$. Then $P(\gamma)\cap M_\gamma=P(\gamma)\cap M'\subseteq P(\gamma)\cap N'=P(\gamma)\cap N\subseteq\bar{A}^n_\gamma$, contradicting lemma 16.8.

If $\gamma=\gamma^*$ the result is deduced easily. □

Note that $\underline{A}_\gamma\in\bar{A}_\gamma$. If $\gamma\in I$ this is true from the definition: but otherwise let $\underline{A}^*=\pi_{\gamma\gamma^*}(\underline{A}_\gamma)$: then $\underline{A}_\gamma=\underline{J}^{A^*}_\gamma$. Hence $\underline{A}^*=\bigcup_{\gamma\in I}\underline{A}=\underline{A}_{\gamma^*}$.

Let $A^*_\gamma=\{\langle i,\gamma_1\ldots\gamma_n\rangle: i\in\omega\wedge\underline{A}_\gamma\models\phi_i(\gamma_1\ldots\gamma_n)\wedge\gamma_1\ldots\gamma_n<\gamma\}$. Then A^*_γ may be coded as a subset of γ and $A^*_\gamma\in\bar{A}_\gamma$. Hence by lemma 16.9 $A^*_\gamma\in M_\gamma$. $\pi_{\gamma\gamma'}(A^*_\gamma)=A^*_{\gamma'}$, of course.

Let ν_γ be least such that $A^*_\gamma\in J^{U_\gamma}_{\nu_\gamma+1}$. Then $\underline{J}^{U_\gamma}_{\nu_\gamma+1}$ is a mouse. Given a formula ϕ_i with m free variables - ϕ_i enumerates all formulae - and define a set $X_\gamma^{i\delta}\subseteq\gamma$ as follows:

For $x_1<\ldots<x_m<\gamma$ $X^0_{x_1\ldots x_m}=\gamma\setminus x_m$. So $X^0_{x_1\ldots x_m}\in U_\gamma$.

Suppose $X^k_{x_1\ldots x_{m-k}}$ is defined such that $X^k_{x_1\ldots x_{m-k}}\in U_\gamma$ and whenever $y_1\ldots y_{k-1},y_1'\ldots y_{k-1}'\in X^k_{x_1\ldots x_{m-k}}$, $y_1<\ldots<y_{k-1},y_1'<\ldots<y_{k-1}'$ and $\delta'<\delta$

$$\underline{A}_\gamma\models\phi_i(x_1\ldots x_{m-k},y_1\ldots y_{k-1},\delta')\Leftrightarrow\phi_i(x_1\ldots x_{m-k},y_1'\ldots y_{k-1}',\delta')\tag{16.3}$$

Let $Y_{\delta'}^{x_1\ldots x_{m-k-1}}=\{y:\phi_i(x_1\ldots x_{m-k-1},y,y_1\ldots y_{k-2},\delta')\}$ (some $y_1<\ldots$ $\ldots<y_{k-2}\in X^k_{x_1\ldots x_{m-k-1},y}$) if this is in U_γ ; otherwise its complement in γ. Let $Z_{\delta'}^{x_1\ldots x_{m-k-1}}=Y_{\delta'}^{x_1\ldots x_{m-k-1}}\cap\{z:\forall y<z(z\in X^k_{x_1\ldots x_{m-k-1},y})\}$. Suppose $y_1\ldots y_k\in Z_{\delta'}^{x_1\ldots x_{m-k-1}}$, $y_1<\ldots<y_k$. Then $y_1\in Y_{\delta'}^{x_1\ldots x_{m-k-1}}$ and $y_2<\ldots<y_k\in X^k_{x_1\ldots x_{m-k-1},y_1}$ so

$$\underline{A}_\gamma\models\phi_i(x_1\ldots x_{m-k-1},y_1\ldots y_k,\delta')\Rightarrow Y_{\delta'}^{x_1\ldots x_{m-k-1}}=\{y:\phi_i(x_1\ldots\ldots x_{m-k-1},y,y_1\ldots y_{k-2})\}.\tag{16.4}$$

So if $y_1'<\ldots<y_k'<\gamma$ then $\underline{A}_\gamma\models\phi_i(x_1\ldots x_{m-k-1},y_1'\ldots y_k',\delta')$. Also $Z_{\delta'}^{x_1\ldots x_{m-k-1}}\in U_\gamma$. Now let $X^{k+1}_{x_1\ldots x_{m-k-1}}=\bigcap_{\delta'<\delta}Z_{\delta'}^{x_1\ldots x_{m-k-1}}$. So $X^{k+1}_{x_1\ldots x_{m-k-1}}\in U_\gamma$ and satisfies (16.3).

If $k=m$ then $X^m\in U_\gamma$ and for all $\delta'<\delta,y_1<\ldots<y_m\in X^m$, $y_1'<\ldots<y_m'\in X^m$

$$\underline{A}_\gamma\models\phi_i(y_1\ldots y_m,\delta')\Leftrightarrow\phi_i(y_1'\ldots y_m',\delta').\tag{16.5}$$

So let $X_\gamma^{i\delta}=X^m$. Let $X_\gamma^i=\{\delta:\forall\delta'<\delta(\delta\in X_\gamma^{i\delta'})\}$. Then $X^i_\gamma\in U_\gamma$ and $\bigcap_{i\in\omega}X^i_\gamma$ are good indiscernibles for $\underline{A}_\gamma$. Also by the uniformity of the definition $\pi_{\gamma\gamma'}(X^i_\gamma)=X^i_{\gamma'}$.

Let $I'=\bigcap_{i\in\omega}X_\gamma^{i}{}^*$. Then $I\subseteq I'$, for $X^i_\gamma\in U_\gamma$ so $\gamma\in\pi_{\gamma\gamma^*}(X^i_\gamma)=X_\gamma^i{}^*$. I' are good indiscernibles for $\underline{A}_\gamma{}^*$ and hence for $\underline{A}$ as $\underline{A}_\gamma{}^*\prec\underline{A}$. Finally I' were defined entirely from $\underline{J}^{U_\gamma^*}_{\nu_{\gamma^*+1}}$ which is a mouse. So $I'\in K$. We have proved the indiscernibles lemma:

Lemma 16.10. *Suppose* $\underline{A}=\langle K_\kappa,D\cap\kappa^2,A_1\ldots A_N\rangle$, $\underline{A}\models R^+$. *Suppose I is a good set of indiscernibles for* $\underline{A}$ *and* $cf(otp(I))>\omega$. *Then there is* $I'\in K$ *such that* $I\subseteq I'$ *and I' is a good set of indiscernibles for* $\underline{A}$.

THE EMBEDDINGS LEMMA

The indiscernibles lemma is used in the proof of the embeddings lemma, which says that if $j:K\to_e K$ is non-trivial then there is a ρ-model. The proof in chapter 12 will not give the result: we have seen that K may not be κ-minimal for any κ. What we now show is that a sufficiently drastic failure of minimality yields an inner model with a measurable cardinal.

Suppose K had been minimal. Then we should get a normal measure U on some κ in K such that $\langle K,U\rangle$ is iterable. We can assume that U is countably complete, for its ω_1-th iterate will be. This cannot give a mouse for all mice are in K. This enables us to prove the following

Lemma 16.11. *Suppose $\langle K_{\kappa^+},U\rangle$ is amenable, U a normal measure on κ in K and U is countably complete (κ^+ calculated in K). Then U is a normal measure in L[U].*

Proof: Suppose not. So $L[U]\not\subseteq K$. Let α be greatest such that $J^U_\alpha\subseteq K$. Then let $\bar{U}=U\cap J^U_\alpha$. If $\bar{U}$ were a normal measure in $L[\bar{U}]$ then $\bar{U}=U$; for if $a\in L[U]$, $a\subseteq\kappa$ but $a\notin J^U_\alpha$ and β is least with $a\in J^U_{\beta+1}$ then $\Sigma_{\omega}(J^U_\beta)\subseteq L[\bar{U}]$ so $\omega\rho^n_{J^U_\beta}>\kappa$, all $n<\omega$, so $J^U_{\beta+1}=\mathrm{rud}_{U\cap J^U_\beta}(J^U_\beta)$; thus $a\in L[\bar{U}]$ so $a\in J^U_\alpha$. Contradiction! (β must be taken least for any such a in the above so that $J^U_\beta=J^{\bar{U}}_\beta$)

So $\bar{U}$ is not normal in $L[\bar{U}]$. So there is $\beta\geq\alpha$ with $a\subseteq\kappa$, $a\in J^{\bar{U}}_{\beta+1}$, but $a\notin J^{\bar{U}}_\beta$. Take β least such; so $\underline{J}^{\bar{U}}_\beta$ is a premouse. $\underline{J}^{\bar{U}}_\beta$ is iterable because $\bar{U}$ is countably complete. And $\underline{J}^{\bar{U}}_\beta$ is a mouse because $a\in J^{\bar{U}}_{\beta+1}$ $a\notin J^{\bar{U}}_\beta$, $a\subseteq\kappa$ implies $\omega\rho^n_{J^{\bar{U}}_\beta}\leq\kappa$ for some n. So $\underline{J}^{\bar{U}}_\beta\in K$. Hence $\beta<\kappa^+$ so $J^{\bar{U}}_\beta\in K_{\kappa^+}$. But $\langle K_{\kappa^+},U\rangle$ is amenable so $J^U_{\beta+1}\subseteq K$ (for $J^{\bar{U}}_\beta=J^U_\beta$). But $\beta+1>\alpha$; contradiction! □

Lemma 16.11 occupies only a small proportion of this chapter but it is one of the most important results in this book: its argument was one of the original motivations for the definition of K. It is irritating that in order to prove the embeddings lemma it has to be embedded between two slabs of technicalities. But then: in general K is not minimal.

Suppose $j:K\to_e K$ is non-trivial with critical point κ. Let $U=\{X\in K:\kappa\in j(X)\wedge X\subseteq\kappa\}$. As in lemma 12.14 $\langle K_{\kappa^+},U\rangle$ is amenable ($\kappa^+=(\kappa^+)^K$ here and for the rest of this chapter). K is κ-maximal by lemma 14.19. Let $\langle\langle \underline{M}_\alpha\rangle_{\alpha\in On},\langle\pi_{\alpha\beta}\rangle_{\alpha\leq\beta\in On}\rangle$ be a weak iterated ultrapower. If $\underline{M}_\alpha$ is well-founded then $\underline{M}_\alpha=K$. Let ρ be least such that $\underline{M}_\rho$ is not well-founded. If ρ does not exist then letting U' be the ω_1-th measure, $\langle K,U'\rangle$ satisfies the conditions of lemma 16.11 and so there is a ρ-model.

By lemma 12.16 ρ cannot be a successor ordinal. Let $X^*_0=$ $=\{x\in K:0\leq j<\lambda\Rightarrow\pi_{0j}(x)=x\}$ (this is the X of chapter 12). Let $\pi:\underline{M}\tilde{=}K|X^*_0$ where $\underline{M}$ is transitive. If $\underline{M}=K$ then the proof in chapter 12 still works, for $\pi'_{0j}:K\to_e M'_j$ implies that each M'_j is K. So $\underline{M}\neq K$; $\underline{M}=K_N$, say.

We can generalise this a bit. Let $X_i^*=\{x\in K: i\leq j<\rho \Rightarrow \pi_{ij}(x)=x\}$. If there is i such that $\underline{K}\cong\underline{K}|X_i^*$ then the proof of chapter 12 works. So let $\pi_i:\underline{M}_i\cong\underline{K}|X_i^*$, $\underline{M}_i=\underline{K}_{N_i}$. This is what we meant by a sufficiently drastic breakdown of minimality.

A technical lemma is of great help: the argument extends that of lemma 12.16. Let $\kappa_\alpha=\pi_{0\alpha}(\kappa)$.

Lemma 16.12. *Suppose $i,j<\rho$. Then $i+j<\rho$ and $\pi_{0i}\pi_{0j}=\pi_{0i+j}$. Also $\pi_{0i}(\kappa_h)=\kappa_{i+h}$ $(h<j)$*

Proof: Suppose $g\in K$. Then $g=\pi_{0j}(f)(\kappa_{k_1}\ldots\kappa_{k_p})$ $(k_1<\ldots<k_p<j)$ say. Also $x=\pi_{0i}(g)(\kappa_{h_1}\ldots\kappa_{h_q})$ $(h_1<\ldots<h_q<i)$ say. So

$$x=\pi_{0i}\pi_{0j}(f)(\kappa_{h_1}\ldots\kappa_{h_q},\kappa'_{k_1}\ldots\kappa'_{k_p}) \tag{16.6}$$

where $\kappa'_k=\pi_{0i}(\kappa_k)$. Define $\sigma:K\to M_{i+j}$ by

$$\sigma(\pi_{0i}\pi_{0j}(f)(\kappa_{h_1}\ldots\kappa_{h_q},\kappa'_{k_1}\ldots\kappa'_{k_p}))=\pi_{0i+j}(f)(\kappa_{h_1}\ldots\kappa_{h_q},\kappa_{i+k_1}\ldots\kappa_{i+k_p}) \tag{16.7}$$

For all formulae ϕ (with one free variable for simplicity)

$$\begin{aligned}K\models\phi(\pi_{0i}\pi_{0j}(f)(\kappa_{h_1}\ldots\kappa_{h_q},\kappa'_{k_1}\ldots\kappa'_{k_p})) &\Leftrightarrow \\ &\Leftrightarrow \{\langle\xi_1\ldots\xi_q\rangle:\phi(\pi_{0j}(f)(\xi_1\ldots\xi_q,\kappa_{k_1}\ldots\kappa_{k_p}))\}\in U^q \\ &\Leftrightarrow \{\langle\xi_1\ldots\xi_q\rangle:\{\langle\zeta_1\ldots\zeta_p\rangle:\phi(f(\xi_1\ldots\xi_q,\zeta_1\ldots\zeta_p))\}\in U^p\}\in U^q \\ &\Leftrightarrow \{\langle\xi_1\ldots\xi_q,\zeta_1\ldots\zeta_p\rangle:\phi(f(\xi_1\ldots\xi_q,\zeta_1\ldots\zeta_p))\}\in U^{p+q} \\ &\Leftrightarrow \underline{M}_{i+j}\models\phi(\kappa_{h_1}\ldots\kappa_{h_q},\kappa_{i+k_1}\ldots\kappa_{i+k_p})\end{aligned} \tag{16.8}$$

so σ is well-defined, $\sigma:\underline{K}\to_e\underline{M}_{i+j}$. Hence $\sigma:\underline{K}\cong\underline{M}_{i+j}$; so $i+j<\rho$, $\underline{M}_{i+j}=\underline{K}$. So σ is the identity and $\pi_{0i}\pi_{0j}=\pi_{0i+j}$ and $\pi_{0i}(\kappa_h)=\kappa_{i+h}$. □

Lemma 16.13. *Suppose that for all j with $i\leq j<\rho$ $x\in rng(\pi_{ij})$. Then $x\in X_i^*$.*

Proof: Let $x_j=\pi_{ij}^{-1}(x)$. Then if $<'$ is the canonical order of K $j<j' \Rightarrow x_{j'}\leq' x_j$, for $x=\pi_{ij}(x_j)=\pi_{ij'}(x_{j'})$ and $\pi_{jj'}(\pi_{ij}(x_j))\geq'\pi_{ij}(x_j)$, i.e. $\pi_{ij'}(x_j)\geq'\pi_{ij}(x_j)=\pi_{ij'}(x_{j'})$. So for some j $j'\geq j \Rightarrow x_j=x_{j'}$. For any k, $i\leq k<\rho$

$$\begin{aligned}\pi_{ik}(x)&=\pi_{ik}(\pi_{ij}(x_j)) \\ &=\pi_{ik}(\pi_{ii+h}(x_j)) \qquad (j=i+h \text{ say}) \\ &=\pi_{ik+h}(x_j) \\ &=\pi_{ik+h}(x_{k+h}) \\ &=x\end{aligned} \tag{16.9}$$

□

Now pick some τ such that τ is regular, $\tau>(\sup_{i<\rho}\kappa_i)^+$, $\tau>\overline{\overline{N}}_i$, all $i<\rho$.

Lemma 16.14. $\tau \in X_0^*$.

Proof: Suppose $\eta<\tau$. So $\pi_{0i}(\eta)=\{\pi_{0i}(f)(\kappa_{h_1}\ldots\kappa_{h_m}): f:\kappa^m\to\eta, f\in K,$ $h_1<\ldots<h_m<i\}$, so that $\pi_{0i}(\eta)\leq\eta^\kappa.\bar{\bar{I}}=\bar{\eta}.\bar{I}<\tau$ (in K). Thus $\pi_{0i}(\eta)<\tau$.

Also if $f:\kappa^m\to\tau$ then $f:\kappa^m\to\eta$, some $\eta<\tau$ so $\pi_{0i}(\tau)=\sup_{\eta<\tau}\pi_{0i}(\eta)\leq\tau$. Thus $\pi_{0i}(\tau)=\tau$. □

Lemma 16.15. $\bigcup_{i<\rho}(\tau^+\cap X_i^*)$ *is cofinal in* τ^+ *(calculated in K).*

Proof: Let $\delta=\sup\bigcup_{i<\rho}(\tau^+\cap X_i^*)$. $\delta\notin\bigcup_{i<\rho}(\tau^+\cap X_i^*)$, for $X_i^*\prec K$ so $\delta\in X_i^*\Rightarrow\delta+1\in X_i^*$. δ is a limit ordinal.

$\pi_{0i}\pi_{hj}(x)=\pi_{i+hi+j}\pi_{0i}$ when $i,h,j<\rho, h\leq j$. This is a modification of lemma 16.12: suppose $x=\pi_{0h}(f)(\kappa_{k_1}\ldots\kappa_{k_p})$ $(k_1<\ldots<k_p<h)$; then

$$\begin{aligned}\pi_{0i}\pi_{hj}(x)&=\pi_{0i}(\pi_{0j}(f)(\kappa_{k_1}\ldots\kappa_{k_p})) &&\\ &=\pi_{0i}\pi_{0j}(f)(\kappa_{i+k_1}\ldots\kappa_{i+k_p}) && \text{(lemma 16.12)}\\ &=\pi_{0i+j}(f)(\kappa_{i+k_1}\ldots\kappa_{i+k_p}) && \text{(lemma 16.12)}\\ &=\pi_{i+hi+j}\pi_{0i+h}(f)(\kappa_{i+k_1}\ldots\kappa_{i+k_p}) &&\\ &=\pi_{i+hi+j}\pi_{0i}\pi_{0h}(f)(\kappa_{i+k_1}\ldots\kappa_{i+k_p}) && \text{(lemma 16.12)}\\ &=\pi_{i+hi+j}\pi_{0i}(\pi_{0h}(f)(\kappa_{k_1}\ldots\kappa_{k_p})) && \text{(lemma 16.12)}\\ &=\pi_{i+hi+j}\pi_{0i}(x).\end{aligned} \tag{16.10}$$

Hence $\pi_{0i}``X_j^*\subseteq X_{i+j}^*$ $(i,j<\rho)$. For suppose $x\in X_j^*$: then $\pi_{jk}(x)=x$ when $j\leq k<$ $<\rho$. So $\pi_{i+ji+k}\pi_{0i}(x)=\pi_{0i}\pi_{jk}(x)=\pi_{0i}(x)$. That is, $\pi_{0i}(x)\in X_{i+j}^*$.

Suppose $cf_K(\delta)\neq\kappa$. Then $\pi_{0i}(cf_K(\delta))=\sup_{\nu<\gamma}\pi_{0i}(\nu)$, all $i<\rho$, where $\gamma=cf_K(\delta)$. If $cf_K(\delta)<\kappa$, this is clear: otherwise letting $f:\kappa^m\to\delta$ there is $\nu<\gamma$ such that $f:\kappa^m\to\nu$, so $\pi_{0i}(f)(\kappa_{k_1}\ldots\kappa_{k_m})<\pi_{0i}(\nu)$. Take $f\in K$ such that $f:\gamma\to\delta$ cofinally. So

$$\begin{aligned}\pi_{0i}(\delta)&=\sup_{\nu<\mu}\pi_{0i}(f)(\nu) \qquad (\mu=\pi_{0i}(\gamma))\\ &=\sup_{\bar{\nu}<\gamma}\pi_{0i}(f)(\pi_{0i}(\bar{\nu}))\\ &=\sup_{\bar{\nu}<\gamma}\pi_{0i}(f(\bar{\nu})).\end{aligned} \tag{16.11}$$

Given $\bar{\nu}<\gamma$ $f(\bar{\nu})<\delta'\in X_j^*$ (some δ', j) so $\pi_{0i}(f(\bar{\nu}))<\pi_{0i}(\delta')\in X_{i+j}^*$. So $\pi_{0i}(f(\bar{\nu}))<\delta$. Thus $\pi_{0i}(\delta)=\delta$, all $i<\rho$. So $\delta\in X_0^*$. Contradiction! So $cf_K(\delta)=\kappa$.

Let $f\in K$ with $f:\kappa\to\delta$ cofinally. Then

$$\begin{aligned}\pi_{1j}(\delta)&=\sup\pi_{1j}(f)``\kappa\\ &=\sup_{\bar{\nu}<\kappa}\pi_{1j}(f(\bar{\nu})).\end{aligned} \tag{16.12}$$

But $\pi_{0j}(f(\bar{\nu}))=\pi_{1j}(\pi_{01}(f(\bar{\nu}))\geq\pi_{1j}(f(\bar{\nu}))$; and $\pi_{0j}(f(\bar{\nu}))<\delta$ as before so $\pi_{1j}(f(\bar{\nu}))<\delta$. Thus $\pi_{1j}(\delta)=\delta$, so $\delta\in X_1^*$: contradiction! □

Now let $\langle\langle \underline{N}_{ij}\rangle_{j\in On},\langle\pi^i_{jk}\rangle_{j\leq k\in On}\rangle$ be the mouse iteration of $\underline{N}_i$. Let τ^i_j be the critical point of $\underline{N}_{ij}$. Let $C_i=\{\tau^i_j:j<\tau\}$ and let $C'=\bigcap_{i<\rho} C_i$. Let $C=\{\eta\in C':\pi''\eta\subseteq\eta\}$ (recall that $\pi:\underline{K}_{N_0}\cong\underline{K}|X_0^*$).

Lemma 16.16. *$C\subseteq X_0^*$.*

Proof: If $\eta\in C$ then $\pi''\eta\subseteq\eta$; thus given $\gamma<\eta$, $\pi(\gamma)<\eta$. $\pi(\gamma)\in X_0^*$ so $\pi_{0j}(\pi(\gamma))=\pi(\gamma)$. Hence $\pi_{0j}(\gamma)\leq\pi_{0j}(\pi(\gamma))=\pi(\gamma)<\eta$. So $\pi_{0j}''\eta\subseteq\eta$.

Suppose $f\in K$, $f:\gamma\to\eta$ cofinally, $\gamma<\kappa_j$. Then applying π_{jk} $\pi_{jk}(\eta)=\sup_{\gamma'<\gamma}\pi_{jk}(f(\gamma))$. But since $f(\gamma)<\eta$ $f(\gamma)\leq\pi(f(\gamma))\in X_0^*\subseteq X_j^*$ so $\pi_{jk}(f(\gamma))\leq\pi_{jk}(\pi(f(\gamma)))<\eta$. So $\eta\in X_j^*$. But then η is regular in K_{N_j} as $\eta=\tau^j_k$ for some k. $\pi_j:\underline{K}_{N_j}\cong\underline{K}|X_j^*$ and $\pi_j(\eta)\geq\eta$. So $\pi_j(\eta)>\eta$, as η is singular in K. Hence $\eta=\pi_j(\eta')$, some $\eta'<\eta$. On the other hand $\pi(\eta')<\eta$. So $\pi_j(\eta')>\pi(\eta')$. But $X_0^*\subseteq X_j^*$. Contradiction!

So in particular in K $cf(\eta)>\kappa$. It follows as usual that

$\pi_{0j}(\eta)=\sup_{\gamma<\eta}\pi_{0j}(\gamma)=\eta$. □

Hence $\pi_i|C=id|C$.

Lemma 16.17. *C is closed unbounded in τ.*

Proof: Each C_i is closed unbounded in τ as $\tau>\bar{\bar{N}}_i$, and $\tau>\rho$ so C' is closed unbounded in τ.

C is closed, obviously. To get unboundedness it is only necessary to show that $\eta<\tau \Rightarrow \pi''\eta\subseteq\tau$, i.e. that $\tau\cap X_0^*$ is cofinal in τ. But the set $D=\{\eta'<\tau:\forall i<\rho(\pi_{0i}''\eta'\subseteq\eta')\wedge cf(\eta')>\kappa\}$ is certainly cofinal; this suffices by the argument of lemma 16.14. □

Lemma 16.18. *Suppose $A\in P(\tau)\cap K$. There is $\gamma<\tau$ such that $C\setminus\gamma$ is a good set of indiscernibles for $\langle K_\tau,D\cap\tau^2,A\rangle$.*

Proof: $A\in K_{\tau^+}$. Because τ is regular it is easily seen that $\langle K_\tau,D\cap\tau^2,A\rangle\models R^+$. By lemma 16.15 there are $i<\rho$ and $\xi\in rng(\pi_i)$ with $A\in K_\xi$. Suppose $\xi=\pi_i(\bar{\xi})$. Let $\langle\bar{B}_\nu:\nu<\tau\rangle\in K_{N_i}$ enumerate $P(\tau)\cap J^{K}_{\xi}N_i$. Let $\bar{B}=\{\langle\nu,\nu'\rangle:\nu'\in\bar{B}_\nu\}$, $B=\pi_i(\bar{B})$, $B_{\nu'}=\{\nu:\langle\nu,\nu'\rangle\in B\}$. It suffices to show that $C\setminus\gamma$ is good for $\langle K_\tau,D\cap\tau^2,B\rangle$ for some $\gamma>\nu'$ where $A=B_{\nu'}$.

$\overline{B}\in H^K_\delta N_i = H^N_\delta i+1$ $(\delta=\tau^i_{\tau+1})$, so there is $\overline{\gamma}$ with $\overline{B}$ Σ_1-definable in $\underline{N}'_{i\tau+1}$ from $\overline{\gamma}\cup p_{N'_{i\tau+1}}\cup\{\tau^i_\tau\}(\nu<\overline{\gamma}<\tau$, where $\underline{N}'_{i\gamma}=\underline{N}^{n(N_{i\gamma})}_{i\gamma})$. Let $\gamma=\pi_i(\overline{\gamma})$. Then we claim $C\setminus\gamma$ is a good set of indiscernibles for $\langle K_\tau, D\cap\tau^2, B\rangle$.

Let $\underline{A}=\langle K_\tau, D\cap\tau^2, B\rangle$ and $\overline{\underline{A}}=\langle J^K_\tau N_i, D^K N_i\cap\tau^2, \overline{B}\rangle$. Then $\pi_i(\overline{\underline{A}})=\underline{A}$. If $\mu\in C\setminus\gamma$ then we must show $\overline{\underline{A}}_\mu=J^{\overline{A}}_\mu\prec\overline{\underline{A}}$. But given $\vec{\xi}<\mu$

$$\overline{\underline{A}}_\mu\models\phi(\vec{\xi}) \Leftrightarrow N_{i\nu}\models\text{"}\overline{\underline{A}}_\mu\models\phi(\vec{\xi})\text{"} \qquad (\mu=\tau^i_\nu) \tag{16.13}$$

$$\Leftrightarrow N_{i\tau}\models\text{"}\overline{\underline{A}}\models\phi(\vec{\xi})\text{"}\text{ (by }\Sigma_1\text{-definability of B)}$$

$$\Leftrightarrow \overline{\underline{A}}\models\phi(\vec{\xi}).$$

Secondly suppose $\xi_1<\ldots<\xi_m<\mu\leq\zeta_1<\ldots<\zeta_n<\tau, \mu\leq\zeta'_1<\ldots<\zeta'_n<\tau$ and each $\zeta_i,\zeta'_i\in C\setminus\gamma$. Then

$$\overline{\underline{A}}\models\phi(\xi_1\ldots\xi_m,\zeta_1\ldots\zeta_n) \Leftrightarrow \phi(\xi_1\ldots\xi_m,\zeta'_1\ldots\zeta'_n) \tag{16.14}$$

by a similar argument.

But these results transfer to $\underline{A}$, $\underline{A}_\mu$ using $\pi_i|C=\text{id}|C$. □

Define $U\subseteq P(\tau)\cap K$ by $A\in U \Leftrightarrow \exists\gamma<\tau\, C\setminus\gamma\subseteq A$. Then U is a normal measure on τ by lemma 16.18.

Lemma 16.19. *$\langle K_{\tau^+}, U\rangle$ is amenable.*

Proof: It is enough to show that $\tau\leq\xi<\tau^+ \Rightarrow U\cap K_\xi\in K$. Let $\langle A_\nu:\nu<\tau\rangle\in K$ enumerate $P(\tau)\cap K_\xi$. Let $A=\{\langle\nu,\nu'\rangle:\nu\in A_{\nu'}\}$. Pick γ so that $C\setminus\gamma$ is good for $\langle K_\tau, D\cap\tau^2, A\rangle$. By lemma 16.10 there is $I\in K$ with $C\setminus\gamma\subseteq I$, I good for $\langle K_\tau, D\cap\tau^2, A\rangle$. So $U\cap K_\xi=\{A_\nu:\nu<\tau \wedge (\exists\gamma<\tau)(I\setminus\gamma\subseteq A_\nu)\}$. So $U\cap K_\xi\in K$. □

Corollary 16.20. *There is a ρ-model.*

Proof: By lemma 16.11. U is countably complete because $\text{cf}(\tau)>\omega$. □

At last we have proved:

Lemma 16.21. *If there is $j:K\to_e K$ that is non-trivial then there is a ρ-model.*

This complex construction is perhaps necessary: see appendix II.

Exercises

1. Suppose $L[U]$ is a ρ-model with iterated ultrapower $\langle\langle L[U_\alpha]\rangle_{\alpha\in On}, \langle\pi_{\alpha\beta}\rangle_{\alpha\leq\beta\in On}\rangle$. Show that $\pi_{0\alpha}\pi_{0\beta}=\pi_{0+\pi_{0\alpha}(\beta)}$.

2. Strengthen 16.21 to: suppose $j:K\to_e K$ with critical point κ; suppose τ regular and greater than $\max(\kappa,\aleph_1)$. Then there is a ρ-model with critical point τ.

17: APPLICATIONS OF K

There are two important applications of the indiscernibles lemma other than the embeddings lemma. These concern large cardinal properties and ultrafilters: the first is given in detail in this chapter, the second summarised very briefly. Also summarised briefly are the combinatorial properties of K. In part five, once we have proved the covering lemma, we shall obtain further applications to cardinal arithmetic and concerning the #s of inner models that are not of the form K^M.

LARGE CARDINALS

There are no measurable cardinals in K. On the other hand many of the properties implied by measurability are preserved: to discuss these we must introduce the arrow notation.

Definition 17.1. *Suppose $f:[A]^n \to I$. H is f-homogeneous provided that $H \subseteq A$ and and $f''[H]^n = \{i\}$ for some $i \in I$. Suppose $f:[A]^{<\omega} \to I$. H is f-homogeneous provided that H is $f|[A]^n$-homogeneous for all $n<\omega$.*

Definition 17.2. *Suppose λ is a limit ordinal. $\kappa \to (\lambda)^n_I$ means that whenever $f:[\kappa]^n \to I$ there is an f-homogeneous set of order type λ. Similarly with n replaced by $<\omega$. I is omitted when it is 2.*

Definition 17.3. *κ is Ramsey if and only if $\kappa \to (\kappa)^{<\omega}$.*

In the following, observe the similarity with the final argument in the proof of the indiscernibles lemma.

Lemma 17.4. *If κ is measurable then κ is Ramsey.*

Proof: Let U be a normal measure on κ. We shall first prove by induction on n that whenever $f:[\kappa]^n \to 2$ there is an f-homogeneous set in U.

If n=1 then either $\{\alpha : f(\{\alpha\})=0\}$ or $\{\alpha : f(\{\alpha\})=1\}$ is in U so this is clear.

Suppose $n=m+1$, $m \geq 1$ and let $f:[\kappa]^n \to 2$. Define $f_\alpha:[\kappa]^m \to 2$ by

$$f_\alpha(x) = f(x \cup \{\alpha\}) \quad (\alpha < \min x) \qquad (17.1)$$
$$= 0 \quad \text{(otherwise)}.$$

By induction hypothesis there is X_α homogeneous for f_α. Let $\gamma_\alpha = f_\alpha(x)$ for some $x \in [X_\alpha]^m$. Suppose that $Y = \{\alpha : \gamma_\alpha = \gamma\} \in U$ ($\gamma = 0$ or 1). Let $X = \{\beta < \kappa : \forall \alpha < \beta (\beta \in X_\alpha)\}$. Then $X \in U$ as each $X_\alpha \in U$. Furthermore if $\{\alpha_1 \ldots \alpha_n\} \in [X]^n$ and $\alpha_1 < \ldots < \alpha_n$ then $\{\alpha_2 \ldots \alpha_n\} \in [X_{\alpha_1}]^m$ so $f_{\alpha_1}(\{\alpha_2 \ldots \alpha_n\}) = \gamma_{\alpha_1}$, i.e. $f(\{\alpha_1 \ldots \alpha_n\}) = \gamma_{\alpha_1}$. Let $Z = X \cap Y$. Then for $\{\alpha_1 \ldots \alpha_n\} \in [Z]^n$ $f(\{\alpha_1 \ldots \alpha_n\}) = \gamma$. So Z is f-homogeneous and $Z \in U$.

Finally suppose $f : [\kappa]^{<\omega} \to 2$ and let $f_n = f | [\kappa]^n$. Let X_n be f_n-homogeneous ($0 < n < \omega$) and $X_n \in U$. Let $X = \bigcap_{0<n<\omega} X_n$; then $X \in U$ and X is f-homogeneous. □

Definition 17.5. *$\kappa(\alpha)$ denotes the least κ such that $\kappa \to (\alpha)^{<\omega}$, for α a limit ordinal.*

$\kappa(\alpha)$ is clearly a cardinal. Actually the $\kappa(\alpha)$ are special cases of the following partition cardinals.

Definition 17.6. *κ is α-Erdős provided κ is a regular cardinal and whenever $C \subseteq \kappa$ is closed unbounded in κ and $f : [C]^{<\omega} \to \kappa$ satisfies $f(a) < \min(a)$ for all $a \in [C]^{<\omega}$ (such f are called regressive) then there is $D \subseteq C$ with order-type α that is f-homogeneous.*

$\kappa(\alpha)$ is α-Erdős. To see this we first prove:

Lemma 17.7. *If $\kappa \to (\alpha)^{<\omega}$ then $\kappa \to (\alpha)^{<\omega}_{2^{\aleph_0}}$.*

Proof: Let $f : [\kappa]^{<\omega} \to \{f' : f' : \omega \to 2\}$. Let f_n denote $f | [\kappa]^n$ ($n \in \omega$) and let $f_{nk} : [\kappa]^n \to 2$ be defined by

$$f_{nk}(\{\alpha_1 \ldots \alpha_n\}) = f_n(\{\alpha_1 \ldots \alpha_n\})(k). \tag{17.2}$$

Let $\sigma : \omega \times \omega \to \omega$ be the Gödel pairing function. Observe that $\sigma(n,k) \geq n$ for all n. Let $g_{\sigma(n,k)} : [\kappa]^{\sigma(n,k)} \to 2$ be defined by

$$g_{\sigma(n,k)}(\{\alpha_1 \ldots \alpha_{\sigma(n,k)}\}) = f_{nk}(\{\alpha_1 \ldots \alpha_n\}).\ (\alpha_1 < \ldots < \alpha_{\sigma(n,k)}) \tag{17.3}$$

Suppose $g(\{\alpha_1 \ldots \alpha_m\}) = g_m(\{\alpha_1 \ldots \alpha_m\})$ and let $X \subseteq \kappa$ be g-homogeneous of order-type α. Suppose $\{\alpha_1 \ldots \alpha_n\}, \{\beta_1 \ldots \beta_n\} \in [X]^n$. Suppose $f(\{\alpha_1 \ldots \alpha_n\}) \neq f(\{\beta_1 \ldots \beta_n\})$; then for some k $f_{nk}(\{\alpha_1 \ldots \alpha_n\}) \neq f_{nk}(\{\beta_1 \ldots \beta_n\})$ so $g_m(\{\alpha_1 \ldots \alpha_m\}) \neq g_m(\{\beta_1 \ldots \beta_m\})$ for some $\alpha_{n+1} < \ldots < \alpha_m, \beta_{n+1} < \ldots < \beta_m$ chosen in X with $\alpha_{n+1} > \alpha_n$, $\beta_{n+1} > \beta_n$ (α is a limit ordinal) where $m = \sigma(n,k)$. Contradiction! So X is f-homogeneous. □

Lemma 17.8. *Let $\underline{A}$ be a structure with a countable number of relations, functions and constants and with $\kappa \subseteq A$, where $\kappa \to (\alpha)^{<\omega}_{2^{\aleph_0}}$. Then $\underline{A}$ has a set of indiscernibles of order-type α.*

Proof: Let Φ be the set of formulae in the appropriate language.

Let $f(\{\alpha_1 \ldots \alpha_n\}) = \{\phi \in \Phi : \underline{A} \models \phi(\alpha_1 \ldots \alpha_n)\}$ $(\alpha_1 < \ldots < \alpha_n)$. So $f : [\kappa]^{<\omega} \to P(\Phi)$. Let X be f-homogeneous of order type α. Then for $\phi \in \Phi, \alpha_1 < \ldots < \alpha_n$, $\beta_1 < \ldots < \beta_n$ all in X

$$\begin{aligned} \underline{A} \models \phi(\alpha_1 \ldots \alpha_n) &\Leftrightarrow \phi \in f(\{\alpha_1 \ldots \alpha_n\}) \\ &\Leftrightarrow \phi \in f(\{\beta_1 \ldots \beta_n\}) \\ &\Leftrightarrow \underline{A} \models \phi(\beta_1 \ldots \beta_n). \quad \square \end{aligned} \tag{17.4}$$

Lemma 17.9. *$\kappa(\alpha)$ is α-Erdős.*

Proof: Suppose $f : [C]^{<\omega} \to \kappa$ is regressive and $C \subseteq \kappa = \kappa(\alpha)$ is closed in κ (i.e. $\gamma < \kappa$ and $\gamma = \sup(\gamma \cap C) \Rightarrow \gamma \in C$) and of cardinality κ. For $\nu < \kappa$ suppose $h_\nu : [\nu]^{<\omega} \to 2$ has no homogeneous subset of order-type α. Let $f_n : C^n \to \kappa$ and $g_n : \kappa^n \to 2$ be defined by

$$\begin{aligned} f_n(\nu_1 \ldots \nu_n) &= f(\{\nu_1 \ldots \nu_n\}) \\ g_n(\nu_1 \ldots \nu_n) &= h_{\nu_n}(\{\nu_1 \ldots \nu_{n-1}\}). \end{aligned} \tag{17.5}$$

Let $\underline{A}$ be the structure $\langle \kappa, <, C, f_1 \ldots f_n \ldots, g_1 \ldots g_n \ldots \rangle$.

Claim 1 There is a set $X \subseteq C$ of order-type α such that X are indiscernibles for $\underline{A}$.

Proof: Adapt the proof of lemma 17.8 to C by letting $f : [C]^{<\omega} \to P(\Phi)$ be defined as there, and use the fact that $\overline{\overline{C}} = \kappa$.

$\square$(Claim 1).

Pick a set X of indiscernibles for $\underline{A}$ of order-type α with min(X) minimal. Say $X = \{b_\nu : \nu < \alpha\}$. $b_\nu = \sup(b_\nu \cap C)$ for all $\nu < \alpha$; for otherwise let $b'_\nu = \sup(b_\nu \cap C)$, $X' = \{b'_\nu : \nu < \alpha\}$. Then $X' \subseteq C$ as C is closed; X' are indiscernibles for $\underline{A}$, since b'_ν is uniformly definable from b_ν, C; $otp(X') = \alpha$, as $\nu < \nu' \Rightarrow b'_\nu \neq b'_{\nu'}$; and $b'_0 < b_0$ (since for some ν $b'_\nu < b_\nu$). This contradicts the minimality of min(X).

Claim 2 X is f-homogeneous.

Proof: Suppose not. So for some m $f(\{b_{k_1} \ldots b_{k_m}\}) \neq f(\{b_{h_1} \ldots b_{h_m}\})$, some $k_1 < \ldots < k_m < \alpha, h_1 < \ldots < h_m < \alpha$. Pick $p_1 \ldots p_m > k_m, h_m$; so $f(\{b_{k_1} \ldots b_{k_m}\}) \neq f(\{b_{p_1} \ldots b_{p_m}\})$. Hence $f(\{b_1 \ldots b_m\}) \neq f(\{b_{m+1} \ldots b_{2m}\})$, i.e. $f_m(b_1 \ldots b_m) \neq f_m(b_{m+1} \ldots b_{2m})$. Let $d_\nu = f_m(b_{(m+1)\nu} \ldots b_{(m+1)\nu+m-1})$ $(\nu < \alpha)$. d_ν is defined for all $\nu < \alpha$ since α is a limit ordinal. If $\nu < \mu$ then $d_\nu \neq d_\mu$, so $d_\nu < d_\mu$ by well-foundedness.

Let $c_\nu = \min(C \setminus (d_\nu + 1))$. If $c_1 < c_0$ then $d_1 < d_0$; so $c_0 \leq c_1$. If $c_0 < c_1$ then let $X'' = \{c_\nu : \nu < \alpha\}$. $X'' \subseteq C$; X'' are indiscernibles for $\underline{A}$ as X are; $otp(X'') = \alpha$. But $c_0 = \min(C \setminus (d_0 + 1))$, and $d_0 = f(\{b_0 \ldots b_{m-1}\})$ so $d_0 < b_0$ as f is regressive. But $b_0 = \sup(C \cap b_0)$ so $b_0 > c_0$, contradicting minimality.

So $c_0 = c_1$, and so $c_\nu = c_\mu$ for all $\nu, \mu < \alpha$. That implies $d_\nu < c_0$ for all $\nu<\alpha$. Now suppose $h_{c_0}(\{d_1\ldots d_{n-1}\})\neq h_{c_0}(\{d_n\ldots d_{2n-2}\})$. Thus $g_n(d_1\ldots d_{n-1},c_0)\neq g_n(d_n\ldots d_{2n-2},c_0)$ and so by indiscernibility $\{g_n(d_1\ldots d_{n-1},c_0),g_n(d_n\ldots d_{2n-2},c_0),g_n(d_{2n-1}\ldots d_{3n-3},c_0)\}$ has three members. But $g_n:\kappa^n\to 2$! So $h_{c_0}(\{d_1\ldots d_{n-1}\})=h_{c_0}(\{d_n\ldots d_{2n-2}\})$ and it follows that $\{d_\nu:\nu<\alpha\}$ is h_{c_0}-homogeneous. Contradiction! □(Claim 2)

Finally $\kappa(\alpha)$ is regular. For suppose it were singular, of cofinality γ, say. Let $\kappa(\alpha)=\sup\{\kappa_i:i<\gamma\}$. Suppose $f_i:[\kappa_i]^{<\omega}\to 2$ has no homogeneous subset of order-type α. Let $f:[\kappa]^{<\omega}\to\kappa$ be defined by

$$f(\{\alpha_1\ldots\alpha_{2n}\})=0 \text{ if } f_i(\{\alpha_1\ldots\alpha_n\})=f_i(\{\alpha_{n+1}\ldots\alpha_{2n}\}) \tag{17.6}$$

$$\text{where } \kappa_i \text{ is least with } \kappa_i>\alpha_1$$

$$=1 \text{ otherwise}$$

$$f(\{\alpha_1\ldots\alpha_{2n+1}\})=\text{the least } i \text{ such that } \alpha_1<\kappa_i.$$

f is regressive on $[\kappa\smallsetminus\gamma]^{<\omega}$ so we may take $X\subseteq\kappa\smallsetminus\gamma$ of order-type α f-homogeneous. Now for all $\alpha\in X$ $f(\{\alpha\})=i$, say. So $X\subseteq\kappa_i$. Also if $\{\alpha_1\ldots\alpha_{2n}\}\in[X]^{2n}$ then $f_i(\{\alpha_1\ldots\alpha_n\})=f_i(\{\alpha_{n+1}\ldots\alpha_{2n}\})$ as $f_i:[\kappa_i]^{<\omega}\to 2$ so X is f_i-homogeneous. Contradiction! □

Since κ Ramsey $\Rightarrow \kappa=\kappa(\kappa)$ this implies

<u>Corollary 17.10</u>. *κ is Ramsey if and only if κ is κ-Erdős.*

κ α-Erdős is a large cardinal property. For example:

<u>Lemma 17.11</u>. *If κ is α-Erdős then κ is strongly inaccessible.*

Proof: We must show κ to be a strong limit. So suppose $\beta<\kappa$ but $2^\beta\geq\kappa$. Let $\tau:\kappa\to\{f:\ f:\beta\to 2\}$ be one-one. Let $f(\{\alpha_1,\alpha_2\})$ be the least α such that $\tau(\alpha_1)(\alpha)\neq\tau(\alpha_2)(\alpha)$. So $f:[\kappa\smallsetminus\beta]^2\to\kappa$ is regressive. Let $X\subseteq\kappa\smallsetminus\beta$ be f-homogeneous of order-type α. Take $\alpha_1,\alpha_2,\alpha_3\in X$. Let $\alpha=f(\{\alpha_1,\alpha_2\})$. Then $\tau(\alpha_3)(\alpha)=\tau(\alpha_1)(\alpha)$ or $\tau(\alpha_3)(\alpha)=\tau(\alpha_2)(\alpha)$. Contradiction! □

<u>Lemma 17.12</u>. *Suppose κ is a regular cardinal. κ is α-Erdős if and only if every transitive model $\underline{A}\models R^+$ with $On_A=\kappa$ has a good set of indiscernibles of order-type α.*

Proof: First of all suppose κ is α-Erdős. Let $\underline{A}_\gamma$ denote $\underline{J}_\gamma^A$. By the reflection principle there is a closed unbounded $C\subseteq\kappa$ such that $\underline{A}_\gamma\prec\underline{A}$ for $\gamma\in C$. Let g be the Gödel pairing function; we may assume that each γ in C is closed under g. Now define $f:[C]^{<\omega}\to\kappa$ by

$$f(\{\alpha_1\ldots\alpha_{2n}\})=\text{ the least }\lambda<\kappa\text{ such that for some }\delta_1<\ldots \tag{17.7}$$

$$\ldots<\delta_k<\alpha_1,\ m<\omega,\ \lambda=g(m,\delta_1\ldots\delta_k)\text{ and}$$

$$\underline{A}\not\models\phi_m(\delta_1\ldots\delta_k,\alpha_1\ldots\alpha_n)\leftrightarrow\phi_m(\delta_1\ldots\delta_k,\alpha_{n+1}\ldots\alpha_{2n})$$

$=0$ if there is none.

Then f is regressive. Let $X\subseteq C$ be homogeneous of order-type α. Then if $\alpha_1\ldots\alpha_n,\alpha_{n+1}\ldots\alpha_{2n},\alpha_{2n+1}\ldots\alpha_{3n}$ are in X and $f(\{\alpha_1\ldots\alpha_{2n}\})\neq 0$ then letting $g(m,\delta_1\ldots\delta_k)=f(\{\alpha_1\ldots\alpha_{2n}\})$,

$$\underline{A}\not\models\phi_m(\delta_1\ldots\delta_k,\alpha_1\ldots\alpha_n) \Leftrightarrow \phi_m(\delta_1\ldots\delta_k,\alpha_{n+1}\ldots\alpha_{2n}). \tag{17.8}$$

So either

$$\underline{A}\models\phi_m(\delta_1\ldots\delta_k,\alpha_1\ldots\alpha_n) \Leftrightarrow \phi_m(\delta_1\ldots\delta_k,\alpha_{2n+1}\ldots\alpha_{3n})$$

or $\underline{A}\models\phi_m(\delta_1\ldots\delta_k,\alpha_{n+1}\ldots\alpha_{2n}) \Leftrightarrow \phi_m(\delta_1\ldots\delta_k,\alpha_{2n+1}\ldots\alpha_{3n})$.

Contradiction! So $f(\{\alpha_1\ldots\alpha_{2n}\})=0$ and it follows that X are good indiscernibles for $\underline{A}$.

Conversely suppose $f:[C]^{<\omega}\to\kappa$ is regressive. Let $f_n(\alpha_1\ldots\alpha_n)=$ $=f(\{\alpha_1\ldots\alpha_n\})$ $(\alpha_1<\ldots<\alpha_n)$, let $B=\{\langle n,\alpha_1\ldots\alpha_n,\gamma\rangle : f_n(\alpha_1\ldots\alpha_n)=\gamma\}$ and let $\underline{A}=\underline{J}_\kappa^B$. Let I be good indiscernibles for $\underline{A}$ of order-type α. Then given $n\in\omega$ let $\alpha_1\ldots\alpha_n\in I$, $\alpha_1<\ldots<\alpha_n$ and let $\gamma=f_n(\alpha_1\ldots\alpha_n)$. Pick $\alpha_1'<\ldots<\alpha_n'\in I$. Then $\gamma<\alpha_1$ so

$$\underline{A}\models\langle n,\alpha_1\ldots\alpha_n,\gamma\rangle\in B \Rightarrow \underline{A}\models\langle n,\alpha_1'\ldots\alpha_n',\gamma\rangle\in B.$$

Hence I is f-homogeneous. □

We could have obtained a stronger result than this; for example $\kappa\subseteq A$ suffices, and R^+ was not really necessary. But lemma 17.12 suffices for our main result, which follows.

Lemma 17.13. *Suppose κ is α-Erdős, where $cf(\alpha)>\omega$. Then κ is α-Erdős in K.*

Proof: Suppose $A_1\ldots A_N\in K$. Let $\underline{A}=\underline{J}_\kappa^{\vec{A}}$. Then $A\subseteq K_\kappa$ and $\langle \underline{K}_\kappa,\vec{A}\rangle\models R^+$. Since κ is α-Erdős there is a good set of indiscernibles I of order-type α. By lemma 16.10 there is a good set of indiscernibles $I'\supseteq I$ for $\langle \underline{K}_\kappa,\vec{A}\rangle$, $I'\in K$. $otp(I')\geq\alpha$ and I' are good indiscernibles for $\underline{A}$. This suffices by lemma 17.12. □

Corollary 17.14. *If κ is Ramsey then κ is Ramsey in K.*

Lemma 17.15. *Suppose κ is α-Erdős, $\alpha\geq\omega_1$. Then $V\neq L$.*

Proof: Let I be good indiscernibles of order-type α for J_κ. Let X be the smallest elementary substructure of J_κ containing I. Say $\sigma:J_\beta\cong J_\kappa|X$. Then $\beta\geq\omega_1$, as $\alpha\geq\omega_1$. So $P(\omega)\cap L\subseteq J_\beta$. So $P(\omega)\cap L\subseteq X$, as $\sigma|\omega+1=id|\omega+1$.

Suppose t is a term of L and suppose $t^{J_\kappa}(\alpha_1\ldots\alpha_n)\subseteq\omega,\alpha_1\ldots\alpha_n\in I$. Take $\alpha_1'<\ldots<\alpha_n'\in I$. Then $t^{J_\kappa}(\alpha_1'\ldots\alpha_n')\subseteq\omega$ and for all $m\in\omega$

$$m\in t^{J_\kappa}(\alpha_1'\ldots\alpha_n') \Leftrightarrow m\in t^{J_\kappa}(\alpha_1\ldots\alpha_n). \tag{17.9}$$

So each term gives at most one subset of ω. Hence $P(\omega)\cap L$ is countable, so $V\neq L$. □

Corollary 17.16. *Suppose κ is α-Erdős, $cf(\alpha)>\omega$. Then $0^{\#}$ exists.*

Proof: Otherwise K=L. □

Much stronger results than this can easily be obtained the same way. In particular the # sequence can be iterated ω_1 times.

Lemma 17.17. *Suppose κ is α-Erdős, $cf(\alpha)>\omega$. Then there is an inner model M such that κ is α-Erdős in M and for any inner model N in which κ is α-Erdős $M\subseteq N$.*

Proof: Suppose for some sharplike $\underline{N}$ $K_{\underline{N}}\models\kappa$ is α-Erdős; then let $M=K_{\underline{N}}$ where $\underline{N}$ is <-least such. Otherwise set M=K. Suppose M' is an inner model, $M'\models\kappa$ is α-Erdős . Then $K^{M'}\models\kappa$ is α-Erdős. If $K^{M'}=K$ then $K^{M'}\supseteq M$. Otherwise say $K^{M'}=K_{\underline{N}'}$. Then by minimal choice of $\underline{N}$ $\underline{N}'>\underline{N}$ or $\underline{N}'=\underline{N}$. So $M=K_{\underline{N}}\subseteq K_{\underline{N}'}\subseteq M'$. □

It is also true if $cf(\alpha)>\omega$ that $\kappa\to(\alpha)^{<\omega}\Rightarrow K\models\kappa\to(\alpha)^{<\omega}$. For $\kappa\to(\alpha)^{<\omega}\Rightarrow$ $\Rightarrow\kappa\geq\kappa(\alpha)$. $\kappa(\alpha)$ is α-Erdős in K so $K\models\kappa(\alpha)\to(\alpha)^{<\omega}$. Hence $K\models\kappa\to(\alpha)^{<\omega}$. Lemma 17.17 can be repeated with $\kappa\to(\alpha)^{<\omega}$ in place of κ α-Erdős.

SLITHY TOVES

The indiscernibles lemma is also used to prove results in the theory of ultrafilters. The briefest summary is given here.

Definition 17.18. *(i) An ultrafilter on a regular cardinal κ is uniform if and only if all elements of U have cardinality κ.*

(ii) An ultrafilter U is (γ,κ)-regular if there is a family $\{M_x:x\in S\}$ of subsets of κ such that $\overline{\overline{M}}_x<\gamma$ for all x and such that $\{x:\gamma\in M_x\}\in U$ for all $\gamma<\kappa$.

Lemma 17.19. *Suppose there are no ρ-models. Let κ be any uncountable cardinal. Then every uniform ultrafilter on κ^+ is (κ,κ^+) regular.*

Regularity is a form of non-normality.

Definition 17.20. *An ultrafilter U on a regular cardinal κ is weakly normal if and only if U is uniform and whenever $X\in U$ and $f:X\to\kappa$ is regressive then for some $\gamma<\kappa$ $\{\xi:f(\xi)=\gamma\}\in U$.*

Lemma 17.21. *Suppose there are no ρ-models. Let κ be a regular cardinal, $\kappa=\sup_{\gamma<\kappa}2^{\gamma}$. Then there is no weakly normal ultrafilter on κ.*

It is conjectured that the restriction on κ is unnecessary.

COMBINATORIAL RESULTS

Almost all of the combinatorial results commonly used in L can be proved in K. We mention only two.

Definition 17.22. *$\diamondsuit$ is the statement: there exists a sequence of sets $S_\alpha, \alpha<\omega_1$, such that for every $X\subseteq\omega_1$ the set*

$$\{\alpha<\omega_1 : X\cap\alpha=S_\alpha\}$$

is stationary.

Lemma 17.23. $K\models\diamondsuit$

The proof is not that hard and the reader who knows the proof in L is encouraged to work it out. The same cannot be said of the next principle.

Definition 17.24. *$\square_\kappa$ is the statement: there is a sequence C_λ defined for limit $\lambda<\kappa^+$ such that*

(i) C_λ is closed unbounded in λ;

(ii) if $cf(\lambda)<\kappa$ then $\overline{\overline{C_\lambda}}<\kappa$;

(iii) if γ is a limit point of C_λ then $C_\gamma=\gamma\cap C_\lambda$.

Lemma 17.25. $K\models\square_\kappa$ *for all* κ.

They are only a couple of examples: stronger $\square$ principles and morasses are available as well. We shall not go into the applications of the principles, which are wide-ranging and not restricted to set theory.

PART FIVE: THE COVERING LEMMA

The applications of K in chapter 17 were discovered after the invention of mice, whereas K was designed with the covering lemma in mind. Much of the motivation for this part was discussed in the introduction. The historical notes to chapters 18 and 19 discuss the connection between the proof given here and the proof originally discovered.

Chapter 18 opens with a summary of the proof. The reader is strongly urged to consult the proof for L at this point: for although a proof for L is embedded in our proof it can be substantially simplified. As we have said, we are concerned not to abridge proofs to such an extent that important techniques are lost, provided that it seems that those techniques show off the structure of K in an important way. It may well be that fine-structure could be done without in this proof; we have explained why it is used here in the introduction. Chapter 18 is mainly occupied with a technical property called goodness; the reader who is unfamiliar with covering lemma proofs would be wise to take goodness on faith.

Chapter 19 proves the covering lemma. Up to the case where C_N is infinite the proof is very similar to the proof in L, but a mucky argument is needed to settle this final case. The muck intensifies in chapter 20 when we extract a Prikry sequence from this final case and obtain a covering lemma on the assumption $-0^{\dagger}$.

The pace relaxes in chapter 21 and there is no fine-structure in the discussion of SCH or patterns of indiscernibles that it contains.

The covering claim for M is the statement: if $X \subseteq On$ is uncountable then there is $Y \in M$ with $X \subseteq Y$ and $X = Y$. In chapter 18 and 19 the term 'covering claim' denotes the covering claim for K. From its failure we deduce the existence of a ρ-model.

This very technical chapter begins the proof of the covering lemma. What we should like to prove is

Suppose $\pi:K_{\bar\tau}\to_{\Sigma_1}K_\tau$ cofinally and $\underline{M}$ is a mouse such that $\forall\gamma<\bar\tau\, P(\gamma)\cap M\subseteq P(\gamma)\cap K_{\bar\tau}$. Then there is a mouse $\underline{M}'$ and $\pi':\underline{M}\to_{\Sigma_n}\underline{M}'$ such that $\pi'\supseteq\pi|M\cap K_{\bar\tau}$ $(n=n(\underline{M})+1)$. (18.1)

(18.1) is not true in general, but a watered down version is. Various closure conditions must be imposed upon $K_{\bar\tau}$; also we prefer to use n completions rather that general Σ_{n+1} maps. But let us see how (18.1) might have been used in a proof of the covering lemma.

Recall that the covering claim says: suppose $X\subseteq On$ is uncountable. There is $Y\supseteq X$ with $Y\in K$ and $\bar{\bar{X}}=\bar{\bar{Y}}$. Suppose this failed, then. Take X with $\tau=\sup X$ minimal for such failure. By minimality τ is a cardinal in K. Take $Y\supseteq X$ such that $\bar{\bar{X}}=\bar{\bar{Y}}$ and $Y\prec K_\tau$. Let $\pi:K_{\bar\tau}\cong K_\tau|Y$ (here a closure property is being used). There are two cases:

(a) $\bar\tau$ is a cardinal in K.

Then every mouse has the property (18.1). Furthermore the various extension maps fit nicely together so that there is $\pi':K\to_{\Sigma_1}K$ with $\pi'\supseteq\pi$. But π' is non-trivial since π is, so by lemma 16.21 there is a ρ-model.

(b) $\bar\tau$ is not a cardinal in K.

Then there is a mouse $\underline{M}$ in which $\bar\tau$ is a cardinal but over which there is a definable map of $\gamma<\bar\tau$ onto $\bar\tau$. If the map is Σ_{n+1}-definable then $\omega\rho_M^{n+1}<\bar\tau$. Suppose κ is the critical point of $\underline{M}$; and for the sake of argument suppose $\omega\rho_M^n>\kappa$, so $n=n(\underline{M})$. The order-type of C_M is no greater than ω. Suppose it is less, i.e. that C_M is finite. We know that

$$M^n=h_{M^n}(C_M\cup\omega\rho_M^{n+1}\cup p_M^{n+1}). \tag{18.2}$$

Let $\pi':M^n\to_{\Sigma_1}M'^n$ be as in (18.1) and let

$$Z=h_{M'^n}(\pi'(p_M^{n+1})\cup\pi'(C_M)\cup\sup\pi''\omega\rho_M^{n+1}) \tag{18.3}$$

so $Z\in K$. In K, $\bar{\bar{Z}}<\tau$; and $Y\cap\tau\subseteq Z$ as $\bar\tau\subseteq M^n$. So $X\subseteq Z$. Then the minimal choice of τ enables us to deduce that for some Z' $\bar{\bar{Z'}}=\bar{\bar{X}}$, $Z'\in K$ and $X\subseteq Z'$.

There are lots of details to be filled in. Firstly all the

closure conditions: these are dealt with in this chapter. Secondly the case where $\kappa \leq \omega\rho_M^n$; this is easier than the case we have discussed. Thirdly $otp(C_M)$ may be ω. Then C_M is not necessarily in M and we only get a covering lemma in $L[\pi'"C_M]$. However, as we shall see in the next chapter, this model has an inner model with a measurable cardinal.

This should be enough to make the structure of the argument clear. This chapter and chapter 19 prove the covering lemma. A more powerful covering lemma occupies chapter 20 and chapter 21 gives applications.

The reader with no familiarity with the covering lemma should read the proof for L before going further; then read up to lemma 18.15, skip from there to lemma 18.24 and then go to the next chapter returning to the technicalities later. We shall deal with the case K=L as we go along, but we shall do so rather sketchily, leaving many details to the reader.

Initially we work with general transitive models of R^+.

Definition 18.1 *$K_{\bar{\tau}}$ is a γ-base for $\underline{M} \models R^+$ provided that $\gamma \leq On_M$ and for all $\bar{\gamma} < \gamma$ $P(\bar{\gamma}) \cap M \subseteq K_{\bar{\tau}}$.*

Suppose $\underline{M} = \langle M, \in, A_1 \ldots A_N \rangle$ and suppose $\pi : K_{\bar{\tau}} \to_{\Sigma_0} K_\tau$ cofinally. Suppose γ is G-closed and $K_{\bar{\tau}}$ is a γ-base for $\underline{M}$.

Definition 18.2. *$T^{M,\pi,\gamma} = \{f : f \in M \wedge dom(f) \subseteq \nu < \gamma$, some $\nu\}$.*

$T^{M,\pi,\gamma} = \{\langle f,x \rangle : f \in T^{M,\pi,\gamma} \wedge x \in \pi(dom(f))\}$.

While $\underline{M}, \pi, \gamma$ are fixed we may omit the superscripts.

Define an equivalence relation $\sim$ on T by

$$\langle f,x \rangle \sim \langle g,y \rangle \leftrightarrow \langle x,y \rangle \in \pi(\{\langle s,t \rangle : f(s) = g(t)\}) \tag{18.4}$$

and let $M' = T/\sim$. $[f,x]$ denotes the equivalence class of $\langle f,x \rangle$.

Define

$$[f,x]E[g,y] \leftrightarrow \langle x,y \rangle \in \pi(\{\langle s,t \rangle : f(s) \in g(t)\}) \tag{18.5}$$

$$A'_k(\langle f,x \rangle) \leftrightarrow x \in \pi(\{s : A_k(f(s))\}) \qquad (0 < k \leq N).$$

It is easily checked that these definitions do not depend on the choices of f,g,x,y.

Lemma 18.3. *Suppose ϕ is Σ_0. Then, letting $\underline{M}' = \langle M', E, A'_1 \ldots A'_N \rangle$,*

$$\underline{M}' \models \phi([f_1,x_1] \ldots [f_k,x_k]) \leftrightarrow \langle x_1 \ldots x_k \rangle \in \pi(\{\langle s_1 \ldots s_k \rangle : \underline{M} \models \phi(f_1(s_1) \ldots f_k(s_k))\}).$$

Proof: Like Los theorem. The atomic formulae are given by definition. We use induction on Σ_0 structure.

Case 1 ϕ is $\neg\psi$.

Then $\underline{M}' \models \phi([f_1,x_1] \ldots [f_k,x_k]) \leftrightarrow \underline{M}' \not\models \psi([f_1,x_1] \ldots [f_k,x_k])$

(*) $\Leftrightarrow \langle x_1 \ldots x_k \rangle \notin \pi(\{\langle s_1 \ldots s_k \rangle : \underline{M} \models \psi(f_1(s_1) \ldots f_k(s_k))\})$.

Now $\langle x_1 \ldots x_k \rangle \in \pi(\mathrm{dom}(f_1) \times \ldots \times \mathrm{dom}(f_k))$ and $\mathrm{dom}(f_1) \times \ldots \times \mathrm{dom}(f_k) =$ $= \{\langle s_1 \ldots s_k \rangle : \underline{M} \models \psi(f_1(s_1) \ldots f_k(s_k))\} \cup \{\langle s_1 \ldots s_k \rangle : \underline{M} \models -\psi(f_1(s_1) \ldots f_k(s_k))\}$ so $\pi(\mathrm{dom}(f_1) \times \ldots \times \mathrm{dom}(f_k)) = \pi(\{\langle s_1 \ldots s_k \rangle : \underline{M} \models \psi(f_1(s_1) \ldots f_k(s_k))\}) \cup$ $\cup \pi(\{\langle s_1 \ldots s_k \rangle : \underline{M} \models -\psi(f_1(s_1) \ldots f_k(s_k))\})$. It is of course essential that ψ be Σ_0, otherwise these sets may not exist. Hence (*) is equivalent to $\langle x_1 \ldots x_k \rangle \in \pi(\{\langle s_1 \ldots s_k \rangle : \underline{M} \models \phi(f_1(s_1) \ldots f_k(s_k))\})$

Case 2 ϕ is $\psi \wedge \chi$.

This is left to the reader.

Case 3. ϕ is $(\exists y \in z)\psi$.

Suppose first that $\underline{M}' \models \psi(y', [f_1, x_1] \ldots [f_k, x_k]) \wedge y' \in [h, z]$. Let y' be $[g, y]$. Then

$$\langle y, z, x_1 \ldots x_k \rangle \in \pi(\{\langle t, u, s_1 \ldots s_k \rangle : \underline{M} \models \psi(g(t), f_1(s_1) \ldots f_k(s_k)) \wedge g(t) \in h(u)\}) \tag{18.6}$$

so

$$\langle z, x_1 \ldots x_k \rangle \in \pi(\{\langle u, s_1 \ldots s_k \rangle : \underline{M} \models \exists y \in h(u)\, \psi(y, f_1(s_1) \ldots f_k(s_k))\}) \tag{18.7}$$

For the other direction a bit more work is needed. Suppose

$$\langle z, x_1 \ldots x_k \rangle \in \pi(\{\langle u, s_1 \ldots s_k \rangle : \underline{M} \models \exists y' \in h(u)\, \psi(y', f_1(s_1) \ldots f_k(s_k))\}) \tag{18.8}$$

Let $g(u, s_1 \ldots s_k)$ be the least (in the canonical order) y' such that $y' \in h(u)$ and $\psi(y', f_1(s_1) \ldots f_k(s_k))$ if there is one; 0 otherwise. Then $g \in M$. (This uses the boundedness of the quantifier: if $h \in S^M_\gamma$ then $W_M \cap (S^M_\gamma)^2 \in M$).

$\langle z, x_1 \ldots x_k \rangle \in \pi(\{\langle u, s_1 \ldots s_k \rangle : \underline{M} \models \psi(g(u, s_1 \ldots s_k), f_1(s_1) \ldots f_k(s_k)) \wedge$ $\wedge\, g(u, s_1 \ldots s_k) \in h(u)\})$. Let $g'(\prec u, s_1 \ldots s_k \succ) = g(u, s_1 \ldots s_k)$ where $\prec\succ$ denotes the Gödel pairing function. $g' \in \dot{T}$ because γ is G-closed. If $y = \prec z, x_1 \ldots x_k \succ$ then it is easily seen that

$$\langle y, z, x_1 \ldots x_k \rangle \in \pi(\{\langle t, u, s_1 \ldots s_k \rangle : \underline{M} \models \psi(g'(t), f_1(s_1) \ldots f_k(s_k)) \wedge g'(t) \in h(u)\}) \tag{18.9}$$

so $\underline{M}' \models \psi([g', y], [f_1, x_1] \ldots [f_k, x_k]) \wedge [g', y] \in [h, z]$. □

Let $\pi' : M \to M'$ be defined by $\pi'(x) = [\{\langle 0, x \rangle\}, 0]$.

Lemma 18.4. $\pi' : \underline{M} \to_{\Sigma_0} \underline{M}'$.

Proof: Let ϕ be Σ_0. Then

$$\underline{M}' \models \phi([\{\langle 0, x_1 \rangle\}, 0] \ldots [\{\langle 0, x_k \rangle\}, 0]) \Leftrightarrow \langle 0 \ldots 0 \rangle \in \pi(\{\langle s_1 \ldots s_k \rangle : \underline{M} \models \phi(x_1 \ldots x_k) \wedge s_1 \ldots s_k = 0\}) \tag{18.10}$$

$$\Leftrightarrow \underline{M} \models \phi(x_1 \ldots x_k). \quad \square$$

Lemma 18.5. $\pi' : \underline{M} \to_{\Sigma_0} \underline{M}'$ *cofinally*.

Proof: Given $[f, x]$ $\langle x, 0 \rangle \in \pi(\{\langle s, t \rangle : f(s) \in \mathrm{rng}(f) \wedge t = 0\})$, i.e.

$[f,x]E[\{\langle 0,\mathrm{rng}(f)\rangle\},0]=\pi'(\mathrm{rng}(f))$. □

<u>Corollary 18.6</u>. *$\pi':\underline{M}\to_{\Sigma_1}\underline{M}'$ and $\underline{M}'\models R^+$.*

Now we construct a map $\sigma:\gamma'\to On_{M'}$, where $\gamma'=\sup\pi''\gamma$. Suppose $\alpha<\pi(\beta)$; then $\sigma(\alpha)=[id|\beta,\alpha]$. Clearly the choice of β is irrelevant. σ is order-preserving: for take $\alpha,\alpha'<\pi(\beta)$. Then $\pi(\{\langle s,t\rangle:(id|\beta)(s)\in\in(id|\beta)(t)\})=\pi(\{\langle s,t\rangle\in\beta^2:s\in t\})=\{\langle s,t\rangle\in\pi(\beta)^2:sEt\}$. So $\underline{M}'\models\sigma(\alpha)\in\sigma(\alpha')$.

Suppose $\underline{M}'\models[f,x]\in\sigma(\alpha)$. Then $\langle x,\alpha\rangle\in\pi(\{\langle x',\alpha'\rangle:f(x')\in\alpha'\})$. Assume $f\in K_{\bar{\tau}}$ (otherwise replace f by $f\cap(\mathrm{dom}(f)\times\beta)$, where $\alpha<\pi(\beta)$). Let $\delta=\pi(f)(x)$. We claim $\sigma(\delta)=[f,x]$.

For all x $\langle f(x),x\rangle\in\{\langle s,t\rangle\in\beta^2:s=f(t)\}$ so for all x $\langle\pi(f),x\rangle\in\pi(\{\langle s,t\rangle\in\beta^2:s=f(t)\})$. Hence $[id|\beta,\pi(f)(x)]=[f,x]$ i.e. $\sigma(\delta)==[f,x]$. So $\mathrm{rng}(\sigma)$ is an initial segment of $On_{M'}$. Let $\gamma''=\mathrm{rng}(\sigma)$: so $\sigma:\gamma'\cong\gamma''$. So we may assume that $\underline{M}'$ is standard. We shall not change any notation, though, except for our convention about $\in_{M'}$. σ must now be the identity and $\gamma'=\gamma''$.

<u>Lemma 18.7</u>. $\pi'\supseteq\pi|\gamma$.

Proof: $[\{\langle 0,\delta\rangle\},0]=[id|(\delta+1),\pi(\delta)]$ where $\delta<\gamma$. For $\langle 0,\pi(\delta)\rangle\in\in\pi(\{\langle 0,t\rangle:\delta=t\})$. So $\pi'(\delta)=\sigma\pi(\delta)$. But $\sigma=id|\gamma'$. □

Fix the above terminology by a definition:

<u>Definition 18.8</u>. *(a) $\underline{M}'$ will be called $\underline{M}^{\pi,\gamma}$;*

(b) π' will be called $\pi^{M,\gamma}$.

<u>Definition 18.9</u>. *Suppose $\underline{M}\models RA$ is transitive and p-sound, $p\in PA_n^M$. Suppose $\omega\rho_M^n\geq\gamma$. Then $\pi^{M,\gamma,n}:\underline{M}\to\underline{M}^{\pi,\gamma,n}$ denotes the n-completion of $\pi^{M^{np},\gamma}:M^{np}\to(M^{np})^{\pi,\gamma}$.*

We should say <u>an</u> n-completion, of course, but we may select any standard one. We are interested in whether or not $\underline{M}^{\pi,\gamma,n}$ is well-founded. In general it is not.

<u>Definition 18.10</u>. *$\bar{\tau}$ is π-good provided that $\pi:K_{\bar{\tau}}\to_{\Sigma_0}K_\tau$ cofinally and for all $\underline{M}$ for which $K_{\bar{\tau}}$ is a $\bar{\tau}$-base:*

(i) if $\underline{M}$ is a mouse and $n=n(\underline{M})$ then $\underline{M}^{\pi,\bar{\tau},n}$ <u>is</u> n-iterable provided $\omega\rho_M^n\geq\bar{\tau}$;

(ii) if $\underline{M}$ is a core mouse and $\omega\rho_M^m\geq\bar{\tau}$ then $\underline{M}^{\pi,\tau,m}$ is n-iterable;

(iii) if $\underline{M}\neq\underline{J}_\beta$ and $\omega\rho_M^m\geq\bar{\tau}$ then $\underline{M}^{\pi,\tau,m}$ is well-founded.

(iii) is only needed when K=L. In the above τ was the least τ such that for some uncountable $X\subseteq\tau$ there is no $Y\supseteq X$ with $Y\in K$ and $\bar{\bar{X}}=\bar{\bar{Y}}$. Fix

τ from now on. All roads lead either to a contradiction or to a ρ-model, but not until the end of the next chapter.

Lemma 18.11. *τ is a cardinal in K.*

Proof: Let $\delta=\overline{\overline{\tau}}^K$. Let $f:\delta\leftrightarrow\tau$. If $\delta<\tau$ then given $X\subseteq\tau$ there is $Y'\supseteq f^{-1}$"X with $\overline{\overline{Y'}}=\overline{\overline{X}}$ and $Y'\in K$ (assuming $f\in K$). So let $Y=f$"Y'. Then $X\subseteq Y$ and $\overline{\overline{X}}=\overline{\overline{Y}}$ and $Y\in K$. Contradiction! □

Lemma 18.12. *Suppose $X\subseteq\tau$, $X\subseteq Y$, $Y\in K$ and $\overline{\overline{Y}}<\tau$ in K. Then there is $Y'\in K$ with $X\subseteq Y'$ and $\overline{\overline{Y'}}=\overline{\overline{X}}$.*

Proof: Let $\delta=\overline{\overline{Y}}^K$. Let $f:\delta\leftrightarrow Y$, $f\in K$. Let $\hat{X}=f^{-1}$"X. There is $\hat{Y}\in K$ with $\hat{Y}\subseteq\delta$ $\overline{\overline{\hat{X}}}=\overline{\overline{\hat{Y}}}$, $\hat{X}\subseteq\hat{Y}$, because $\delta<\tau$. Let $Y'=f$"$\hat{Y}$. □

The result intended to replace (18.1) is:

Suppose $X\subseteq\tau$. Then there is $\bar{\tau}>\omega$ and $\pi:K_{\bar{\tau}}\to_{\Sigma_0}K_\tau$ cofinally (18.11) such that

(i) $Y\subseteq\pi$"$\bar{\tau}$;

(ii) the cardinality of $\bar{\tau}$ does not exceed $\max(Y,\aleph_1)$;

(iii) $\bar{\tau}$ is π-good.

Even (i) and (ii) require a little care: just as K may not be minimal, so $Y \prec K_\tau$ need not imply that $Y\cong K_{\bar{\tau}}$ for some $\bar{\tau}$.

Definition 18.13. *A set $Y\subseteq\tau$ is ω-closed in τ provided that whenever $\langle\alpha_i:i\in\omega\rangle$ are such that $\alpha_i\in Y$ for all $i\in\omega$ and $\sup_{i\in\omega}\alpha_i<\tau$ then $\sup_{i\in\omega}\alpha_i\in Y$.*

Lemma 18.14. *Suppose $Y\prec K_\tau$, $Y\cap\tau$ is ω-closed and cofinal in τ, and $Y\cap\omega_2$ is transitive. Then there are $\bar{\tau},\pi$ such that $\pi:\underline{K}_{\bar{\tau}}\cong\underline{K}_\tau|Y$.*

Proof: Clearly $\omega_1\subseteq Y$. Suppose $\pi:\langle\bar{K},\bar{D}\rangle\cong\underline{K}_\tau|Y$, $\bar{K}$ transitive. As $\omega_1\subseteq\bar{K}$ $\langle\bar{K},\bar{D}\rangle\models$"$\underline{M}$ is a mouse" $\Rightarrow$ $\underline{M}$ is a mouse. Assume $K_\tau\neq J_\tau$ otherwise the result is trivial. So $\bar{K}=\bigcup\{\underline{M}\in\bar{K}:\ \underline{M}$ is a mouse$\}$. So $\bar{K}\subseteq K_{\bar{\tau}}$.

Suppose $\underline{N}\in K_{\bar{\tau}}$ is a mouse with κ critical and $\mathrm{otp}(C_N)=\omega$. Suppose $\kappa\notin\omega_2\cap Y$.

Claim κ is singular in $\underline{\bar{K}}$.

Proof: Let $\kappa'=\sup\pi$"C_N. $\kappa'\in Y$ by ω-closure, so $\kappa'=\pi(\kappa)$. It suffices to show $\underline{K}_\tau\models$"$\kappa'$ is singular". If $\kappa'<\omega_2$ then $\kappa'\in\omega_2\cap Y$, so $\kappa\in\omega_2\cap Y$. So $\kappa'\geq\omega_2$. Now $\kappa'<\tau$ so there is $B\in K$ with $B\supseteq\pi$"C_N and $\overline{\overline{B}}=\aleph_1$. And $\overline{\overline{B}}^K<\kappa'$ so κ' is singular in K; hence in K_τ as τ is a K-cardinal. □(Claim)

Let $\underline{M}\in\bar{K}$ be a mouse with critical point above κ such that $\underline{M}\models\kappa$ singular. Let $\underline{M}',\underline{N}'$ be comparable mouse iterates of $\underline{M},\underline{N}$ respectively. If $M'\subseteq N'$ then $\underline{N}'\models\kappa$ singular so $\underline{N}\models\kappa$ singular. Hence $N'\in M'$. So $A_N^{n(N)+1}\subseteq(\omega\times(\kappa)^{<\omega})\in\bar{K}$, so $\underline{N}\in\bar{K}$.

If $\underline{N}\in K_{\bar{\tau}}$ is a mouse with κ critical and $\kappa\in\omega_2\cap Y$ then $\pi(\kappa)=\kappa$ and

$\pi^{-1}{}''C_{\underline{N}}=C_{\overline{N}}$ so $\underline{N}\in\overline{K}$. Hence $K_{\overline{\tau}}\subseteq\overline{K}$. □

<u>Lemma 18.15</u>. *Suppose $X\subseteq\tau$ is cofinal. There is $Y\supseteq X$ such that*

(i) $Y\prec K_\tau$;

(ii) $Y\cap\tau$ is ω-closed in τ;

(iii) $Y\cap\omega_2$ is transitive;

(iv) $\overline{\overline{Y}}=\max(\overline{\overline{X}},\aleph_1)$.

Proof: Let Z^* denote some $Z'\supseteq Z$ such that $\overline{\overline{Z'}}=\overline{\overline{Z}}$ and $Z'\prec K$. Define

$$X_0=X$$
$$X_{\alpha+1}=(X_\alpha\cup X'_\alpha\cup\bigcup(X_\alpha\cap\omega_2))^*$$
$$X_\lambda=\bigcup_{\alpha<\lambda}X_\alpha\quad(\lambda\text{ a limit})$$

where $X'=\{\sup\alpha_i:\langle\alpha_i:i\in\omega\rangle\in(X_\alpha)^\omega\}\cap\tau$. Let $Y=X_{\omega_1}$. Then $Y\prec K_\tau$. As to cardinality, $\overline{\overline{X'_\alpha}}\leq\overline{\overline{X_\alpha}}$ and if $\overline{\overline{X_\alpha}}\leq\aleph_1$ then $\overline{\overline{\bigcup(X_\alpha\cap\omega_2)}}\leq\aleph_1$ so we can show inductively that $\overline{\overline{X_\lambda}}\leq\max(\overline{\overline{X}},\aleph_1)$, $\overline{\overline{Y}}\leq\overline{\overline{X}}+\aleph_1$.

If $\{\alpha_i:i\in\omega\}\subseteq Y\cap\tau$ then $\{\alpha_i:i\in\omega\}\subseteq X_\alpha\cap\tau$, some $\alpha<\omega_1$, so $\sup\{\alpha_i:i\in\omega\}\in$ $\in X'_\alpha\subseteq X_{\alpha+1}\subseteq Y$. If $\delta\in\beta\in Y\cap\omega_2$ and $\beta\in X_\alpha$, say, then $\delta\in\bigcup(X_\alpha\cap\omega_2)$, so $\delta\in X_{\alpha+1}$. □

This gives (i) and (ii) of (18.11). The bulk of the rest of the chapter is devoted to getting $\overline{\tau}$ to be π-good. Suppose, then, that $\pi:K_{\overline{\tau}}\rightarrow_{\Sigma_0}K_\tau$ cofinally. Suppose $\underline{M}$ is a mouse but that $\underline{M}^{\pi,\gamma}$ is not well-founded - this is the simplest case, and we shall see that once we can handle this we can handle any relation E^+_n and therefore get the full result. Suppose $n(\underline{N})=0$. There are $\langle f_i,x_i\rangle\in T^{M,\pi,\gamma}$ with $[f_{i+1},x_{i+1}]E[f_i,x_i]$ $(i\in\omega)$, where E is as in (18.5). It cannot be the case that $x_i\in\text{rng}(\pi)$ for all i: for

$$[f_{i+1},\pi(\overline{x}_{i+1})]E[f_i,\pi(\overline{x}_i)]\Rightarrow\langle\pi(\overline{x}_{i+1}),\pi(\overline{x}_i)\rangle\in\pi(\{\langle u,v\rangle:f_{i+1}(u)\in f_i(v)\})$$
$$\Rightarrow\langle\overline{x}_{i+1},\overline{x}_i\rangle\in\{\langle u,v\rangle:f_{i+1}(u)\in f_i(v)\}$$
$$\Rightarrow f_{i+1}(\overline{x}_{i+1})\in f_i(\overline{x}_i).$$

But this contradicts the axiom of foundation. Our strategy is as follows: start with $X_0=X$. Define X_α inductively such that X_α is ω-closed in τ, $X_\alpha\cap\omega_2$ is transitive, $X_\alpha\prec K_\tau$ and whenever $\pi_\alpha:\underline{K}_{\tau_\alpha}\cong\underline{K}_\tau|X_\alpha$ and $\underline{M}$ is a mouse for which K_{τ_α} is a γ-base but $M^{\pi_\alpha,\gamma}$ is not well-founded add some sequence with the above property, $\{x_i:i<\omega\}$ to $X_{\alpha+1}$ (add the elements, that is). Unions are taken at limits. Let $Y=X_{\omega_1}$, $\pi:\underline{K}_{\overline{\tau}}\cong\underline{K}_\tau|Y$. Suppose $\underline{M}$ is a mouse for which $K_{\overline{\tau}}$ is a γ-base; but suppose $M^{\pi,\gamma}$ is not well-founded. Say f_i,x_i are as above. By a condensation argument we obtain for some α $\overline{\underline{M}}$ for which K_{τ_α} is a $\overline{\gamma}$-base, say, with each $\pi(f_i)$ in $\text{rng}(\pi_\alpha)$ (this cannot literally be the case, but we shall see how to arrange things) and, letting

$\overline{f}_i$ denote $\pi_\alpha^{-1}(\pi(f_i))$ then for all $i<\omega$ $[\overline{f}_{i+1},x_{i+1}]E_{\overline{M}}[\overline{f}_i,x_i]$. So the x_i must all have been added at stage α of the construction and hence are in Y. But we have seen that this is impossible. There are several complications, though: firstly we need to deal with general rudimentary relations and not just $\in$; secondly we cannot really add all the x_i sequences because this would increase cardinality too much, so that we have to pick out nice ones that will be preserved under our maps; and thirdly, since f_i is not necessarily in $K_{\overline{\tau}}$ we have to add apparatus to code it. All this accounts for the complexity of the next definition. Note in particular that we want to ensure that $\underline{M}=\bigcup\{rng(f_i):i\in\omega\}$ with notation as before; this is to make the condensation argument work.

<u>Definition 18.16</u>. *Suppose $\underline{M}$ is a transitive model of R^+. Suppose $\pi:K_{\overline{\tau}}\to_{\Sigma_1} K_\tau$ cofinally and $K_{\overline{\tau}}$ is a γ-base for $\underline{M}$ where γ is G-closed. Let T be some relation on M such that $\langle M,T\rangle$ is amenable. Let R be a relation rudimentary on $\underline{M}$ and let R' have the same rudimentary definition over $\underline{M}^{\pi,\gamma}$. A sequence $\langle\langle f_i,x_i\rangle:i\in\omega\rangle$ with each $\langle f_i,x_i\rangle\in T^{M,\pi,\gamma}$ is R-vicios for $\underline{M},\pi,\gamma,T$ provided:*

(i) for all $i<\omega$ $[f_{i+1},x_{i+1}]R'[f_i,x_i]$;

(ii) if $\mu,\nu\in dom(f_i)$ and $0<k\leq 17+N$ then $\langle k,\mu,\nu\rangle\in dom(f_{i+1})$ and $f_{i+1}(\langle k,\mu,\nu\rangle)=$ $=F_k(f_i(\mu),f_i(\nu))$.

(iii) $f_{i+1}(0)=\langle\delta_i,f_i,S^M_{\mu_i}\rangle$ where:

(a) $\delta_i<\delta_{i+1}<\gamma$ and δ_i is closed under the Gödel pairing function;

(b) $dom(f_i)\subseteq\delta_i\subseteq rng(f_i)$;

(c) $\mu_i<\mu_{i+1}$; $T\cap S^M_{\mu_i}\in S^M_{\mu_{i+1}}$;

(d) $f_i\in S^M_{\mu_i}$;

(e) $\sup_{i<\omega}\delta_i=\gamma$.

<u>Lemma 18.17</u>. *Suppose, with the notation of definition 18.16, that R' is not well-founded. Then there is a sequence which is R-vicious for $\underline{M},\pi,\gamma'$ (some $\gamma'\leq\gamma$)*

Proof: There is a sequence $\langle\langle f_i,x_i\rangle:i\in\omega\rangle$that satisfies (i). We aim to get $\langle\langle f'_i,x'_i\rangle:i\in\omega\rangle$ satisfying (i)-(iii). $\overline{\tau}$ is G-closed. Pick $\delta_0>\sup(dom(f_0))$ with δ_0 closed under the Gödel pairing function. Let $dom(f'_0)=\{\langle 1,\nu\rangle:\nu\in dom(f_0)\}\cup\{\langle 2,\nu'\rangle:\nu'<\delta_0\}$. Let $f'_0(\langle 1,\nu\rangle)=f_0(\nu)$, $f'_0(\langle 2,\nu'\rangle)=\nu'$. Pick μ_0 with $f'_0\in S^M_{\mu_0}$. Let $x'_0=\langle 1,x_0\rangle$. (i)-(iii) are easily checked.

Suppose, then, that δ_i,μ_i,f'_i,x'_i are defined. Let $dom(f'_i)=$ $=\{\langle k,\mu,\nu\rangle:\ \mu,\nu\in dom(f'_i)\wedge 0<k\leq 17+N\}\cup\{\langle 17+N+1,\nu,0\rangle:\ \nu\in dom(f_{i+1})\}\cup$ $\cup\{\langle 17+N+2,\nu,0\rangle:\ \nu<\delta_{i+1}\}$, where δ_{i+1} is picked greater than δ_i and $\sup(dom(f_{i+1}))$ and closed under the Gödel pairing function.

Let

$f'_{i+1}(0)=\langle\delta_i, f'_i, S^M_{\mu_i}\rangle$

$f'_{i+1}(\langle k,\mu,\nu\rangle)=F_k(f'_i(\mu),f'_i(\nu))$ $\quad (0<k\leq 17+N)$

$f'_{i+1}(\langle 17+N+1,\nu,0\rangle)=f_{i+1}(\nu)$

$f'_{i+1}(\langle 17+N+2,\nu,0\rangle)=\nu.$

Pick μ_{i+1} such that $\mu_i, f'_{i+1}, T\cap S^M_{\mu_i} \in S^M_{\mu_{i+1}}$. Let $x'_{i+1}= 17+N+1, x_{i+1}, 0$. Set $\gamma'=\sup_{i\in\omega}\delta_i$. □

Now suppose that R is defined by a uniform rudimentary definition on $\underline{M}^m$ whenever $\underline{M}$ is a mouse with $n(\underline{N})=n$. (m,n) is called the type of R. If R is defined on $\underline{M}^m$ when $M=J_\beta$ then the type of R is m. Suppose R is of type (m,n) and $\pi: K_{\bar\tau} \to_{\Sigma_0} K_\tau$ cofinally. Suppose that for some $\underline{M}$ a mouse with $n(\underline{M})=n$, $K_{\bar\tau}$ a γ-base for $\underline{M}^m$, the critical point of $\underline{M}$ not less than γ R' in $(\underline{M}^m)^{\pi,\gamma}$ is not well-founded. Call this property $P(\underline{M},\pi,\gamma)$. Let $\underline{M}$ be the <-least core mouse some mouse iterate of which satisfies $P(\underline{M}_\alpha,\pi,\gamma)$ for some γ.
A sequence is vicious for $\underline{M}^m,\pi,\gamma$ iff it is vicious for $\underline{M}^m,\pi,\gamma,T^m_M$.

Definition 18.18. *(i) $\underline{M}_\pi$ is the least mouse iterate of $\underline{M}$ satisfying $P(\underline{M}_\pi,\pi,\gamma)$ for some γ.*

(ii) Let γ be least such that $P(\underline{M}_\pi,\pi,\gamma)$. The canonical R-vicious sequence for π is defined inductively as follows:

(a) Suppose $\langle f_j,x_j\rangle$ is defined for $j<i$. f_i is least in the canonical order of $\underline{M}_\pi$ such that there is an R-vicious sequence $\langle\langle f'_i,x'_i\rangle : i\in\omega\rangle$ for $\underline{M}^m_\pi,\pi,\gamma$ with $f'_j=f_j$, $x'_j=x_j (j<i)$ and $f'_i=f_i$.

(b) Suppose $\langle f_j,x_j\rangle$ is defined for $j<i$ and f_i is defined. x_i is least such that there is an R-vicious sequence $\langle\langle f'_i,x'_i\rangle : i\in\omega\rangle$ for $\underline{M}^m_\pi,\pi,\gamma$ with $f'_j=f_j$, $x'_j=x_j$ for $j\leq i$.

Lemma 18.19. *Suppose there are $\underline{M},\gamma$ such that $P(\underline{M},\pi,\gamma)$. Then there is a unique canonical R-vicious sequence for π which is a vicious sequence for $\underline{M}^m_\pi,\pi,\gamma'$ (some $\gamma'\leq\gamma$).*

Proof: Uniqueness is clear. Existence is also clear. The only point in the definition of R-viciousness that is in doubt is (iii) and this is easily seen to hold for $\gamma'=\sup\delta_i$. □

We mention now two slight modifications that we sometimes need. First of all, if the type of R is (n,n) then we add to the definition of viciousness the requirement that $f_0(0)=p^{n+1}_M$. This is clearly possible. Secondly if the type of R is n then instead of taking $\underline{M}$ as above we let $\underline{M}=\underline{J}_\beta$ where β is least such that for some γ $P'(J_\beta,\pi,\gamma)$, where $P'(\underline{M},\pi,\gamma)$ is the statement that $K_{\bar\tau}$ is a γ-base for $\underline{M}^n$ and R' in $(\underline{M}^n)^{\pi,\gamma}$ is not well-founded. Definition 18.18 is redone with P' in place of P.

Lemma 18.20. *Suppose $\langle\langle f_i,x_i\rangle : i\in\omega\rangle$ is the canonical R-vicious sequence for π. Then $M^m_\pi = \bigcup_{i\in\omega} rng(f_i)$, where R is of type (m,n) (or m).*

Proof: Let $X=\bigcup_{i\in\omega} rng(f_i)$. Let $\gamma=\sup\delta_i$ (notation as in definition 18.16), so that $\langle\langle f_i,x_i\rangle : i\in\omega\rangle$ is vicious for $\underline{M}^m_\pi,\underline{\pi},\gamma$. Since $\delta_i\subseteq rng(f_i)$, $\gamma\subseteq X$. Let $\omega\mu=\sup_{i\in\omega}\mu_i$. Since $rng(f_i)\subseteq S^{M^m_\pi}_{\mu_{i+1}}$, $X\subseteq J^{M^m}_\mu$.

Claim $\underline{M}^m|X\prec_{\Sigma_1} J^{M^m}_\mu$. $\qquad(\underline{M}=\underline{M}_\pi)$.

Proof: First of all observe that X is rud closed. It follows that if ϕ is Σ_0 and $x,\vec{t}\in X$ then $\{y\in x:\phi(y,\vec{t})\}\in X$. Hence it suffices to show that $x\in X$, $x\neq\phi \Rightarrow x\cap X\neq\phi$ to get the result for Σ_0. Suppose $x\in S^{M^m}_{\mu_i}$; the canonical order of $S^{M^m}_{\mu_i}$ is in X so $x\cap X\neq\phi$.

For Σ_1, $J^{M^m}_\mu\models\exists y\phi(y,\vec{t}) \Rightarrow (\exists i)(\exists y\in S^{M^m}_{\mu_i})\phi(y,\vec{t})$. $\qquad\square$(Claim)

Let $\sigma:\overline{\underline{M}}'\tilde{=}\underline{M}^m|X$; so $\sigma:\overline{\underline{M}}'\to_{\Sigma_1}\underline{J}^{M^m}_\mu$. $\langle J^{M^m}_\mu,T^m_M\rangle$ is amenable; letting $x\in\overline{T} \Leftrightarrow \sigma(x)\in T^m_{M'}$, then, $\langle\overline{\underline{M}}',\overline{T}\rangle\models MC_m$. Let $\overline{\underline{M}}$ be m-sound with $\overline{\underline{M}}^m=\overline{\underline{M}}'$ and $\tilde{\sigma}:\overline{\underline{M}}\to_{\Sigma_1}\underline{M}$ such that $\tilde{\sigma}\supseteq\sigma$ and $\sigma(p^k_{\overline{M}})=p^k_M$ $(k\leq m)$. Then clearly $\overline{\underline{M}}$ is a mouse with $n(\overline{\underline{M}})=n$. Let $\overline{\kappa}=\tilde{\sigma}^{-1}(\kappa)$ be the critical point of $\overline{\underline{M}}$. If a is a bounded subset of γ then since $\kappa\geq\gamma$, $\overline{\kappa}\geq\gamma$, so $\tilde{\sigma}(a)=a$. Hence $K_{\overline{\tau}}$ is a γ-base for $\overline{\underline{M}}$. Each f_i is in X; let $\overline{f}_i=\tilde{\sigma}^{-1}(f_i)$. $\delta_i=\sigma^{-1}(\delta_i)$ and it follows easily that $\langle\langle\overline{f}_i,x_i\rangle : i\in\omega\rangle$ is an R-vicious sequence for $\overline{\underline{M}}^m,\pi,\gamma$. Iterate $\overline{\underline{M}},\underline{M}$ to comparable $\overline{\underline{N}},\underline{N}$. If $N\in\overline{N}$ then there is a Σ_0 non-cofinal map of $\overline{\underline{M}}$ to $\overline{\underline{N}}$. But $P(\overline{\underline{M}},\pi,\gamma)$, so $\overline{N}\in N$ is impossible. Thus $\underline{N}=\overline{\underline{N}}$. If $\underline{M}\neq\overline{\underline{M}}$ then $\overline{\underline{M}}$ is a mouse iterate of $\underline{M}$, since $P(\overline{\underline{M}},\pi,\gamma)$. But this is impossible since $\overline{\kappa}\leq\kappa$. So $\overline{\underline{M}}=\underline{M}$. But then $\overline{f}_i\leq f_i$ in the canonical order of $\overline{\underline{M}}$ so $\overline{f}_i=f_i$, all $i\in\omega$. $\overline{M}^m=\bigcup_{i\in\omega} rng(\overline{f}_i)$; hence $M^m=\bigcup_{i\in\omega} rng(f_i)$. $\qquad\square$

The preceding lemma is a simple form of the next one, which is the essential property of vicious sequences that we need.

Lemma 18.21. *Suppose $Y,Y'\prec K_\tau$, $Y\subseteq Y'$, $\pi:K_{\overline{\tau}}\tilde{=}K_\tau|Y$, $\pi':K_{\overline{\tau}'}\tilde{=}K_\tau|Y'$ and let $\hat{\pi}=\pi'^{-1}\pi$; so $\hat{\pi}:K_{\overline{\tau}}\to_e K_{\overline{\tau}'}$. Suppose $\langle\langle f_i,x_i\rangle : i\in\omega\rangle$ is the canonical R-vicious sequence for π' and let*

$b_i=dom(f_i)$;

$h_i=f_i\cap\delta_i^2$.

If $\pi'(b_i),\pi'(h_i)\in Y$ for all $i\in\omega$ then there are $\overline{f}_i$ such that $\langle\langle\overline{f}_i,x_i\rangle : i\in\omega\rangle$ is *the canonical R-vicious sequence for* π.

Proof: Let $\bar{b}_i = \Uparrow^{-1}(b_i)$, $\bar{h}_i = \Uparrow^{-1}(h_i)$. Let $\underline{M}=\underline{M}_{\pi'}$. Let (m,n) be the type of R. Let $Z = \bigcup_{i\in\omega} f_i "\Uparrow"\bar{b}_i$.

Claim 1 Z is rud closed in $\underline{M}^m$.

Proof: Given $s,t\in Z, k\leq 17+N$ with $k>0$, suppose $s=f_i(\Uparrow(\bar{s}))$ and $t=f_j(\Uparrow(\bar{t}))$. Suppose $i\leq j$. We can show that there is $s'\in\bar{b}_{i+1}$ such that $s=f_{i+1}(\Uparrow(s'))$; in fact picking $\tilde{t}$ such that $\phi=f_i(\Uparrow(\tilde{t}))$

$$\begin{aligned} f_{i+1}(\langle 3,\Uparrow(\bar{s}),\Uparrow(\tilde{t})\rangle) &= F_3(f_i(\Uparrow(\bar{s})),f_i(\Uparrow(\tilde{t}))) \\ &= F_3(f_i(\Uparrow(\bar{s})),\phi) \\ &= f_i(\Uparrow(\bar{s})) \\ &= s. \end{aligned} \tag{18.12}$$

So let s' be $\langle 3,\bar{s},\tilde{t}\rangle$.

Using this result we can show by induction that there is s" in $\bar{b}_j$ such that $s=f_j(\Uparrow(s''))$. So

$$\begin{aligned} f_{j+1}(\Uparrow(\langle k,s'',\bar{t}\rangle)) &= f_{j+1}(\langle k,\Uparrow(s''),\Uparrow(\bar{t})\rangle) \\ &= F_k(f_j(\Uparrow(s'')),f_j(\Uparrow(\bar{t}))) \\ &= F_k(s,t) \end{aligned} \tag{18.13}$$

□(Claim 1)

Claim 2 $S^{\underline{M}^m}_{\mu_i}\in Z$ for all i.

Proof: It suffices by claim 1 to show $f_{i+1}(0)\in X$. But $0\in b_{i+1} \Rightarrow 0\in\bar{b}_{i+1}$ so this is clear. □(Claim 2)

Claim 3 $\omega\rho^m_M = \sup_{i\in\omega}\mu_i$.

Proof: By lemma 18.20. □(Claim 3)

Claim 4 $\underline{M}^m|Z \prec_{\Sigma_1} \underline{M}^m$.

Proof: Just as in the claim in lemma 18.20, using claims 1-3.

□(Claim 4)

Let $\sigma:\bar{\underline{M}}'\cong\underline{M}^m|X$ with $\bar{M}'$ transitive; so $\sigma:\bar{\underline{M}}'\to_{\Sigma_1}\underline{M}^m$. By supposition $\underline{M}$ is m-sound; let $\tilde{\sigma}:\bar{\underline{M}}\to_{\Sigma_1}\underline{M}$ such that $\tilde{\sigma}\supseteq\sigma$, $\bar{\underline{M}}$ is m-sound, $\bar{\underline{M}}^m=\bar{\underline{M}}'$ and $\tilde{\sigma}(p^{\bar{M}}_k)=$ $=p^M_k$ for $k\leq m$.

Claim 5 $\bar{\underline{M}}$ is a mouse; and a core mouse if m>n.

Proof: $\bar{\underline{M}}$ is n-iterable so it is only necessary to check that it is critical. Its critical point $\bar{\kappa}$ must be $\tilde{\sigma}^{-1}(\kappa)$ so if m>n this is clear. Suppose m=n. Given $x\in\bar{M}^n$ there are $i\in\omega,\vec{\gamma}<\kappa$ such that $\sigma(x)=$ $=h_M n(i,\langle\vec{\gamma},p^{n+1}_M\rangle)$. So

$$\underline{M}^n \models \exists i\in\omega\exists\vec{\gamma}<\kappa(\sigma(x)=h_M n(i,\langle\vec{\gamma},p^{n+1}_M\rangle)). \tag{18.14}$$

But $p^{n+1}_M = f_0(0)\in Z$ so

$$\bar{\underline{M}}^n \models \exists i\in\omega\exists\vec{\gamma}<\bar{\kappa}(x=h_{\bar{M}} n(i,\langle\vec{\gamma},\sigma^{-1}(p^{n+1}_M)\rangle)). \tag{18.15}$$

so $\omega\rho^{n+1}_{\overline{M}} \leq \overline{\kappa}$. □(Claim 5)

<u>Claim 6</u> $Z\cap\gamma = \mathrm{rng}(\pitchfork)\cap\gamma$ $(\gamma = \sup\{\delta_i : i\in\omega\})$

Proof: Suppose $\nu = \pitchfork(\overline{\nu})$ and $\nu<\gamma$. Suppose $\nu<\delta_i$, $\delta_i = \pitchfork(\overline{\delta}_i)$ ($\delta_i \in \mathrm{rng}(\pitchfork)$ as $\delta_i = \mathrm{rng}(h_i)$). So $\overline{\nu}<\overline{\delta}_i$. Say $\overline{\nu} = \overline{h}_i(\overline{\xi})$, $\overline{\xi}\in\mathrm{dom}(\overline{h}_i)$. Let $\xi = \pitchfork(\overline{\xi})$; then $\nu = h_i(\xi) = f_i(\xi)\in Z$.

Conversely suppose $\nu\in Z\cap\gamma$. By induction as in claim 1 we may assume $\nu = f_i(\xi)$, $\nu<\delta_i$, and $\xi = \pitchfork(\overline{\xi})$. So $\nu = h_i(\xi)$, and so $\nu = \pitchfork(\overline{h}_i(\overline{\xi}))$.

□(Claim 6)

Let $\overline{\gamma} = \sup \overline{\delta}_i$.

<u>Claim 7</u> $\sigma|\overline{\gamma} = \pitchfork|\overline{\gamma}$.

Proof: Immediate from claim 6. □(Claim 7)

<u>Claim 8</u> If $a\in Z$, $a\subseteq\gamma'<\gamma$ then $a\in\mathrm{rng}(\pitchfork)$.

Proof: Assume $a = f_i(\nu)$, $\nu = \pitchfork(\overline{\nu})$ and $a\subseteq\delta_i$. Then

$$f_i(\xi)\in f_i(\nu) \leftrightarrow \{f_i(\xi)\}\smallsetminus f_i(\nu) = \phi \qquad (18.16)$$
$$\leftrightarrow F_3(F_1(f_i(\xi), f_i(\xi)), f_i(\nu)) = \phi$$
$$\leftrightarrow F_3(f_{i+1}(\langle 1,\xi,\xi\rangle), f_{i+1}(\langle 3,\nu,\pitchfork(\tilde{t})\rangle)) = \phi \quad (\tilde{t} \text{ as in C1})$$
$$\leftrightarrow f_{i+2}(\langle 3,\langle 1,\xi,\xi\rangle,\langle 3,\nu,\pitchfork(\tilde{t})\rangle\rangle) = \phi$$
$$\leftrightarrow h_{i+2}(\langle 3,\langle 1,\xi,\xi\rangle,\langle 3,\nu,\pitchfork(\tilde{t})\rangle\rangle) = \phi.$$

Let $\overline{a} = \{\overline{h}_i(\xi) : \overline{h}_{i+2}(\langle 3,\langle 1,\overline{\xi},\overline{\xi}\rangle,\langle 3,\overline{\nu},\tilde{t}\rangle\rangle) = \phi\}$. Then $f_i(\xi)\in\pitchfork(\overline{a}) \leftrightarrow$
$\leftrightarrow h_{i+2}(\langle 3,\langle 1,\xi,\xi\rangle,\langle 3,\nu,\pitchfork(\tilde{t})\rangle\rangle) = \phi \leftrightarrow f_i(\xi)\in a$; so $a = \pitchfork(\overline{a})$. □(Claim 8)

<u>Claim 9</u> If $a\in Z$, $a\subseteq\gamma'<\gamma$ then $\pitchfork^{-1}(a) = \sigma^{-1}(a)$.

Proof: $\sigma^{-1}(a) = \sigma^{-1}``(a\cap(Z\cap\gamma)) = \pitchfork^{-1}``(a\cap(\mathrm{rng}(\pitchfork)\cap\gamma)) = \pitchfork^{-1}(a)$. □(Claim 9)

<u>Claim 10</u> $K_{\overline{\tau}}$ is a $\overline{\gamma}$-base for $\overline{\underline{M}}^m$.

Proof: Suppose $a\subseteq\gamma'<\overline{\gamma}$. Then $\sigma(a)\in M$, so $\sigma(a)\in K_{\overline{\tau}'}$; and $\sigma(a) = \pitchfork(a)$. So $a\in K_{\overline{\tau}}$. □(Claim 10)

<u>Claim 11</u> $\overline{\gamma}$ is G-closed.

Proof: Each $\overline{\delta}_i$ is closed under the Gödel pairing function.

□(Claim 11)

Let $\overline{f}_i = \sigma^{-1}(f_i)$. Then $\mathrm{dom}(\overline{f}_i) = \sigma^{-1}(\mathrm{dom}(f_i)) = \sigma^{-1}(b_i) = \pitchfork^{-1}(b_i) = \overline{b}_i$.
Also $\overline{h}_i = \overline{f}_i\cap\overline{\delta}_i^{\,2}$.

<u>Claim 12</u> $\langle\langle\overline{f}_i, x_i\rangle : i\in\omega\rangle$ is an R-vicious sequence for $\overline{\underline{M}}^m, \pi, \overline{\gamma}$.

Proof: Since $f_i, f_{i+1}\in Z$ and Z is rud closed $\{\langle s,t\rangle : \underline{M}^m \models f_{i+1}(s) R f_i(t)\}\in$
$\in Z$ so by claim 9 (which clearly can be extended to subsets of γ^2)
$\pitchfork^{-1}(\{\langle s,t\rangle : \underline{M}^m\models f_{i+1}(s) R f_i(t)\}) = \sigma^{-1}(\{\langle s,t\rangle : \underline{M}^m\models f_{i+1}(s) R f_i(t)\}) =$
$= \{\langle s,t\rangle : \overline{\underline{M}}^m\models \overline{f}_{i+1}(s) R\overline{f}_i(t)\}$. Hence

$$(\overline{\underline{M}}^m)^{\pi,\overline{\gamma}}\models[\overline{f}_{i+1}, x_{i+1}]R'[\overline{f}_i, x_i] \leftrightarrow \langle x_{i+1}, x_i\rangle\in\pi(\{\langle s,t\rangle : \qquad (18.17)$$
$$: \overline{\underline{M}}^m\models\overline{f}_{i+1}(s) R\overline{f}_i(t)\})$$
$$\leftrightarrow \langle x_{i+1}, x_i\rangle\in\pi'\pitchfork(\{\langle s,t\rangle : \overline{\underline{M}}^m\models\overline{f}_{i+1}(s) R\overline{f}_i(t)\})$$
$$\leftrightarrow \langle x_{i+1}, x_i\rangle\in\pi'(\{\langle s,t\rangle : \underline{M}^m\models f_{i+1}(s) R f_i(t)\})$$
$$\leftrightarrow (\underline{M}^m)^{\pi',\gamma}\models[f_{i+1}, x_{i+1}]R'[f_i, x_i].$$

The verification of the other clauses is left to the reader.

□(Claim 12)

It follows that $\underline{M}_\pi$ is defined. Let $\overline{\underline{N}}$ be $\underline{M}_\pi$ and let $\underline{Q},\underline{\Omega}'$ be comparable mouse iterates of $\overline{\underline{M}},\overline{\underline{N}}$ respectively. $Q\in Q'$ is impossible by the <-minimality of core($\overline{\underline{N}}$). Suppose $\Omega'\in\Omega$. Let $\underline{\tilde{\Omega}},\underline{\tilde{\Omega}}',\underline{\tilde{\Omega}}''$ be comparable mouse iterates of $\overline{\underline{M}},\overline{\underline{N}},\underline{M}$ respectively. So $\tilde{\Omega}'\in\tilde{\Omega}$. And there is a map $\bar{\sigma}:\underline{\tilde{\Omega}}\to_{\Sigma_1}\underline{\tilde{\Omega}}''$ such that $\bar{\sigma}\tilde{\pi}=\tilde{\pi}''\tilde{\sigma}$, where $\tilde{\pi},\tilde{\pi}''$ are the respective iteration maps. But then $\bar{\sigma}(\underline{\tilde{\Omega}}')$contradicts the <-minimality of core($\underline{M}$). Hence $\Omega=Q'$ so core($\overline{\underline{M}}$)=core($\overline{\underline{N}}$). By minimality again $\overline{\underline{M}}$ is a mouse iterate of $\overline{\underline{N}}$.

<u>Claim 13</u> $\overline{\underline{M}}=\overline{\underline{N}}$

Proof: Suppose not. Let $\bar{k}$ be the mouse iteration map from $\overline{\underline{N}}$ to $\overline{\underline{M}}$ and let κ' be the critical point of $\overline{\underline{N}}$, so $\bar{k}(\kappa')=\bar{\kappa}$, $\bar{k}|\kappa'=\mathrm{id}|\kappa'$. Unless m=n the result is trivial since both $\overline{\underline{N}}$ and $\overline{\underline{M}}$ are core mice. So $\bar{k}''\overline{\underline{N}}^n=h^n_{\overline{M}}(\kappa'\cup p^{n+1}_{\overline{M}})$. Let $p=\sigma(p^{n+1}_{\overline{M}})$, $\varrho=\sup\sigma''\kappa'$ and let $Y=h^n_M(\varrho\cup p)$. So $\sigma\bar{k}''\overline{\underline{N}}^n\subseteq Y$. And $Y\prec_{\Sigma_1}\underline{M}^n$, so we may let $k:\underline{N}'\tilde{=}\underline{M}^n|Y$, with N' transitive; then $k:\underline{N}'\to_{\Sigma_1}\underline{M}^n$ and there is a mouse $\underline{N}$ with $\underline{N}^n=\underline{N}'$, $n(\underline{N})=n$. Then letting $\hat{\theta}=k^{-1}\sigma\bar{k}$, $\hat{\theta}:\overline{\underline{N}}^n\to_{\Sigma_1}\underline{N}^n$. The critical point of $\underline{N}$ is $\hat{\theta}(\kappa')$ and $K_{\bar{\tau}}$, is a γ-base for $\underline{N}^n$. Iterate $\underline{N}$, $\underline{M}$ to comparable $\underline{Q},\underline{\Omega}'$. $Q'\in\Omega$ is impossible because we should get a Σ_0 non-cofinal map of $\underline{N}$ to $\underline{\Omega}$; $Q\in Q'$ is impossible, because if$\langle\langle g_i,y_i\rangle:i\in\omega\rangle$ were R-vicious for $\overline{\underline{N}}^n,\pi,\delta$ then $\langle\langle\hat{\theta}(g_i),y_i\rangle:i\in\omega\rangle$ would be R-vicious for $\underline{N}^n,\pi',\delta'$ (some δ') contradicting the minimality of core($\underline{M}$); so $Q=\Omega'$. By minimality again $\underline{N}$ must be a mouse iterate of $\underline{M}$; but $k:\underline{N}^n\to_{\Sigma_1}\underline{M}^n$ so $\underline{N}=\underline{M}$. So $\underline{M}^n=h^n_M(\varrho\cup p)$. It follows that $\overline{\underline{M}}^n=h^n_{\overline{M}}(\kappa'\cup p^{n+1}_{\overline{M}})$, so $\overline{\underline{M}}^n=\bar{k}''\overline{\underline{N}}^n$. But $\kappa'\notin\bar{k}''\overline{\underline{N}}^n$. Contradicticn! □(Claim 13).

Hence $\langle\langle\bar{f}_i,x_i\rangle:i\in\omega\rangle$ is R-vicious for $\underline{M}^m,\pi,\bar{\gamma}$. Suppose it were not the canonical R-vicious sequence for π. Then there would be an R-vicious sequence for $\underline{M}^m,\pi,\gamma'$ such that either $\gamma'<\bar{\gamma}$ or for some i the sequence $\langle\langle g_i,y_i\rangle:i\in\omega\rangle$ would satisfy $j<i\Rightarrow f_j=g_j,x_j=y_j$ but either $g_i<f_i$ in the canonical order of $\underline{M}^m$, or $g_i=f_i$ and $y_i<x_i$. But in any of these cases $\langle\langle\sigma(g_i),y_i\rangle:i\in\omega\rangle$ contradicts the canonical status of $\langle\langle f_i,x_i\rangle:i\in\omega\rangle$. □

Lemmas 18.20 and 18.21 are proved for the case where R is of type (m,n). The reader is left to do the proofs for R of type n; all that needs to be done is to replace the iterate-and-compare arguments with direct comparison.

Given $X\subseteq\tau$ define a set Y as follows:

Let Z* denote some $Z'\supseteq Z$ with $\bar{\bar{Z}}'=\bar{\bar{Z}}+\aleph_1$, $Z'\cap\tau$ ω-closed in τ, $Z'\cap\omega_2$ transitive and $Z'\prec\underline{K}_\tau$.

(i) $X_0=X^*$.

(ii) Suppose X_α is defined and $\pi_\alpha : \underline{K}_{\tau_\alpha} \cong \underline{K}_\tau | X$. For each (m,n) such that, for some m-sound mouse with critical point not less than τ_α for which K_{τ_α} is a γ-base, $(E^m_n)'$ in $(M^m)^{\pi_\alpha,\gamma}$ is not well-founded let $\langle\langle f^{mn}_i, x^{mn}_i\rangle : i\in\omega\rangle$ be the canonical E^m_n-vicious sequence for π_α. For each n such that there is some $\underline{M}=\underline{J}_\beta$ for which K_{τ_α} is a γ-base and $(\widetilde{E}_n)'$ in $(\underline{M}^m)^{\pi_\alpha,\gamma}$ is not well-founded let $\langle\langle f^n_i, x^n_i\rangle : i\in\omega\rangle$ be the canonical $\widetilde{E}_n$-vicious sequence for π_α($\widetilde{E}_n$ is defined inductively by

$\widetilde{E}_0 = \in$;

$\langle i,x\rangle \widetilde{E}_{n+1} \langle j,y\rangle \leftrightarrow \langle s(\phi), \langle x,y\rangle\rangle \in A^{n+1}_M$

where ϕ is the formula $h(i,\langle x, p^{n+1}_M\rangle)\widetilde{E}_n h(j,\langle y, p^{n+1}_M\rangle))$. Then $X_{\alpha+1} = (X_\alpha \cup \{x^{mn}_i : x^{mn}_i \text{ defined}\} \cup \{x^n_i : x^n_i \text{ defined}\})^*$.

(iii) $X_\lambda = (\bigcup_{\alpha<\lambda} X_\alpha)^*$ (λ a limit).

Let $Y = \bigcup_{\alpha<\omega_1} X_\alpha$. Then $Y\cap\tau$ is ω-closed and cofinal in τ(provided X is cofinal in τ), $Y\cap\omega_2$ is transitive and $\overline{\overline{Y}} = \overline{\overline{X}} + \aleph_1$. Let $\pi : \underline{K}_{\overline{\tau}} \cong \underline{K}_\tau | Y$.

<u>Lemma 18.22</u>. *$\overline{\tau}$ is π-good.*

Proof: Suppose not. That is, there is (say) a mouse which is m-sound $\underline{M}$ with $\omega\rho^m_{\underline{M}} \geq \overline{\tau}$ but $\underline{M}^{\pi,\overline{\tau},m}$ is not $n(\underline{M})$-iterable. Therefore $(E^m_n)'$ in $(\underline{M}^m)^{\pi,\overline{\tau}}$ is not well-founded. $K_{\overline{\tau}}$ is a $\overline{\tau}$-base for $\underline{M}$ and $\omega\rho^m_{\underline{M}} \geq \overline{\tau}$ imply that $K_{\overline{\tau}}$ is a $\overline{\tau}$-base for $\underline{M}^m$. Let $\langle\langle f_i, x_i\rangle : i\in\omega\rangle$ be the canonical E^m_n-vicious sequence for π. Let b_i and h_i be as in lemma 18.21 and pick α such that for all $i<\omega$ $\pi(b_i), \pi(h_i)$ are in X_α. Then there are by lemma 18.21 $\overline{f}_i$ such that $\langle\langle \overline{f}_i, x_i\rangle : i\in\omega\rangle$ is the canonical E^m_n-vicious sequence for π_α. It follows that for all $i\in\omega$ $x_i \in Y$. But let $\overline{x}_i = \pi^{-1}(x_i)$. It follows that $\underline{M}^m_\pi \models f_{i+1}(\overline{x}_{i+1}) E^m_n f_i(\overline{x}_i)$. But $\underline{M}_\pi$ is a a mouse. Contradiction! □

In summary:

<u>Lemma 18.23</u>. *Suppose $X \subseteq \tau$ is cofinal in τ. Then there is $\overline{\tau} > \omega$, $\pi : K_{\overline{\tau}} \to_e K_\tau$ such that*

(i) $X \subseteq \pi''\overline{\tau}$;

(ii) cardinality of $\overline{\tau} \leq \max(\overline{\overline{X}}, \aleph_1)$;

(iii) $\overline{\tau}$ is π-good.

Finally we apply this to get

<u>Lemma 18.24</u>. *Suppose* $\pi: K_{\bar\tau} \to_{\Sigma_1} K_\tau$ *and* $\bar\tau$ *is* π*-good. Suppose that* $\bar\tau$ *is a cardinal in* K. *Then there is* $\tilde\pi: K \to_{\Sigma_1} K$ *with* $\tilde\pi \supseteq \pi$.

Proof: Suppose $\kappa > \tau$. Then $K_{\bar\tau}$ is a $\bar\tau$-base for Ω_κ. Let $\pi^{(\kappa)} = \pi^{\Omega_\kappa, \bar\tau}$, $\underline{Q}^* = \underline{Q}^{\pi, \bar\tau}$. Suppose $\underline{Q}^* = \underline{J}^{F^*}_{\theta^*}$ with $\kappa^* = \pi^{(\kappa)}(\kappa)$. Then $\pi^{(\kappa)} | \kappa : \kappa \to \kappa^*$ cofinally. For given $\pi^{(\kappa)}(f)(x) < \kappa^*$ with $x < \tau$, $\text{dom}(f) \subseteq \bar\gamma < \bar\tau$ and $\text{rng}(f) \subseteq \kappa$ there is $\gamma < \kappa$ with $\text{rng}(f) \subseteq \gamma$ (we are assuming κ regular, otherwise Ω_κ is not defined), so $\pi^{(\kappa)}(f)(x) < \pi^{(\kappa)}(\gamma)$. ($\pi^{(\kappa)}(f)(x) = [f, x]$, of course). But if $\gamma < \kappa$ then in K

$$\begin{aligned} \overline{\overline{\pi^{(\kappa)}(\gamma)}} &\le \overline{\overline{\{f \in K : f : a \to \gamma,\ a \subseteq \gamma' < \tau\} \times \tau}} \\ &\le (\overline{\overline{\gamma}}^{\bar\tau}) \overline{\overline{\tau}} \\ &< \kappa \end{aligned} \tag{18.18}$$

since $K \models \text{GCH}$. So $\pi^{(\kappa)}(\gamma) < \kappa$. Thus $\kappa^* = \kappa$.

$\theta^* \ge \theta$ and $\pi^{(\kappa)} : \Omega_\kappa \to Q^*$ cofinally. Given $\alpha < \theta$ with $\text{core}(J^F_\alpha \kappa) = \underline{N}$ and $\underline{N} \in K_\kappa$ let $\underline{N}^* = \pi^{(\kappa)}(\underline{N})$. Say $\underline{N}^* = \underline{J}^U_{\alpha^*}$. $\underline{N}^*$ is a mouse as $\underline{N}$ is; let $\underline{N}^*_\kappa$ be its κth mouse iterate. Let $C^* = C_{N^*}$. Then C^* is closed unbounded in κ. And so $N^* = J^F_{\alpha'} \kappa$, some $\alpha' < \theta^*$. So $\underline{\Omega}^* = \underline{J}^F_{\theta^*}$. If $\theta^* > \theta$ let α be such that $\alpha' > \theta^*$. Then $\omega\rho^n_{J^F_{\theta^*}} \kappa < \kappa$ for some n, so $\alpha' < \theta$; hence $\theta = \theta^*$.

Let $\pi^\kappa = \pi^{(\kappa)} | K_\kappa$. $K_\kappa = H^{\Omega}_{\kappa} \kappa$ by lemma 14.13 so $\pi^\kappa : K_\kappa \to_{\Sigma_1} K_\kappa$ and $\pi^\kappa \supseteq \pi$.

<u>Claim</u> If $\kappa < \kappa'$, κ, κ' regular, then $\pi^\kappa \subseteq \pi^{\kappa'}$.

Proof: Let $\pi' = \pi^{\kappa'} | K_\kappa$. Then $\pi' : K_\kappa \to_{\Sigma_1} K_\kappa$. Any $x \in K_\kappa$ can be written in the form $\pi^\kappa(f)(\gamma)$ for $f \in \Omega_\kappa$ with $\text{dom}(f) \subseteq \gamma' < \bar\tau$, $\text{rng}(f) \subseteq K_\kappa$, $\gamma < \tau$. So $f \in K_\kappa$. Let $\sigma(\pi^\kappa(f)(\gamma)) = \pi'(f)(\gamma)$. Then σ is well-defined and $\sigma : \underline{K}_\kappa \to_{\Sigma_0} \underline{K}_\kappa$; for given ϕ Σ_0

$$\begin{aligned} &\underline{K}_\kappa \models \phi(\pi'(f_1)(\gamma_1) \ldots \pi'(f_n)(\gamma_n)) \Leftrightarrow \\ &\Leftrightarrow \underline{Q}_{\kappa'} \models \text{"}\underline{K}_\kappa \models \phi(\pi^{(\kappa')}(f_1)(\gamma_1) \ldots \pi^{(\kappa')}(f_n)(\gamma_n))\text{"} \\ &\Leftrightarrow \langle \gamma_1 \ldots \gamma_n \rangle \in \pi(\{\langle \delta_1 \ldots \delta_n \rangle : \underline{K}_\kappa \models \phi(f_1(\delta_1) \ldots f_n(\delta_n))\}) \\ &\Leftrightarrow \underline{Q}_\kappa \models \text{"}\underline{K}_\kappa \models \phi(\pi^{(\kappa)}(f_1)(\gamma_1) \ldots \pi^{(\kappa)}(f_n)(\gamma_n))\text{"} \\ &\Leftrightarrow \underline{K}_\kappa \models \phi(\pi^\kappa(f_1)(\gamma_1) \ldots \pi^\kappa(f_n)(\gamma_n)). \end{aligned} \tag{18.19}$$

So $\sigma : \underline{K}_\kappa \cong \underline{K}_\kappa$, hence σ is $\text{id} | K_\kappa$. So $\pi^\kappa(f)(\gamma) = \pi'(f)(\gamma)$ and so $\pi^\kappa = \pi'$. Thus $\pi^\kappa \subseteq \pi^{\kappa'}$. □(Claim)

Let $\pi = \bigcup\{\pi^\kappa : \kappa \text{ regular} > \tau\}$. □

<u>Exercises</u>

1. $K_{\bar\tau}$ is a γ-base for $\underline{M}$, $n=n(\underline{M})$. Show that the αth mouse iterate of $\underline{M}^{\pi,\gamma,n}$ is $\underline{M}_{\alpha}^{\pi,\gamma,n}$ where $\underline{M}_{\alpha}$ is the αth mouse iterate of $\underline{M}$.

2. Show that $C_{M^{\pi,\gamma,n}} \setminus \gamma$ is countable.

3. Obtain lemma 18.24 for the case K=L.

4. Suppose $Y_i \prec K_\tau$, each $i<\omega_1$, and $Y = \bigcup_{i<\omega_1} Y_i$. For each $i<\omega_1$ there is τ_i and there is π_i such that $\pi_i : \underline{K}_{\tau_i} \cong \underline{K}_\tau | Y_i$, and τ_i is π_i-good. Show that there are $\pi, \bar\tau$ such that $\pi : \underline{K}_{\bar\tau} \cong \underline{K}_\tau | Y$ and $\bar\tau$ is π-good.

19: THE COVERING LEMMA

Recall that we are assuming that the covering claim fails and that τ is least containing some uncountable X that fails to be covered. Our aim is to get a ρ-model.

Let $\bar{\tau}$ be such that

(i) $\pi: K_{\bar{\tau}} \to_{\Sigma_1} K_\tau$;

(ii) cardinal of $\bar{\tau} \leq \bar{\bar{X}}$;

(iii) $\bar{\tau}$ is π-good;

(iv) $X \subseteq \text{rng}(\pi)$.

This is possible by lemma 18.23. Note that $\bar{\bar{X}} < \tau$, otherwise τ itself would cover X.

Lemma 19.1 *$0^{\#}$ exists.*

Proof: Otherwise K=L. If $\bar{\tau}$ were a cardinal in L then there would be $\tilde{\pi} \supseteq \pi$, $\tilde{\pi}: L \to_{\Sigma_1} L$, $\pi \neq id|L$ so that by the result of chapter 12 $0^{\#}$ exists. Otherwise there is $\beta \geq \bar{\tau}$ with $\Sigma_\omega(J_\beta) \cap P(\gamma) \nsubseteq J_{\bar{\tau}}$ (where γ is the L-cardinality of $\bar{\tau}$). Let β be least such. So for some m $\omega\rho^{m+1}_{J_\beta} < \bar{\tau}$; take m least. Let $\underline{J}_{\beta'} = (\underline{J}_\beta)^{\pi,\bar{\tau},m}$, $\tilde{\pi} = \pi^{J_\beta,\bar{\tau},m}$. $(J_\beta)^m = h_{(J_\beta)^m}(\omega\rho^{m+1}_{J_\beta} \cup p^{m+1}_{J_\beta})$. So $\tilde{\pi}''(J_\beta)^m = h_{J^m_{\beta'}}(\pi''\omega\rho^{m+1}_{J_\beta} \cup \tilde{\pi}(p^{m+1}_{J_\beta}))$. Let $\tilde{\rho} = \sup \pi''\omega\rho^{m+1}_{J_\beta}$, and let Y= $= h_{(J_{\beta'})^m}(\tilde{\rho} \cup \tilde{\pi}(p^{m+1}_{J_\beta}))$. So $X \subseteq Y$ and $Y \in L$. Furthermore $L \models \bar{\bar{Y}} < \tau$, because $\tilde{\rho} < \tau$ and τ is a cardinal in L. By lemma 18.12 there is $Y' \in L$ with $X \subseteq Y'$ and $\bar{\bar{X}} = \bar{\bar{Y}}'$. Contradiction! □

Hence $K = \bigcup\{M: \underline{M} \text{ is a mouse}\}$. The proof of lemma 19.1 foreshadows the proofs of the simpler cases of the covering lemma for K (and gives, of course, the covering lemma for L).

Lemma 19.2. *Suppose $\bar{\tau}$ is a cardinal in K. Then there is a ρ-model whose critical point is less than τ.*

Proof: By lemma 18.24 and lemma 16.21 there is a ρ-model. There is τ' regular with $\bar{\tau} < \tau' < \tau$, $\omega_1 < \tau'$; in fact $\bar{\bar{\tau}}^+$ will do, for $\tau \geq \bar{\bar{\tau}}^+$ (as $\bar{\bar{X}} < \tau$) and $\tau = \bar{\bar{\tau}}^+$ is impossible as τ is singular (X is cofinal in it). Also $\bar{\bar{\tau}}^+ > \omega_1$ as $\bar{\bar{X}} \geq \omega_1$. So the restriction on the critical point follows from chapter 16 exercise 2. □

From now on suppose that $\bar{\tau}$ is not a cardinal in K; if $\gamma = (\bar{\bar{\tau}})^K$ then

$P(\gamma)\cap K\not\subseteq K_{\bar\tau}$. So there is $\underline{N}$, a mouse with critical point $\geq\bar\tau$ such that $P(\gamma)\cap N\not\subseteq K_{\bar\tau}$. Let β be least such that $P(\gamma)\cap J^{N}_{\beta+1}\not\subseteq K_{\bar\tau}$; then by the usual argument $P(\gamma)\cap\Sigma_{\omega}(J^{N}_{\beta})\not\subseteq K_{\bar\tau}$. Clearly $\underline{J}^{N}_{\beta}$ is a mouse. Let $P(\underline{N},\bar\tau)$ be the property: $\underline{N}$ is a mouse with critical point not less than $\bar\tau$ and for some $\gamma<\tau$ $P(\gamma)\cap\Sigma_{\omega}(\underline{N})\not\subseteq K_{\bar\tau}$. So $\exists\underline{N}P(\underline{N},\bar\tau)$.

Definition 19.3. *Let $\underline{M}$ be the <-least core mouse some mouse iterate of which satisfies $P(\underline{M}_{\alpha},\bar\tau)$. Then $\underline{N}_{\bar\tau}$ denotes the least mouse iterate of $\underline{M}$ that satisfies $P(\underline{M}_{\alpha},\bar\tau)$.*

Lemma 19.4. *$K_{\bar\tau}$ is a $\bar\tau$-base for $\underline{N}_{\bar\tau}$.*
Proof: By the <-minimality of core$(\underline{N}_{\bar\tau})$. □

Lemma 19.5. *$C_{\underline{N}_{\bar\tau}}\subseteq\bar\tau$.*
Proof: Otherwise let $C_{\underline{N}_{\bar\tau}}=\{\kappa_{\alpha}:\alpha<\beta\}$ and suppose $\kappa_{\alpha}\notin\bar\tau$. Let $\hat{\underline{N}}$ be the αth mouse iterate of core$(\underline{N}_{\bar\tau})$. Then $P(\hat{\underline{N}},\bar\tau)$. □

Lemma 19.6. *otp$(C_{\underline{N}_{\bar\tau}})\leq\omega$.*
Proof: Otherwise let $\bar{\underline{N}}$ be the ωth mouse iterate of core$(\underline{N}_{\bar\tau})$; then $\bar{\underline{N}}\in K_{\bar\tau}$. But $K_{\bar\tau}\models$"for all α the αth mouse iterate of core$(\underline{N}_{\bar\tau})$ exists". Suppose $C_{\underline{N}_{\bar\tau}}=\{\kappa_{\alpha}:\alpha<\beta\}$. Suppose $a\subseteq\gamma<\bar\tau$, $a\in\Sigma_{\omega}(\underline{N}_{\bar\tau})\setminus K_{\bar\tau}$. Pick α such that $\gamma<\kappa_{\alpha}<\bar\tau$; so letting $\hat{\underline{N}}$ be the αth mouse iterate of core$(\underline{N}_{\bar\tau})$, $a\in\Sigma_{\omega}(\hat{\underline{N}})$. But $\hat{\underline{N}}\in K_{\bar\tau}$, so $a\in K_{\bar\tau}$: contradiction! □

Fix $\underline{N}=\underline{N}_{\bar\tau}$, let $n=n(\underline{N})$ and let m be least such that $\omega\rho^{m+1}_{N}<\bar\tau$. So $m\geq n$. Let κ be the critical point of $\underline{N}$.

Lemma 19.7. *m=n.*
Proof: Suppose m>n. Then $\underline{N}$ is a core mouse; for by lemma 19.5 $C_{N}\subseteq\bar\tau$; but $\omega\rho^{n+1}_{N}\geq\bar\tau$ and $C_{N}\cap\omega\rho^{n+1}_{N}=\phi$ so $C_{N}=\phi$. Let $\underline{M}=\underline{N}^{\pi,\bar\tau,m}$, $\tilde\pi=\pi^{N,\bar\tau,m}$. $\underline{M}$ is n-iterable since $\bar\tau$ is π-good. Also $\omega\rho^{n}_{M}>\tilde\pi(\kappa)\geq\omega\rho^{n+1}_{M}$ so $\underline{M}$ is a mouse. $N^{m}=h_{N^{m}}(\omega\rho^{m+1}_{N}\cup p^{n+1}_{N})$, so $\tilde\pi"N^{m}=h_{M^{m}}(\pi"\omega\rho^{m+1}_{N}\cup\tilde\pi(p^{m+1}_{N}))$ Let $\tilde\rho=\sup\pi"\omega\rho^{m+1}_{N}$, $Y=h_{M^{m}}(\tilde\rho\cup\tilde\pi(p^{m+1}_{N}))$. Since $\underline{M}\in K$, $Y\in K$. $X\subseteq Y$ and $\bar{\bar{Y}}<\tau$, so by lemma 18.12 there is $Y'\in K$ such that $X\subseteq Y'$ and $\bar{\bar{X}}=\bar{\bar{Y}}'$. Contradiction! □

Lemma 19.8 *otp$(C_{N})=\omega$.*
Proof: Suppose C_{N} were finite. Let $\underline{M}=\underline{N}^{\pi,\bar\tau,n}$, $\tilde\pi=\pi^{N,\bar\tau,n}$. Then $\underline{M}$ is n-iterable because $\bar\tau$ is π-good. Every x in M^{n} is of the form $\tilde\pi(f)(\gamma)$ for some $\gamma<\tau$. Suppose $f=h_{N^{n}}(i,\langle\bar\gamma,p^{n+1}_{N}\rangle)$ $(i<\omega,\bar\gamma<\bar\tau)$. Then x= $=(h_{M^{n}}(i,\langle\pi(\bar\gamma),\tilde\pi(p^{n+1}_{N})\rangle))(\gamma)$. So $M^{n}=h_{M^{n}}(\tau\cup\tilde\pi(p^{n+1}_{N}))$; and so $\omega\rho^{n+1}_{M}\leq\tau\leq\tilde\pi(\kappa)$. Hence $\underline{M}$ is a mouse.

$N^n = h_N n(\omega\rho_N^{n+1} \cup p_N^{n+1} \cup C_N)$ so $\pi"N^n = h_M n(\pi"\omega\rho_N^{n+1} \cup \tilde{\pi}(p_N^{n+1}) \cup \pi(C_N))$. Let $\tilde{\rho} = \sup \pi"\omega\rho_N^{n+1}$, $Y = h_M n(\tilde{\rho} \cup \tilde{\pi}(p_N^{n+1}) \cup \pi(C_N))$. Since $\underline{M} \in K$ and C_N is finite $Y \in K$. $X \subseteq Y$ and $\overline{\overline{Y}} < \tau$, so by lemma 18.12 there is $Y' \in K$ such that $X \subseteq Y'$ and $\overline{\overline{X}} = \overline{\overline{Y}}'$. Contradiction! □

Corollary 19.9. $\overline{\tau} = \kappa$.
Proof: $\kappa \geq \overline{\tau} \geq \sup C_N = \kappa$. □

Observe that lemma 19.8 would have given an outright contradiction but for the fact that if $otp(C_N) = \omega$ then $\pi"C_N$ may not be in K. Anyway we could get a covering set in $L[\pi"C_N]$, although this is of limited use since $\underline{N}$ may have depended on the choice of X. In chapter 20 we obtain a single C that is $\pi"C_N$ sufficiently often and prove a covering lemma for L[C]. In this chapter we are only trying to get a ρ-model.

The idea of the construction is that whenever $\overline{\tau}$ is $\underline{\pi}$-good we define a mouse $N^{\overline{\tau}}$ with critical point τ as $N_{\overline{\tau}}^{\pi,\tau,n}$. Let $C^{\overline{\tau}} = \pi"C_{N_{\overline{\tau}}}$, so $otp(C^{\overline{\tau}}) = \omega$. If $\underline{N}^{\overline{\tau}} = \underline{J}_\alpha^{U^{\overline{\tau}}}$ we find that $X \in U^{\overline{\tau}} \leftrightarrow \exists\gamma < \tau C^{\overline{\tau}} \setminus \gamma \subseteq X$. Subject to a restriction to be stated all the $\underline{N}^{\overline{\tau}}$ are comparable. So define $U = \bigcup_{\overline{\tau} < \tau} U^{\overline{\tau}}$. For each $x \in K_{\tau^+}$ we show that there is $\overline{\tau}$ with $x \in N^{\overline{\tau}}$. So U is a normal measure on τ in K_{τ^+} and $\langle K_{\tau^+}, U\rangle$ is amenable. Unfortunately U is not countably complete but anyway we deduce that U is a normal measure on τ in L[U].

Two small modifications are called for. Firstly, although $N_{\overline{\tau}}$ depends only on $\overline{\tau}$ $N^{\overline{\tau}}$ depends on π, which is not uniquely determined by $\overline{\tau}$, so we must index our $\underline{N}$ by sets $Y = rng(\pi)$ rather than by ordinals. Secondly we want $Y \prec K_{\tau^+}$ rather than $Y \prec K_\tau$.

Lemma 19.10. *Suppose $Y \subseteq K_{\tau^+}$ (τ^+ calculated in K here and throughout the chapter) $Y \cap \tau$ cofinal in τ and $\overline{\overline{Y}} < \tau$. Then there is a transitive $\overline{K}$ and $\pi: \overline{K} \to_{\Sigma_1} K_{\tau^+}$ such that*

(i) $Y \subseteq rng(\pi)$;

(ii) cardinality $\overline{K} \leq \max(Y, \aleph_1)$;

(iii) if $\overline{\tau} = \pi^{-1}(\tau)$ then $H_{\overline{\tau}}^{\overline{K}} = K_{\overline{\tau}}$ and $\overline{\tau}$ is $\pi|K_{\overline{\tau}}$-good.

Proof: There is $Y^* \supseteq Y$ such that

(i) $Y^* \prec K_{\tau^+}$;

(ii) $Y^* \cap \tau$ is ω-closed in τ;

(iii) $Y^* \cap \omega_2$ is transitive;

(iv) $\overline{\overline{Y^*}} = \max(Y, \aleph_1)$.

The proof of this is just like that of lemma 18.15. Suppose $\pi: \overline{K} \cong Y^*$ with $\overline{K}$ transitive. There is no hope of proving $\overline{K}$ to be of the form

K_β since the proof of lemma 18.14 depended on the minimality of τ. But letting $Z=Y^*\cap K_\tau$ then $Z\cap\tau$ is ω-closed in τ, $\overline{Z\cap\omega_2}$ is transitive and $Z\prec K_\tau$ so $\pi^{-1}{}''Z=K_{\bar\tau}$, say. Since $K_\tau=H^{K}_{\tau}{}_{\tau^+}$, $K_{\bar\tau}=H^{\bar K}_{\bar\tau}$; and $\bar\tau=\pi^{-1}(\tau)$. The construction of $\bar K$ with π $\bar\tau$-good is just as in the previous chapter.□

Let $C=\{Y:Y\supseteq X$ and for some $\bar K,\pi:\bar K\to_{\Sigma_1} K_{\tau^+}$ $Y=\mathrm{rng}(\pi)$, $\bar K$ satisfies (iii) of lemma 19.10 and $\bar{\bar Y}=\bar{\bar X}\}$. Let τ_Y denote $\bar\tau$, π_Y π, N_Y $N_{\bar\tau}$. By lemma 19.8 $\mathrm{otp}(C_{N_Y})=\omega$ and by corollary 19.9 τ_Y is the critical point of N_Y. Let $N'_Y=N_Y{}^{\pi_Y,\tau_Y,n(N_Y)}$ and $C_Y=\pi_Y{}''C_{N_Y}$. If $Y,Y'\in C$ and $Y\subseteq Y'$ let $\pi_{YY'}=\pi^{-1}_{Y'}\pi_Y$.

<u>Lemma 19.11</u>. *Suppose $Y,Y'\in C$, $Y\subseteq Y'$ and $C_{Y'}\setminus\gamma'\subseteq Y$, some $\gamma'<\tau$. Then $C_{Y'}\setminus\gamma\subseteq C_Y$ for some $\gamma<\tau$.*

Proof: Let $\underline{M}=\underline{N}_Y^{\pi_{YY'},\tau_Y,n(N_Y)}$, $\hat\pi=\pi^{N_Y,\tau_Y,n(N_Y)}_{YY'}$. $\underline{M}$ is easily seen to be a mouse ($n(N_Y)$-iterable because if $\langle\langle f_i,x_i\rangle:i\in\omega\rangle$ were vicious then $\langle\langle\pi_Y(f_i),\pi_{Y'}(x_i)\rangle:i\in\omega\rangle$ would be vicious for N'_Y; and a mouse by the argument of lemma 19.8). Let $\underline{\Omega},\underline{\Omega}'$ be comparable mouse iterates of $\underline{M},\underline{N}_{Y'}$ respectively.

$Q'\notin Q$; for $C_{N_Y}=C_{\Omega'}\cap\tau\in\Sigma_\omega(\underline{Q}')$, so if $Q'\in Q$ we should have $C_{N_Y}\in\Omega$, so $C_{N_Y}\in M$ and $\underline{M}\models\mathrm{cf}(\tau_{Y'})=\omega$. But $\tau_{Y'}$ is the critical point of $\underline{M}$.

So $Q\subseteq Q'$. Let $\pi_{M\Omega}$ be the mouse iteration map of $\underline{M}$ to $\underline{\Omega}$. Let $p^*=$ $=p^{n(N_Y)+1}_{N_Y}$. Pick $\gamma>\gamma'$, $\gamma<\tau_{Y'}$, such that either $\Omega=Q'$ or

$$\{Q\}\cup\pi_{MQ}\hat\pi(p^*)\subseteq h_{\Omega'}{}^{n(Q')}(\gamma\cup(C_{Q'}\setminus\tau_{Y'})\cup p^{n(Q')+1}_{\Omega'}) \qquad (19.1)$$

(remember that $Q\in Q'\Rightarrow Q\in\Omega'^{\,n(Q')}$).

Suppose, for the sake of contradiction, that $\eta>\gamma$ and $\eta\in C_{N_{Y'}}$ but $\pi_{Y'}(\eta)\notin C_Y$. But $\eta\in\mathrm{rng}(\pi_{YY'})$ so we may let $\eta=\pi_{YY'}(\bar\eta)$; then $\bar\eta\notin C_{N_Y}$. So $\bar\eta\in h_{N_Y}{}^{n(N_Y)}(\bar\eta\cup p^*)$. Say $\bar\eta=h_{N_Y}{}^{n(N_Y)}(i,\langle\vec\delta,p^*\rangle)$. So

$$\eta=h_M{}^{n(N_Y)}(i,\langle\pi_{YY'}(\vec\delta),\hat\pi(p^*)\rangle). \qquad (19.2)$$

Hence

$$\eta=h_Q{}^{n(N_Y)}(i,\langle\pi_{YY'}(\vec\delta),\pi_{M\Omega}\hat\pi(p^*)\rangle) \qquad (19.3)$$

so that by (19.1) there are j, $\vec\delta<\gamma$ such that

$$\eta=h_{Q'}{}^{n(Q')}(j,\langle\vec\delta,p^{n(Q')+1}_{Q'}\rangle). \qquad (19.4)$$

Take $\eta'\in C_{N_{Y'}}$ with $\eta'>\eta$ and, using the fact that $C_{N_{Y'}}=C_{\Omega'}\cap\tau_{Y'}$ and $C_{Q'}\setminus\eta$ are Σ_1 indiscernibles for $\langle Q'^{\,n(Q')},p^{n(Q')+1}_{Q'},\delta\rangle_{\delta<\eta}$ we get

$$\eta'=h_{Q'}{}^{n(Q')}(j,\langle\vec\delta,p^{n(Q')+1}_{Q'}\rangle). \qquad (19.5)$$

By (19.4) $\eta=\eta'$. Contradiction! So $C_{Y'}\setminus\pi_{Y'}(\gamma)\subseteq C_Y$. □

<u>Lemma 19.12</u>. *Suppose* $Y_i \in C$ *for* $i<\omega_1$ *and* $i<j \Rightarrow Y_i \subseteq Y_j$. *Then* $\bigcup_{i<\omega_1} Y_i \in C$.

Proof: Each $Y_i \prec K_\tau^+$, so $\bigcup_{i<\omega_1} Y_i \prec K_\tau^+$ by lemma 6.8. If $\{\alpha_i : i\in\omega\} \subseteq \bigcup_{i<\omega_1} Y_i$ then for some $i<\omega_1$ $\{\alpha_i : i\in\omega\} \subseteq Y_i$, so $\bigcup_{i<\omega_1} Y_i \cap \tau$ is ω-closed in τ. $\omega_2 \cap \bigcup_{i<\omega_1} Y_i$ is clearly transitive. Also since X is uncountable $\overline{\overline{\bigcup_{i<\omega_1} Y_i}} = \overline{\overline{X}}$. Suppose $\bar{\pi} : \underline{K}_{\bar{\tau}} \cong \underline{K}_\tau | (\bigcup_{i<\omega_1} Y_i \cap K_\tau)$. It remains to show that $\bar{\tau}$ is $\bar{\pi}$-good. But this follows from chapter 18 exercise 4. □

We shall say that Y is fine provided that

$$(\forall Y' \in C)(Y' \supseteq Y \Rightarrow (\exists Y'' \in C)(Y'' \supseteq Y' \wedge C_{Y''} \subseteq Y)). \tag{19.6}$$

<u>Lemma 19.13</u>. *There is a fine* $Y\in C$.

Proof: Suppose not. Define a sequence as follows:

Y_0 is arbitrary;

Given $Y_\alpha \in C$, Y_α is not fine. Hence there is $Y' \supseteq Y_\alpha$, $Y' \in C$ such that $(\forall Y'' \in C)(Y'' \supseteq Y' \Rightarrow C_{Y''} \nsubseteq Y_\alpha)$. Let $Y_{\alpha+1}$ be some such Y'.

If λ is a limit then pick $Y_\lambda \supseteq \bigcup_{\alpha<\lambda} Y_\alpha$ with $Y_\lambda \in C$.

Let $Y = \bigcup_{\alpha<\omega_1} Y_\alpha$; then $Y\in C$ by lemma 19.12. C_Y is countable so $C_Y \subseteq Y_\alpha$, say, $\alpha<\omega_1$. But $Y \supseteq Y_{\alpha+1}$ and $Y \in C$ so $C_Y \nsubseteq Y_\alpha$. Contradiction! □

Fix some fine $Y\in C$ and let $C' = \{Y' \in C : Y' \supseteq Y \wedge C_{Y'} \setminus \gamma \subseteq Y$, some $\gamma<\tau\}$. Then

<u>Lemma 19.14</u>. *For all* $Y' \in C$ *there is* $Y'' \in C'$ *such that* $Y'' \supseteq Y'$.

Proof: Say $Y' \cup Y \subseteq \hat{Y}$, $\hat{Y} \in C$. Then there is $Y'' \in C$, $Y'' \supseteq \hat{Y}$, $C_{Y''} \subseteq Y$. So $Y'' \in C'$ $Y'' \supseteq Y'$. □

<u>Lemma 19.15</u>. *Suppose* $\underline{N}'_Y = \underline{J}^{U_Y}_{\alpha_Y}$. *Then for* $\bar{X} \in P(\tau) \cap N'_Y$, $\bar{X} \in U_Y \Leftrightarrow \exists\gamma<\tau (C_Y \setminus \gamma \subseteq \bar{X})$.

Proof: Let $\underline{N}^* = \underline{N}'^{\,n(N'_Y)}_Y$. Fix $\bar{X}$. There is $\hat{X} \in N'_Y$ such that $\hat{X} \in \mathrm{rng}(\pi_Y)$, $\hat{X} = \langle \hat{X}_\alpha : \alpha<\tau \rangle$ and $\bar{X} = \hat{X}_\alpha$, some $\alpha<\tau$; for suppose $\bar{X} \in S^{N'_Y}_{\pi(\beta)}$; then there is $X^* = \langle X^*_\alpha : \alpha<\tau_Y \rangle$ enumerating $P(\tau_Y) \cap S^{N_Y}_\beta$, by acceptability: let $\hat{X} = \pi_Y(X^*)$.

Let $p^* = p^{n(N_Y)+1}_{N_Y}$, and pick δ such that $X^* \in h^{n(N_Y)}_{N_Y}(\delta \cup p^*)$, $\delta<\tau_Y$. Take δ with $\pi_Y(\delta)>\alpha$ without loss of generality. Then for $\eta \in C_Y$, $\eta > \pi_Y(\delta)$, $\eta = \pi_Y(\bar{\eta})$, say,

$$\underline{N}_Y \models \forall\alpha<\delta(\bar{\eta} \in X^*_\alpha \Leftrightarrow X^*_\alpha \in U_{N_Y}). \tag{19.7}$$

So

$$\underline{N}'_Y \models \forall\alpha<\pi_Y(\delta)(\eta \in \hat{X}_\alpha \Leftrightarrow \hat{X}_\alpha \in U_Y). \tag{19.8}$$

Thus if $\bar{X} \in U_Y$ then $C_Y \setminus \pi_Y(\delta) \subseteq \bar{X}$.

If $\bar{X} \notin U_Y$ but $C_Y \setminus \gamma \subseteq \bar{X}$ then $\tau \setminus \bar{X} \in U_Y$ so $C_Y \setminus \gamma' \subseteq \tau \setminus \bar{X}$, some $\gamma'<\tau$. So $C_Y \setminus (\max(\gamma,\gamma')) = \phi$. Contradiction! □

<u>Lemma 19.16</u>. *Suppose* $Y' \subseteq Y''$, $Y', Y'' \in C$. *Then for any mouse* $\underline{M}$ *for which* $K_{\tau_{Y'}}$ *is a* $\tau_{Y'}$*-base and* $m \geq n(\underline{M})$, $\omega\rho_{\underline{M}}^{m} \geq \tau_{Y'}$

$$\underline{M}^{\pi_{Y'}, \tau_{Y'}, m} = (\underline{M}^{\pi_{Y'Y''}, \tau_{Y'}, m})^{\pi_{Y''}, \tau_{Y''}, m}.$$

Proof: Suppose $\underline{M}_1 = \underline{M}^m$, $\underline{M}_2 = (\underline{M}_1)^{\pi_{Y'Y''}, \tau_{Y'}}$. Suppose $a \subseteq \gamma < \tau_{Y''}$, $a \in M_2$. Then $a = \pi_{Y'Y''}(f)(\delta)$, say, with $\delta < \tau_{Y''}$. Assume $\gamma, \delta < \pi_{Y'Y''}(\bar{\gamma})$, $\mathrm{dom}(f) \subseteq \bar{\gamma}$, and $\mathrm{rng}(f) \subseteq P(\bar{\gamma})$. Then $f \in K_{\tau_{Y'}}$ so $\pi_{Y'Y''}(f) \in K_{\tau_{Y''}}$ and so $a \in K_{\tau_{Y''}}$. Thus $K_{\tau_{Y''}}$ is a $\tau_{Y''}$-base for M_2. Let $\underline{M}_3 = (\underline{M}_2)^{\pi_{Y''}, \tau_{Y''}}$.

If $f \in M_2$ then $f = [g, \delta]$ with $\langle g, \delta \rangle \in T^{M_1, \pi_{Y'Y''}}$. For such g, δ and $\delta' < \tau$ define $\sigma([[g,\delta],\delta']) = \pi_{Y'}(g)(\pi_{Y''}(\delta))(\delta')$. Then for ϕ Σ_0

$$\underline{M}_3 \models \phi([[g_1,\delta_1],\delta_1'] \ldots [[g_n,\delta_n],\delta_n']) \leftrightarrow \qquad (19.9)$$

$$\leftrightarrow \langle \delta_1' \ldots \delta_n' \rangle \in \pi_{Y''}(\{\langle \gamma_1' \ldots \gamma_n' \rangle : \underline{M}_2 \models \phi([g_1,\delta_1](\gamma_1') \ldots [g_n,\delta_n](\gamma_n'))\})$$

$$\leftrightarrow \langle \delta_1' \ldots \delta_n' \rangle \in \pi_{Y''}(\{\langle \gamma_1' \ldots \gamma_n' \rangle : \langle \delta_1 \ldots \delta_n \rangle \in \pi_{Y'Y''}(\{\langle \gamma_1 \ldots \gamma_n \rangle : \underline{M}_1 \models \phi(g_1(\gamma_1)(\gamma_1') \ldots g_n(\gamma_n)(\gamma_n'))\})\})$$

$$\leftrightarrow \langle \pi_{Y''}(\delta_1) \ldots \pi_{Y''}(\delta_n) \rangle \in \pi_{Y'}(\{\langle \gamma_1 \ldots \gamma_n \rangle : \underline{M}_1 \models \phi(g_1(\gamma_1)(\delta_1') \ldots \ldots g_n(\gamma_n)(\delta_n'))\})$$

$$\leftrightarrow \underline{M}_1^{\pi_{Y'}, \tau_{Y'}} \models \phi(\pi_{Y'}(g_1)(\pi_{Y''}(\delta_1))(\delta_1') \ldots \pi_{Y'}(g_n)(\pi_{Y''}(\delta_n))(\delta_n')).$$

So σ is well-defined and $\sigma : \underline{M}_3 \to_{\Sigma_0} \underline{M}_1^{\pi_{Y'Y''}, \tau_{Y'}}$. It is clearly surjective so it is an isomorphism. □

<u>Lemma 19.17</u>. *If* $Y', Y'' \in C'$ *then* $U_{Y'} \cap N'_{Y''} = U_{Y''} \cap N'_{Y'}$.

Proof: Take $\hat{Y} \in C'$ with $\hat{Y} \subseteq Y' \cup Y''$. Then $C_{\hat{Y}} \setminus \gamma' \subseteq Y \subseteq Y'$ (some $\gamma' < \tau$) so by lemma 19.11 there is $\gamma < \tau$ such that $C_{\hat{Y}} \setminus \gamma \subseteq C_{Y'}$.

Suppose $\bar{X} \in U_{Y'} \cap N'_{\hat{Y}}$. Then by lemma 19.15 there is $\gamma_1 < \tau$ such that $\gamma_1 > \gamma$, $C_{Y'} \setminus \gamma' \subseteq \bar{X}$. Hence $C_{\hat{Y}} \setminus \gamma_1 \subseteq \bar{X}$, so $\bar{X} \in U_{\hat{Y}}$.

<u>Claim</u> $N'_{Y'} \subseteq N'_{\hat{Y}}$.

Proof: By lemma 19.16 it is sufficient to show $\underline{M} = \underline{N}_{Y'}^{\pi_{Y'\hat{Y}}, \tau_{Y'}, n(N_{Y'})} \subseteq \underline{N}_{\hat{Y}}$. But suppose $\underline{M} = \underline{J}_{\alpha}^{U}$. Then for $\bar{X} \in P(\tau_{\hat{Y}}) \cap M$, $\bar{X} \in U \leftrightarrow \pi_{Y'\hat{Y}}{}''C_{N_{Y'}} \setminus \gamma \subseteq \bar{X}$ for some $\gamma < \tau_{Y'}$ by lemma 19.15. So $\underline{M}$, $\underline{N}_{\hat{Y}}$ are comparable. But if $\underline{N}_{\hat{Y}} = \underline{J}_{\beta}^{U}$, for some $\beta < \alpha$ then $C_{N_{\hat{Y}}} \in M$ so $\underline{M} \models \mathrm{cf}(\tau_{\hat{Y}}) = \omega$. Contradiction! So $M \subseteq N_Y$.

□(Claim)

Hence $U_{Y'} \subseteq U_{\hat{Y}}$. Similarly $U_{Y''} \subseteq U_{\hat{Y}}$. This gives the result. □

<u>Corollary 19.18</u>. *If* $Y', Y'' \in C'$ *then there are* U, α', α'' *such that* $\underline{N}'_{Y'} = \underline{J}_{\alpha'}^{U}$, $\underline{N}'_{Y''} = \underline{J}_{\alpha''}^{U}$.

Let $U = \bigcup_{Y' \in C'} U_{Y'}$.

Lemma 19.19. *Suppose $x \in K_{\tau^+}$, $x \subseteq \tau$. Then there is $Y' \in C'$ such that $x \in N'_{Y'}$.*
Proof: There is $Y' \in C'$ with $x \in Y'$. Say $\pi : \overline{K} \cong Y'$, $\overline{K}$ transitive. So $K_{\tau_{Y'}} = H^{\overline{K}}_{\tau_{Y'}}$ and $\pi_{Y'} = \pi | K_{\tau_{Y'}}$. Let $\overline{x} = \pi^{-1}(x)$; $\overline{x} \in \overline{K}$.
Claim $\overline{x} \in N_{Y'}$.
Proof: $\overline{K} \models \exists \underline{M}(\underline{M}$ is a mouse, $\overline{x} \in M$ and the critical point of $\underline{M}$ is not less than $\tau_{Y'}$); and $\omega_1 \in \overline{K}$. Let $\underline{M}$ be such a mouse. Let $\underline{Q}, \underline{\Omega}'$ be comparable mouse iterates of $\underline{M}, \underline{N}_{Y'}$. Suppose $\Omega' \in Q$. There is $a \in \Sigma_\omega(\underline{N}_{Y'}) \setminus K_{\tau_{Y'}}$, $a \subseteq \gamma < \tau_{Y'}$: so $a \in \Omega$, hence $a \in M$. So $a \in \overline{K} \setminus K_{\tau_{Y'}}$. But $K_{\tau_{Y'}} = H^{\overline{K}}_{\tau_{Y'}}$. So $Q \subseteq Q'$. It follows that $\overline{x} \in N_{Y'}$. □(Claim)
Let $\hat{\pi} = \pi_{Y'}^{N_{Y'}, \tau_{Y'}, n(N_{Y'})}$. Then for $\gamma < \tau$

$$\begin{aligned} \gamma \in \hat{\pi}(\overline{x}) &\leftrightarrow \gamma \in \hat{\pi}(\overline{x} \cap \delta) && (\text{some } \delta < \tau_{Y'}) \\ &\leftrightarrow \gamma \in \pi(\overline{x} \cap \delta) && (\text{as } \hat{\pi} \supseteq \pi_{Y'}) \\ &\leftrightarrow \gamma \in \pi(\overline{x}) \\ &\leftrightarrow \gamma \in x. \end{aligned} \tag{19.10}$$

So $x = \hat{\pi}(\overline{x}) \in N'_{Y'}$. □

Corollary 19.20. *U is a normal measure on τ in K and $\langle K_{\tau^+}, U \rangle$ is amenable.*
Proof: If $x \in K_{\tau^+}$, $x \subseteq P(\tau)$ then, coding x as a subset of τ, $x \in N_{Y'}$ for some $Y' \in C'$. So $x \cap U = x \cap U_{N_{Y'}} \in K_{\tau^+}$. □

Were U countably complete this would suffice by lemma 16.11. But it plainly is not. The following result enables us to get a ρ-model anyway.

Lemma 19.21. *$cf(\tau^+) > \omega$.*
Proof: Otherwise take $\{Y_i : i \in \omega\}$ with $\sup_{i\in\omega} On \cap N'_{Y_i} = \tau^+$. Let x_{i+1} code all the subsets of τ in N'_{Y_i}; we may assume $x_{i+1} \in N'_{Y_{i+1}}$. There is $Y \prec K_{\tau^+}$ with $Y \in C'$ and $x_i \in Y$, all $i \in \omega$. As in lemma 19.19 $x_i \in N'_Y$, all $i \in \omega$. So $P(\tau) \cap K \subseteq N'_Y$ and N'_Y is a mouse. Contradiction! □

The only facts used from now on are that U is a normal measure on τ in K, that $\langle K_{\tau^+}, U \rangle$ is amenable and that $cf(\tau^+) > \omega$.

If $\gamma < \tau^+$ then $J^U_\gamma \in K_{\tau^+}$ by amenability. On the other hand $J^U_{\tau^+} \notin K$, otherwise $L[U] \subseteq K$ is a ρ-model. So $J^U_{\tau^+} \subseteq K$ but $J^U_{\tau^+} \notin K$.

Suppose U is not normal in L[U]. Then there is $\delta \geq \tau^+$ with $P(\tau) \cap J^U_{\delta+1} \nsubseteq K$, since $P(\tau) \cap L[U] \neq P(\tau) \cap K$. Take δ least $\geq \tau^+$ such that $P(\tau) \cap J^U_{\delta+1} \nsubseteq J^U_\delta$. Then $\Sigma_\omega(\underline{J}^U_\delta) \cap P(\tau) \nsubseteq J^U_\delta$ and $\underline{J}^U_\delta$ is a premouse. Let $\underline{N} = \underline{J}^U_\delta$. If U were countably complete we should have a contradiction. In the present case we also have a contradiction, but it takes more work. The argument to the end of lemma 19.25 will show that $\underline{N}$ is a mouse.

Lemma 19.22. $(\tau^+)^N=\tau^+$.

Proof: Suppose $\tau<\gamma<\tau^+$. Then $K_{\tau^+}\models\overline{\overline{\gamma}}=\tau$. If $\underline{J}^U_{\tau^+}\models$"$\gamma$ is a cardinal" then $U\cap J^U_{\tau^+}\subseteq J^U_\gamma\in K$, so $J^U_{\tau^+}\in K$. Hence $\underline{N}\models\overline{\overline{\gamma}}=\tau$.

If τ^+ were not a cardinal in N there would be a well-order of τ of order-type τ^+ in N, hence in K. □

Lemma 19.23. *$\underline{N}$ is acceptable.*

Proof: $\underline{J}^U_{\tau^+}$ is acceptable, as it is a limit of mice. An easy induction on $\omega\beta\in On_N$ gives the result. □

$\omega\rho^{n+1}_N\leq\tau$, some $n<\omega$. Let $n=n(\underline{N})$, $\rho=\rho^n_N$, $A=A^n_N$. $\omega\rho\geq\tau^+$, of course.

Lemma 19.24. *$cf(\omega\rho)>\omega$.*

Proof: Suppose $cf(\omega\rho)=\omega$; say $\omega\rho=\sup_{i<\omega}\rho_i$. Let H be Σ_0 such that

$$y=h_{N^n}(i,x)\leftrightarrow\exists zH(z,i,x,y). \qquad (19.11)$$

Let $t=\langle(t)_0,(t)_1\rangle$ be the Gödel pairing function on τ. Define f_i with $\mathrm{dom}(f_i)\subseteq\tau$ by

$$y=f_i(t)\leftrightarrow\exists z\in S^{N^n}_{\rho_i}H(z,(t)_0,\langle(t)_1,p^{n+1}_N\rangle,y)\wedge(t)_0\in\omega_n\wedge \wedge\, y,t\in S^{N^n}_{\rho_i}. \qquad (19.12)$$

We may assume $p^{n+1}_N\in S^{N^n}_{\rho_0}$. Let $\alpha_i=\sup(\mathrm{rng}(f_i)\cap\tau^+)$. Since $J^{N^n}_{\tau^+}=\bigcup_{i\in\omega}\mathrm{rng}(f_i)$ $\sup_{i\in\omega}\alpha_i=\tau^+$. But $cf(\tau^+)>\omega$ so $\tau^+=\alpha_i$, say. So f_i maps a subset of τ cofinally into τ^+. But $f_i\in N$ and $\underline{N}\models\tau^+$ is regular. Contradiction! □

Let $I=\{\nu<\rho:\tau<\nu$ and $p^{n+1}_N\in J^U_\nu$ and $J^U_\nu=J^{U,A}_\nu\}$. Then I is cofinal in ρ (which is a limit ordinal by lemma 19.24); for given $\overline{\nu}<\rho$ let $\nu_0=\overline{\nu}+1$ and take ν_{i+1} least such that $S^{U,A}_{\nu_i}\subseteq S^U_{\nu_{i+1}}$ (this exists because $J^U_\rho=J^{U,A}_\rho$). Letting $\omega\nu=\sup_{i<\omega}\nu_i$, $\nu\in I$.

For $\nu\in I$ let $\underline{N}^\nu=\underline{J}^{U,A}_\nu$ and let $X_\nu=h_{N^\nu}(\tau\cup p^{n+1}_N)$. Let $\sigma^\nu:\overline{\underline{N}}^\nu\cong\underline{N}^\nu|X_\nu$, where $\overline{N}^\nu$ is transitive. Say $\overline{\underline{N}}^\nu=\underline{J}^{\overline{U},\overline{A}}_{\mu_\nu}$; then for $X\in P(\tau)\cap\overline{N}^\nu$, $\sigma^\nu(X)=X$ so $X\in\overline{U}\leftrightarrow X\in U$. Thus $\overline{\underline{N}}_\nu=\underline{J}^{U,\overline{A}}_{\mu_\nu}$. Let $\underline{M}^\nu$ be such that for some $p\in PA^{M^\nu}_n$ $\underline{M}^{\nu p}=\overline{\underline{N}}$, $\underline{M}^\nu$ is p-sound and standard.

Lemma 19.25. *$\underline{M}^\nu\in K$.*

Proof: $U\cap M^\nu=U\cap\overline{N}^\nu$, so $M^\nu=J^U_{\alpha_\nu}$, say. $\alpha_\nu\leq\delta$ as U is normal in $\underline{M}$. But $\mu_\nu=$ $=\rho^n_M\nu\leq\nu<\rho=\rho^n_N$ so $\alpha_\nu<\delta$. $J^U_{\alpha_\nu+1}\models\overline{\overline{\alpha_\nu}}=\tau$, but in $\underline{N}$ τ^+ is a cardinal so $\alpha_\nu<\tau^+$.

It follows that $\underline{J}^{U}_{\alpha_\nu} \in K$. □

In K, U is countably complete, so $\underline{M}^\nu$ is a mouse. Suppose $\underline{N}$ were not a mouse. Then $(E^+_n)_{\underline{N}}$ is not well-founded. Let $\{x_i : i\in\omega\}$ be such that $\underline{N}^n \models x_{i+1} E^+_n x_i$, all $i\in\omega$. Since $N^{\underline{n}} = \bigcup_{\nu\in I} X_\nu$ there is $\nu\in I$ with $x_i \in X_\nu$, all $i\in\omega$. Let $\bar{x}_i = \sigma^{\nu-1}(x_i)$. Then $\underline{\bar{N}}^{\nu n} \models \bar{x}_{i+1} E^+_n \bar{x}_i$, all $i\in\omega$. But $\underline{M}^\nu$ is a mouse and $n(\underline{M}^\nu) = n$. Contradiction!

So $\underline{N}$ is a mouse and so $\underline{N}\in K$. Hence $J^U_{\tau^+} \in K$. Contradiction! So U must be normal in L[U].

Hence we have proved the covering lemma:

Lemma 19.26. *Suppose there is no ρ-model. Then if $X \sqsubseteq On$ is uncountable there is $Y \in K$ such that $Y \sqsupseteq X$ and $\overline{\overline{X}} = \overline{\overline{Y}}$.*

Observe that although the proof made lavish use of the axiom of choice, we could obtain lemma 19.26 in ZF, since given X we may argue in L[X].

The assumption that X is uncountable is necessary: see model 1 in appendix I.

Exercises

1. Prove the converse of lemma 19.26.

2. Let $\pi : K_{\bar{\tau}} \to_{\Sigma_1} K_\tau$. Show that $K_{\bar{\tau}} = H^{N_{\bar{\tau}}}_{\bar{\tau}}$ (notation as in this chapter).

We want to continue the discussion of chapter 19 for a bit. So we are still supposing that the covering property fails for some set $X \subseteq On$ with minimal sup τ and that X is uncountable. So there is an inner model with a measurable cardinal. We are interested particularly in the case just considered where we get a ρ-model whose critical point is τ. This involves assuming that $N_{\bar{\tau}}$ exists whenever $\pi: K_{\bar{\tau}} \to_{\Sigma_1} K_\tau$, $\bar{\tau}<\tau$ and $\bar{\tau}$ is π-good. Let L[U] be the ρ-model we have just constructed: note that $V \neq L[U]$, as $cf(\tau)=\omega$; indeed the covering lemma for L[U] fails.

<u>Definition 20.1</u> *Suppose L[U] is a ρ-model with critical point τ. C is a Prikry sequence for U provided that whenever $X \in P(\tau) \cap L[U]$, $X \in U \leftrightarrow \exists \gamma<\tau (C \setminus \gamma \subseteq X)$; and otp(C)= $=\omega$.*

For example, if $\langle \langle L[U_\alpha] \rangle_{\alpha \in On}, \langle \pi_{\alpha\beta} \rangle_{\alpha \leq \beta \in On} \rangle$ is the iterated ultrapower of L[U] and κ_α is the critical point of $L[U_\alpha]$ then $\{\kappa_n : n \in \omega\}$ is a Prikry sequence for $L[U_\omega]$. Usually we shall be interested in the minimal ρ-model: model 3 in appendix I shows that this may carry a Prikry sequence.

If M is an inner model and $C \in M$ then $U \in M$: in particular L[C]= =L[U,C].

The aim of this chapter is to show that, letting C' be as in chapter 19, there is $C'' \subseteq C'$ such that for all $Y \in C'$ there is $Y' \in C''$ with $Y \subseteq Y'$ and for each $Y \in C''$, C_Y is a Prikry sequence for U (C_Y is always a 'Prikry sequence' for $U \cap N_Y'$, of course). To do this it is sufficient to stabilise C_Y so that for some γ $Y, Y' \in C'' \to C_Y \setminus \gamma =$ $=C_{Y'} \setminus \gamma$. It looks as though a simple appeal to cofinality should do the trick, but it does not.

Suppose that $\langle Y_\alpha' : \alpha<\omega_1 \rangle$ is a sequence satisfying $\alpha \leq \beta \to Y_\alpha \subseteq Y_\beta$ and $Y_\alpha \in C'$. Let $\pi_\alpha : \underline{K}_{\tau_\alpha} \cong \underline{K}_\tau | Y_\alpha \cap K_\tau$, $\underline{N}_\alpha = \underline{N}_{\tau_\alpha}$, $\underline{N}_\alpha' = \underline{N}_{\tau_\alpha}^{\pi_\alpha, \tau_\alpha, n(N_\alpha')}$. Let μ_α denote $(\tau_\alpha^+)^{N_\alpha}$ (or On_{N_α} if $\underline{N}_\alpha \models \tau_\alpha$ is the largest cardinal) and μ_α' denote $(\tau^+)^{N_\alpha'}$ (or $On_{N_\alpha'}$ if $\underline{N}_\alpha' \models \tau$ is the largest cardinal). Then $\mu_\alpha' =$ $=\sup\{\omega\gamma+\omega : \omega\gamma \in N_\alpha' \wedge P(\tau) \cap J_\gamma^{N_\alpha'} \neq P(\tau) \cap J_{\gamma+1}^{N_\alpha'}\}$. Assume that $\alpha<\beta \to \mu_\alpha' < \mu_\beta'$ and that $P(\tau) \cap rng(\pi_\alpha^{N_\alpha, \tau_\alpha, n(N_\alpha)}) \subseteq Y_\alpha$.

Let $Y=\bigcup_{\alpha<\omega_1} Y_\alpha$; so $Y\in C$. In fact $Y\in C'$, for clearly $C_Y\subseteq Y_\alpha$ some $\alpha<\omega_1$ so for some $\gamma<\tau$ $C_Y\setminus\gamma\subseteq C_{Y_\alpha}$. Let $\pi:\underline{K}_{\bar\tau}\cong\underline{K}_\tau|Y\cap K_\tau$, $\underline{\bar N}=\underline{N}_{\bar\tau}$, $\underline{\bar N}'=\underline{\bar N}^{\pi,\bar\tau,n(\bar N)}$. Let μ denote $(\bar\tau^+)^{\bar N}$ and μ' $(\tau^+)^{{}^{\alpha}N}$ (with the usual convention about largest cardinals).

<u>Lemma 20.2</u>. *For* $\alpha<\omega_1$ $\mu'_\alpha<\mu'$.

Proof: Let $\hat\pi=\pi^{-1}\pi_\alpha$. Then $\underline{N}'_\alpha=((\underline{N}_\alpha)^{\hat\pi,\tau_\alpha,n(N_\alpha)})^{\pi,\bar\tau,n(N_\alpha)}$. So there is a Σ_0 map from $\underline{N}'_\alpha$ to $\underline{\bar N}'$; but these are comparable, so for some $\omega\beta\leq On_{\bar N'}$, $\underline{N}'=J_\beta^{\bar N'}$. Thus $\mu'_\alpha\leq\mu'$. But $\mu'_\alpha<\mu'_{\alpha+1}\leq\mu'$ so $\mu'_\alpha<\mu'$. □

<u>Lemma 20.3</u>. $P(\tau)\cap rng(\pi^{\bar N,\bar\tau,n(\bar N)})\subseteq Y$.

Proof: Let $\tilde\pi:\underline{\bar K}\cong\underline{K}_{\tau^+}|Y$, so $\pi=\tilde\pi|K_{\bar\tau}$. Suppose $a\in\bar N$, $a\subseteq\bar\tau$. Let $\underline{M}=core(\underline{\bar N})$ and let $\langle\langle\underline{M}_\alpha\rangle_{\alpha\in On},\langle\pi_{\alpha\beta}\rangle_{\alpha\leq\beta\in On}\rangle$ be its mouse iteration. So $a=\pi_{n\omega}(\bar a)$, some $n\in\omega,\bar a$. Let κ_α denote the critical point of $\underline{M}_\alpha$. So $\bar a\subseteq\kappa_n$. $\bar a=a\cap\kappa_n\in\bar N$ so $\bar a\in K_{\bar\tau}$. Hence $\pi(\bar a)\in Y$, say $\pi(\bar a)\in Y_\alpha$. Suppose $\pi(\bar a)=\pi_\alpha(\bar b)$. Let $\underline{M}'=core(\underline{N}_\alpha)$ and let $\langle\langle\underline{M}'_\alpha\rangle_{\alpha\in On},\langle\pi'_{\alpha\beta}\rangle_{\alpha\leq\beta\in On}\rangle$ be its mouse iteration. We may assume $\pi(\kappa_n)\in\pi_\alpha{}''C_N$; suppose $\pi(\kappa_n)=\pi_\alpha(\kappa'_m)$ where κ'_m is the critical point of $\underline{M}'_m$. $\bar b\in N_\alpha$ so $\bar b\in M'_m$: let $b=\pi'_{m\omega}(\bar b)$. So $b\subseteq\tau_\alpha$ and $b\in\underline{N}_\alpha$. $\pi_\alpha{}^{N_\alpha,\tau_\alpha,n(N_\alpha)}(b)\in$ $\in Y_\alpha\subseteq Y$ so it is sufficient to show $\pi^{N_\alpha,\tau_\alpha,n(N_\alpha)}(b)=\pi^{\bar N,\bar\tau,n(\bar N)}(a)$. Let $\hat\pi=(\pi^{-1}\pi_\alpha)^{N_\alpha,\tau_\alpha,n(N_\alpha)}$: it suffices to show $\hat\pi(b)=a$.

Suppose $\bar b=h_{M'_m}{}^{n(N_\alpha)}(i,\langle\vec\gamma,p_{M'_m}^{n(N_\alpha)+1}\rangle)$ $(\vec\gamma<\kappa_m)$. We may assume $\hat\pi(\vec\gamma)$, $\hat\pi(p_N^{n(N_\alpha)+1})\in rng(\pi_{n\omega})$. So

$$\begin{aligned}\hat\pi(b)&=\hat\pi(\pi'_{m\omega}(\bar b))\\&=\hat\pi(h_{N_\alpha}{}^{n(N_\alpha)}(i,\langle\vec\gamma,p_{N_\alpha}^{n(N_\alpha)+1}\rangle))\\&=h_{\bar N}{}^{n(\bar N)}(i,\langle\hat\pi(\vec\gamma),\hat\pi(p_{N_\alpha}^{n(N_\alpha)+1})\rangle)\\&\in rng(\pi_{n\omega}).\end{aligned}\tag{20.1}$$

Say $\hat\pi(b)=\pi_{n\omega}(a')$. Then for $\delta<\kappa_n$

$$\begin{aligned}\delta\in a'&\Leftrightarrow\delta\in\hat\pi(b)\\&\Leftrightarrow\delta\in\hat\pi(\pi'_{m\omega}(\bar b))\\&\Leftrightarrow\delta\in\hat\pi(\bar b)\\&\Leftrightarrow\delta\in\bar a.\end{aligned}\tag{20.2}$$

So $\bar a=a'$; hence $\hat\pi(b)=\pi_{n\omega}(\bar a)=a$. □

<u>Lemma 20.4</u>. $\mu'=\sup_{\alpha<\omega_1}\mu'_\alpha$.

Proof: Suppose not. Then there is $a\in P(\tau)\cap rng(\pi^{\bar N,\bar\tau,n(\bar N)})$ such that $a\notin N'_\alpha$, all $\alpha<\omega_1$. By lemma 20.3 $a\in Y$ so $a\in Y_\alpha$, some α. Let $\hat\pi_\alpha:\underline{\bar K}_\alpha\cong\underline{K}_{\tau^+}|Y_\alpha$ and suppose $a=\hat\pi_\alpha(\bar a)$. Let $\hat a=\pi^{N_\alpha,\tau_\alpha,n(N_\alpha)}(\bar a)$. For $\delta<\pi_\alpha(\gamma)<\tau$

$\delta\in\hat{a} \Leftrightarrow \delta\in\pi_\alpha(\bar{a}\cap\gamma)$
$\Leftrightarrow \delta\in a.$

Thus $a=\hat{a}\in N'_\alpha$. Contradiction! □

Corollary 20.5. *$cf(\mu')>\omega$.*

Now let $\rho=\rho^n_{\bar{N}'}$. $\omega\rho\geq\mu'>\tau$ and $\bar{N}'^{n(\bar{N}')}=h_{\bar{N}',n(\bar{N}')}(\tau\cup p^{n(\bar{N}')+1}_{\bar{N}'})$ so just as in lemma 19.24 $cf(\omega\rho)>\omega$. Hence ρ is a limit ordinal. Let $\tilde{p}=$ $=\pi^{\bar{N},\bar{\tau},n(\bar{N})}(p^{n(\bar{N})+1}_{\bar{N}})$; so $\bar{N}'^{n(\bar{N}')}=h_{\bar{N}',n(\underline{N}')}(\tau\cup\tilde{p})$. Let $\tilde{p}_\alpha=\pi^{N_\alpha,\tau_\alpha,n(N_\alpha)}_\alpha$ $(p^{n(N_\alpha)+1}_{N_\alpha})$.

For $\nu<\rho$ such that $\tau<\nu$ and $p^{n(\bar{N}')+1}_{\bar{N}'}\in J^{\bar{N}',n(\bar{N}')}_\nu$ let $X_\nu=$ $=h_{J^{\bar{N}',n(\bar{N}')}_\nu}(\tau\cup\tilde{p})$. As ρ is a limit ordinal, $\bar{N}'^{n(\bar{N}')}=\bigcup_{\nu<\rho} X_\nu$. Let $\sigma_\nu:\underline{Q}_\nu\cong\underline{\bar{N}}'^{n(\bar{N}')}|X_\nu$.

Lemma 20.6. *$\underline{Q}_\nu\in J^{\bar{N}'}_{\mu'}$.*

Proof: $X_\nu\in\bar{N}'$ and $\underline{\bar{N}}'\models\bar{\bar{X}}_\nu\leq\tau$. So $\underline{Q}_\nu\in H^{\bar{N}'}_{\mu'}=J^{\bar{N}'}_{\mu'}$. □

Lemma 20.7. *$J^{\bar{N}'}_{\mu'}\subseteq\bigcup_{\nu<\rho} Q_\nu$.*

Proof: Suppose that $a\in J^{\bar{N}'}_{\mu'}$ and $a\in X_\nu$. Then since $X_\nu\models\bar{\bar{a}}\leq\tau$, $a\subseteq X_\nu$. So $X_\nu\cap J^{\bar{N}'}_{\mu'}$ is transitive. Thus $\sigma_\nu^{-1}(a)=a$, so $a\in Q_\nu$. □

The purpose of this construction is shown by the next lemma.

Lemma 20.8. *There is $\alpha<\omega_1$, $\gamma<\tau$ such that $C_{Y_\alpha}\setminus\gamma=C_Y\setminus\gamma$.*

Proof: First of all we may assume that there is $\gamma<\tau$ such that for all $\alpha<\omega_1$ $C_Y\setminus\gamma\subseteq C_{Y_\alpha}$. Otherwise suppose that for all γ $X_\gamma=\{\alpha<\omega_1:C_Y\setminus\gamma\subseteq C_{Y_\alpha}\}$ is bounded in ω_1. Let $x_\gamma=\sup X_\gamma$: let $\langle\gamma_i:i<\omega\rangle$ be cofinal in τ. Then $\langle x_{\gamma_i}:i<\omega\rangle$ is cofinal in ω_1. Contradiction!

Secondly we may assume that for all $\nu\in I$ there is α such that $\sigma_\nu^{-1}(\tilde{p}),Q_\nu\in h_{N'_\alpha,n(N'_\alpha)}((\gamma\cap rng(\pi))\cup\tilde{p})$.

If the lemma fails then for each α there is ξ_α with $\xi_\alpha\in C_{Y_\alpha}\setminus\gamma$ but $\xi_\alpha\notin C_Y$. But supposing, as we may, that for all α $C_{Y_\alpha}\setminus\gamma\subseteq C_{Y_0}$, there are only ω possible values for ξ_α. So we may assume $\xi_\alpha=\xi$ all $\alpha<\omega_1$.

Let $D=(C_Y\setminus\xi)\cup\{\xi\}$. Suppose $\xi=\pi(\bar{\xi})$. Then there are $\vec{\delta}<\bar{\xi}$ and $i\in\omega$ such that

$$\underline{\bar{N}}^{n(\bar{N})}\models\bar{\xi}=h(i,\langle\vec{\delta},p^{n(\bar{N})+1}_{\bar{N}}\rangle) \quad (20.4)$$

so

$$\underline{\bar{N}}'^{n(\bar{N})}\models\xi=h(i,\langle\pi(\vec{\delta}),\tilde{p}\rangle). \quad (20.5)$$

(20.5) is a Σ_1 property of $\xi,\pi(\vec{\delta}),\tilde{p}$; call it $\phi(\xi,\pi(\vec{\delta}),\tilde{p})$. There is $\nu\in I$ such that $\underline{J}_\nu^{\overline{N}',n(\overline{N}')}\models\phi(\xi,\pi(\vec{\delta}),\tilde{p})$; so

$$\underline{J}_\nu^{\overline{N}',n(\overline{N}')}|X_\nu\models\phi(\xi,\pi(\vec{\delta}),\tilde{p}) \tag{20.6}$$

so that

$$\underline{Q}_\nu\models\phi(\xi,\pi(\vec{\delta}),\sigma_\nu^{-1}(\tilde{p})). \tag{20.7}$$

But $\underline{Q}_\nu\in J_{\mu'}^{\overline{N}'}$, so by lemma 20.4 $\underline{Q}_\nu\in J_{\mu'_\alpha}^{\overline{N}'}$, say. Assume $\sigma_\nu^{-1}(\tilde{p}),\underline{Q}_\nu\in$ $\in h_{N',n(N')}((\gamma\cap\text{rng}(\pi_\alpha))\cup\tilde{p}_\alpha)$. So

$$\underline{N}'_\alpha\models\text{"}\underline{Q}_\nu\models\phi(\xi,\pi(\vec{\delta}),\sigma_\nu^{-1}(\tilde{p}))\text{"} \tag{20.8}$$

is a Σ_1 property of $\xi,\pi(\vec{\delta}),\vec{\zeta}\in\gamma\cap\text{rng}(\pi_\alpha),\tilde{p}_\alpha$: call it $\psi(\xi,\pi(\vec{\delta}),\pi_\alpha(\vec{\overline{\zeta}}),\tilde{p}_\alpha)$ Assume $\pi(\vec{\delta})=\pi_\alpha(\vec{\overline{\delta}}),\xi=\pi_\alpha(\hat{\overline{\xi}})$. So

$$\underline{N}_\alpha\models\psi(\hat{\overline{\xi}},\vec{\overline{\delta}},\vec{\overline{\zeta}},p_{N_\alpha}^{n(N'_\alpha)+1}). \tag{20.9}$$

It follows, since $\hat{\overline{\xi}}\in C_N$ that if we take $\overline{\xi}'\in C_{N_\alpha}$, $\overline{\xi}'>\hat{\overline{\xi}}$ then

$$\underline{N}_\alpha\models\psi(\overline{\xi}',\vec{\overline{\delta}},\vec{\overline{\zeta}},p_{N_\alpha}^{n(N'_\alpha)+1}). \tag{20.10}$$

Reversing the argument (20.5)-(20.10)

$$\underline{\overline{N}}'^{\,n(\overline{N}')}\models\pi_\alpha(\overline{\xi}')=h(i,\langle\pi(\vec{\delta}),\tilde{p}\rangle). \tag{20.11}$$

So by (20.5) $\xi=\pi_\alpha(\overline{\xi}')$. So $\hat{\overline{\xi}}=\overline{\xi}'$. Contradiction! □

<u>Lemma 20.9</u>. *There are $\gamma<\tau$ and $Y\in C'$ such that for all $Y'\in C$ there is $Y''\supseteq Y'$ such that $C_{Y''}\smallsetminus\gamma=C_Y\smallsetminus\gamma$.*

Proof: Suppose not. For all $\gamma<\tau$, $Y\in C'$ there is $Y'\in C$ such that for all $Y''\supseteq Y'$ $C_{Y''}\smallsetminus\gamma\neq C_Y\smallsetminus\gamma$.

Pick a sequence $\langle\gamma_i:i\in\omega\rangle$ cofinal in τ. Let $Y_0\in C'$ be arbitrary. Given $Y_i\in C'$ there is $Y'\in C'$ such that $\forall Y''\supseteq Y'$ $C_{Y''}\smallsetminus\gamma_i\neq C_Y\smallsetminus\gamma$. Let $Y_{i+1}\in$ $\in C'$ be such that $Y_i\cup Y'\subseteq Y_{i+1}$. Let $\hat{Y}\in C'$ contain $\bigcup_{i\in\omega}Y_i$. Then whenever $Y^*\supseteq\hat{Y}$, $Y^*\supseteq Y_{i+1}$ so $C_{Y^*}\smallsetminus\gamma_i\neq C_Y\smallsetminus\gamma$. Hence

$$\forall Y\in C'\exists Y'\in C'\forall\gamma\forall Y''\supseteq Y'\ C_{Y''}\smallsetminus\gamma\neq C_Y\smallsetminus\gamma. \tag{20.12}$$

Now define a sequence $\langle Y'_\alpha:\alpha<\omega_1\rangle$ with $Y'_\alpha\in C'$. Given any $Y\in C$ there is $Y^*\in C'$ with $Y^*\supseteq Y$ and $P(\tau)\cap\text{rng}(\pi_{Y^*}^{N_{Y^*},\tau_{Y^*},n(N_{Y^*})})\subseteq Y^*$. This is shown by the argument of lemma 20.3, constructing an ω_1-chain $\langle Z_\alpha:\alpha<\omega_1\rangle$ with $Z_0=Y$, $Z_{\alpha+1}\supseteq P(\tau)\cap\text{rng}(\pi_{Z_\alpha}^{N_{Z_\alpha},\tau_{Z_\alpha},n(N_{Z_\alpha})})$, $Z_\lambda\supseteq\bigcup_{\alpha<\lambda}Z_\alpha$ (λ a limit) with each Z_α in C'. $Y^*=\bigcup_{\alpha<\omega_1}Z_\alpha$ has the desired property.

Let Y be arbitrary and let $Y'_0=Y^*$. Given Y'_α there is $Y'\supseteq Y'_\alpha$ such that $\forall\gamma\forall Y''\supseteq Y'C_{Y''}\smallsetminus\gamma\neq C_{Y'}\smallsetminus\gamma$. We may assume Y' contains a well-order of τ of order-type μ_α. Let $Y'_{\alpha+1}=(Y')^*$. If λ is a limit ordinal let $Y'_\lambda=(\bigcup_{\alpha<\lambda}Y_\alpha)^*$. Let $Y'=\bigcup_{\alpha<\omega_1}Y_\alpha$. By lemma 20.8 $C_{Y'}\smallsetminus\gamma=C_Y\smallsetminus\gamma$ for some $\gamma<\tau$.

Contradiction, as $Y'\supseteq Y_{\alpha+1}$. □

Lemma 20.10. *If Y is as in lemma 20.9 then C_Y is a Prikry sequence for U.*
Proof: Take $x\in P(\tau)\cap K_{\tau^+}$. Suppose $Y'\in C'$, $x\in N_{Y'}$, and $C_{Y'}\setminus\gamma=C_Y\setminus\gamma$. Then $x\in U \Leftrightarrow x\in U_{Y'} \Leftrightarrow \exists\gamma'(C_{Y'}\setminus\gamma'\subseteq x) \Leftrightarrow \exists\gamma'(C_Y\setminus\gamma'\subseteq x)$. □

A NEW COVERING LEMMA

Now we turn to covering properties again. Suppose $L[U]$ is the minimal ρ-model and κ is the critical measure ordinal. Suppose that there is some τ with $X\subseteq\tau$ uncountable but no $Y\in L[U]$ such that $\overline{\overline{X}}=\overline{\overline{Y}}$ and $X\subseteq Y$. Take τ minimal.

Lemma 20.11. *If $0^\dagger$ does not exist then $\kappa=\tau$.*
Proof: Suppose $\tau<\kappa$. Then since $K\subseteq L[U]$ the covering lemma for K fails for $X\subseteq\tau$; so there is a ρ-model whose critical point is not greater than τ. But this contradicts the minimality of κ.

Suppose $\tau>\kappa$. τ is a cardinal in $L[U]$, just as for K. Take $Y\prec \underline{J}^U_\tau$ with $X\subseteq Y$ and $\overline{\overline{X}}=\overline{\overline{Y}}$. Let $\pi:\underline{J}^{\bar U}_{\bar\tau}\cong\underline{J}^U_\tau|Y$. Assume without loss of generality that if $\overline{\overline{X}}\geq\overline{\overline{\kappa}}$ then $\kappa\subseteq X$. We may assume π to be $\bar\tau$-good. Let α be the critical point of π. If $\bar\tau$ were not a cardinal in $L[\bar U]$ then we could follow the strategy of chapter 19 and get a set covering X. So $\bar\tau$ is a cardinal in $L[\bar U]$. Hence there is a map $\hat{\pi}:L[\bar U]\to_e L[U]$ with $\hat{\pi}\supseteq\pi$.

Case 1 $\alpha>\kappa^+$ (calculated in $L[U]$ here and from now on).
Then $\bar U=U$. So $\hat{\pi}:L[U]\to_e L[U]$ is non-trivial, and so by lemma 13.19 $0^\dagger$ exists. Contradiction!

Case 2 $\kappa<\alpha\leq\kappa^+$.
α is a cardinal in $L[\bar U]$; for if $\beta<\alpha$, $f:\beta\to\alpha$ onto, $f\in L[\bar U]$ then $\hat{\pi}(\beta)=\beta$ so $\hat{\pi}(f):\beta\to\hat{\pi}(\alpha)$ onto. Say $\alpha=\hat{\pi}(f)(\gamma)$, $\gamma<\beta$. So $\alpha\in\text{rng}(\hat{\pi})$. Contradiction!

Hence $\hat{\pi}(\alpha)\geq\kappa^+$. If $\hat{\pi}(\alpha)>\kappa^+$ then $\alpha>(\kappa^+)^{L[\bar U]}$. But $\kappa^+\in\text{rng}(\hat{\pi})$, $\alpha\leq\kappa^+$ so this is impossible. So $\hat{\pi}(\alpha)=\kappa^+$. Hence $\alpha=(\kappa^+)^{L[\bar U]}$. So $\bar U=U\cap J^U_\alpha$. But then there is $\beta\geq\alpha$ with $J^{\bar U}_{\beta+1}\cap P(\kappa)\neq J^{\bar U}_\beta\cap P(\kappa)$. Taking β least such $P(\kappa)\cap\Sigma_\omega(\underline{J}^U_\beta)\neq P(\kappa)\cap J^U_\beta$ and $J^{\bar U}_\beta=J^{\bar U}_\beta$. Hence $\bar U$ is not a normal measure in $L[\bar U]$. Contradiction!

Case 3 $\alpha\leq\kappa$.
$\kappa\in\text{rng}(\hat{\pi})$; say $\kappa=\hat{\pi}(\bar\kappa)$. $\bar\kappa=\kappa$ by minimality of κ. If $\overline{\overline{X}}\geq\overline{\overline{\kappa}}$ then $\kappa\subseteq X$ so $\pi|\kappa=\text{id}|\kappa$, so $\alpha=\bar\kappa=\kappa$. But $\pi(\bar\kappa)=\kappa$, so $\pi(\alpha)=\alpha$. Hence $\overline{\overline{X}}<\overline{\overline{\kappa}}$. So $\overline{\overline{Y}}<\overline{\overline{\kappa}}$. So $\bar\tau<\kappa$. But $\bar\kappa<\bar\tau$. Contradiction! □

Corollary 20.12. *If $0^\dagger$ does not exist then there is a Prikry sequence for U.*
Proof: τ is the least point where the covering claim fails over K. The conditions described at the start of this chapter must hold, since otherwise there would be a ρ-model whose critical point was

less than τ. Hence there is a Prikry sequence, by lemma 20.10. □

Call such a sequence C. Suppose the covering property for L[C] fails so that there is τ' with $X \subseteq \tau'$ uncountable but no $Y \in L[C]$ with $\overline{\overline{Y}} = \overline{\overline{X}}$ and $Y \supseteq X$. Take τ' least such. Suppose $0^{\dagger}$ does not exist.

<u>Lemma 20.13</u>. $\tau' = \kappa$.

Proof: If $\tau' < \kappa$ then κ is not the critical measure ordinal. If $\tau' > \kappa$ then just as in lemma 20.11 $0^{\dagger}$ exists. □

<u>Lemma 20.14</u>. *There is $X \subseteq \tau'$ with $otp(X) = \omega$ such that there is no $Y \in L[C]$ with $Y \supseteq X$ and $\overline{\overline{Y}} = \aleph_1$.*

Proof: Suppose not. With X <u>some</u> uncoverable set, suppose $C = \{\gamma_i : i < \omega\}$ and let $X_i = X \cap \gamma_i$. Take $Y_i \in K$ with $\overline{\overline{Y_i}} = \overline{\overline{X_i}} + \aleph_1$ and $X_i \subseteq Y_i$. Let $\delta = \overline{\overline{X}}^{+}$; then $\delta < \kappa$ ($\overline{\overline{X}} < \overline{\overline{\kappa}}$ so $\delta \leq \kappa$: but $cf(\kappa) = \omega$). There is $f \in K$ with $f : \kappa \leftrightarrow H_{\kappa}^{K}$, since $H_{\kappa}^{K} = K_{\kappa}$. Let $\nu_i = f^{-1}(Y_i)$. By hypothesis there is $Z \in L[C]$ with $Z \supseteq \{\nu_i : i \in \omega\}$, $Z \subseteq \kappa$ and $\overline{\overline{Z}} = \aleph_1$. Let $Y = \bigcup \{f(\nu) : \nu \in Z \wedge \overline{\overline{f(\nu)}}^{K} < \delta\}$. Then $Y \in L[C]$. If $\overline{\overline{f(\nu)}}^{K} < \delta$ then $\overline{\overline{f(\nu)}} \leq \overline{\overline{X}}$ so $\overline{\overline{Y}} = \overline{\overline{X}}$. Finally $X \subseteq \bigcup_{i \in \omega} X_i \subseteq \bigcup_{i \in \omega} Y_i \subseteq Y$. Contradiction! □

Now we derive a contradiction by showing

<u>Lemma 20.15</u>. *Suppose $X \subseteq \kappa$, $\overline{\overline{X}} = \omega$. Then there is $Y \in L[C]$ with $\overline{\overline{Y}} = \aleph_1$ and $X \subseteq Y$.*

Proof: This is trivial unless $\kappa = \sup X$. There is $Y' \in C'$ with $X \subseteq Y'$ and $C_{Y'} \setminus \gamma = C \setminus \gamma$ (some γ). Let $\pi_{Y'} : \underline{N}_{Y'}^{n(N_{Y'})} \to_{\Sigma_1} N_{Y'}^{n(N_{Y'})}$, $n = n(N_{Y'})$. Since $N_{Y'}^{n} = h_{N_{Y'}^{n}}(C_{N_{Y'}} \cup \omega\rho_{N_{Y'}}^{n+1} \cup p_{N_{Y'}}^{n+1})$ therefore $\pi_{Y'}{}'' N_{Y'}^{n} = h_{N_{Y'}^{n}}(C_{Y'} \cup \pi_{Y'}{}'' \omega\rho_{N_{Y'}}^{n+1} \cup \pi_{Y'}(p_{N_{Y'}}^{n+1})) \subseteq h_{N_{Y'}^{n}}(C_{Y'} \cup \sup \pi_{Y'}{}'' \omega\rho_{N_{Y'}}^{n+1} \cup \pi_{Y'}(p_{N_{Y'}}^{n+1}))$. Call this last set Y*. Since $N_{Y'}$ is a mouse, $N_{Y'} \in K$; and $C_{Y'} \setminus \gamma = C \setminus \gamma$, $C_{Y'} \cap \gamma$ is finite so $C_{Y'} \in L[C]$. Thus $Y^* \in L[C]$. $X \subseteq Y$ as $X \subseteq rng(\pi_{Y'})$. And $\overline{\overline{Y^*}}^{L[C]} < \kappa$. Let $f : Y^* \leftrightarrow \delta$, $f \in L[C]$, $\delta < \kappa$. Let $X' = f''X$. There is $Z \supseteq X'$, $Z \subseteq \delta$, $Z \in L[C]$ with $\overline{\overline{Z}} = \aleph_1$. Let $Y = f^{-1}{}''Z$. □

Thus we have a more general covering lemma:

<u>Lemma 20.16</u>. *Suppose $0^{\dagger}$ does not exist. Then either the covering property holds over K or it holds over the minimal ρ-model L[U] or there is a Prikry sequence for U such that it holds over L[C].*

A few final observations about Prikry sequences. Let L[U] be the minimal ρ-model and let $\langle \langle L[U_\alpha] \rangle_{\alpha \in On}, \langle \pi_{\alpha\beta} \rangle_{\alpha \leq \beta \in On} \rangle$ be its iterated ultrapower. Let $C = \{\kappa_n : n < \omega\}$. Then C is a Prikry sequence for $L[U_\omega]$. Let $M = L[C]$. Then $L[U_\omega] \subseteq M$, for $U_\omega = \{X \subseteq \kappa_\omega : \exists \gamma < \kappa_\omega (C \setminus \gamma \subseteq X)\}$.

Also $M\subseteq L[U_n]$ for $n\in\omega$: for, given $n\in\omega$, $C\setminus\kappa_n$ is definable in $L[U_{n+1}]$ as the critical points of the iterated ultrapower of V; but $C\cap\kappa_n$ is finite. So for no n is $L[U_n]\subseteq M$, otherwise $U_n\in L[U_{n+1}]$. Hence $M\models L[U_\omega]$ is the minimal ρ-model. Also $M\models 0^\dagger$ does not exist, since $0^\dagger\in$ $\in L[U]$ is impossible (since $0^\dagger$ exists ⇒ the critical measure ordinal is countable). But as $C\in M$ the covering lemma fails over $L[U_\omega]$ in M. This shows that the L[C] part of lemma 20.16 cannot be omitted.

Consideration of $L[U_\gamma]$ where $\gamma=\aleph_{\omega_1}$ shows that the minimal ρ-model was needed. For let $D=\{\kappa_{\aleph_\alpha}:\alpha<\omega_1\}$. Since there could be no Prikry sequence for U_γ $(cf(\kappa_\gamma)>\omega)$ dropping "minimal" in lemma 20.16 would give $D'\in L[U_\gamma]$ with $\bar{\bar{D}}'=\aleph_1$ and $D\subseteq D'$. But $\kappa_\gamma\geq\gamma$ so $L[U_\gamma]\models\bar{\bar{D}}'<\kappa_\gamma$. Yet κ_γ is regular in $L[U_\gamma]$.

Finally although lemma 19.26 could be proved in ZF, lemma 20.16 requires AC. Let $M=\bigcap_{i<\omega^2} L[U_i]$; then $M\models ZF$. Let $\Delta=\{g\in L[U]:g:\omega\to\omega^2$ is cofinal and strictly increasing$\}$. For $g\in\Delta$ let $C_g=\{\kappa_{g(n)}:n\in\omega\}$; let $G=$ $=\{C_g:g\in\Delta\}$. Let $G'=\{C:\exists C'\in\ \exists\lambda<\kappa_{\omega^2}\ C\setminus\lambda=C'\setminus\lambda\}$. Then $G'\in M$.

<u>Lemma 20.17</u>. *G' cannot be well-ordered in M.*

Proof: $L[U_{\omega^2}]\subseteq M$, of course, but $G'\cap L[U_{\omega^2}]=\phi$. Suppose G' could be well-ordered in M, by $<_{G'}$, say. Let $G''=\{C\in G':\forall C'\in G'(C'<_{G'}C\Rightarrow C'\setminus\gamma\neq$ $\neq C\setminus\gamma$ all $\gamma<\kappa_{\omega^2})\}$. There is Δ' in L[U] such that $\Delta'\subseteq\Delta$, $g,g'\in\Delta'\Rightarrow$ $\Rightarrow\forall\gamma<\omega^2\ rng(g)\cap\gamma\neq rng(g')\cap\gamma$ and for all $g\in\Delta$ there is $g'\in\Delta'$ such that $\exists\gamma<\omega^2\ rng(g)\cap\gamma=rng(g')\cap\gamma$.

Let $b:G''\to\omega$ be defined by: b(C) is the least $b<\omega$ such that if γ is the bth member of C then $C\setminus\gamma\subseteq\{\kappa_\alpha:\alpha<\omega^2\}$. b can be coded as a map of $(\omega^2)^\omega$ to ω, and hence is in M. Let $X=\bigcup\{C\setminus\gamma:C\in G''\wedge C\setminus\gamma\subseteq\{\kappa_\alpha:\alpha<\omega^2\}\}$. Then $X\in M$. There is $a:\Delta'\to\omega$ in L[U] such that $\forall g\in C\forall n\geq a(g)(\kappa_{g(n)}\in X)$. Let $K=\{g(m): g\in\Delta'\wedge m\geq a(g)\}$. So $K\in M$. Now suppose K is of order-type ω. Since K is cofinal in ω^2, letting $g:\omega\to K$ be increasing and onto $g\in\Delta$. So letting $g'(n)=g(n)+1$, $g'\in\Delta$, so $C_{g'}\subseteq X$, and hence $rng(g')\subseteq K$. Contradiction! But letting γ be the ωth member of K $K\cap\gamma\in M$. If γ is κ_δ then U_δ can be defined from $K\cap\gamma$. So $L[U_\delta]\subseteq M\subseteq L[U_{\delta+1}]$. Contradiction! □

<u>Lemma 20.18</u>. *Suppose the covering lemma holds over L[C] and for all $\gamma<\kappa$ $a\subseteq\gamma\Rightarrow$ $\Rightarrow a\in L[C]$. Then $P(\kappa)\subseteq L[C]$.*

Proof: (in M). Suppose $a\subseteq\kappa$. Let $C=\{\gamma_i:i\in\omega\}$ and let $a_i=a\cap\gamma_i$. Then each $a_i\in L[C]$. Let $L[C]\models\theta=\sup_{\beta<\kappa} 2^\beta$. Then if $f:\theta\leftrightarrow\bigcup_{\beta<\kappa} P(\beta)$, $f\in L[C]$ let $\nu_i=f^{-1}(a_i)$. So $\nu_i\in On$. Let $Z\supseteq\{\nu_i:i\in\omega\}$, $\bar{\bar{Z}}=\aleph_1$, $Z\in L[C]$. Then in L[C] $\bar{\bar{Z}}<\kappa$ since κ is inaccessible in M. Take $g\in L[C]$, $g:Z\leftrightarrow\bar{\bar{Z}}$. Let $A=\{g(\nu_i):i\in\omega\}$ so $A\in L[C]$. Hence $\{\nu_i:i\in\omega\}\in L[C]$ so $\{a_i:i\in\omega\}\in L[C]$, so $a\in L[C]$. □

But then if $M \not\models$ the covering lemma over L[C] then since $a \subseteq \gamma < \kappa$, $a \in M \Rightarrow$
$\Rightarrow a \in L[U_\alpha]$ (some $\kappa_\alpha > \gamma$) $\Rightarrow a \in L[U_{\omega_1}] \Rightarrow a \in L[C]$. So by lemma 20.18 $G' \subseteq L[C]$
and hence can be well-ordered in M, contradicting lemma 20.17.

Exercises

1. Show that the converse of lemma 20.16 holds.

2. Let L[C], κ be as usual. Show that if $\gamma \geq \kappa$ is an L[C]-cardinal then
$L[C] \models 2^\gamma = \gamma^+$ (actually $L[C] \models$ GCH: show this if you can!)

21: APPLICATIONS OF THE COVERING LEMMA

SINGULAR CARDINALS

Various properties of singular cardinals follow. Lemma 21.1 for example will say that if there is no ρ-model and $\tau\geq\omega_2$ is regular in K then $cf(\tau)=\overline{\overline{\tau}}$. Because no generic extension of the core model contains a ρ-model, this result tells us that it is impossible by forcing to change the cofinality of a cardinal unless we also change its cardinality (unless $\tau<\omega_2$), unless there is a ρ-model. Since we certainly do have a forcing construction to preserve the cardinality of a measurable cardinal while changing its cofinality to ω the converse holds too. The first results of the chapter are best read as saying that certain large cardinals are necessary for certain forcing constructions. Lemma 21.11, on the other hand, makes no mention of inner models and is of significance apart from forcing considerations.

<u>Lemma 21.1</u>. *If there is no ρ-model and $\tau\geq\omega_2$ is regular in K then $cf(\tau)=\overline{\overline{\tau}}$.*
Proof: Suppose $\delta=cf(\tau)$, $\delta<\overline{\overline{\tau}}$. Suppose $X\subseteq\tau$ is cofinal in τ and $\overline{\overline{X}}=\delta$. Let $Y\supseteq X$, $Y\subseteq\tau$, $Y\in K$, $\overline{\overline{Y}}=\overline{\overline{X}}+\aleph_1$. Since $K\models\tau$ regular, $K\models\overline{\overline{Y}}=\tau$. So $\overline{\overline{Y}}=\overline{\overline{\tau}}$. Hence $\overline{\overline{\tau}}=$ $=\delta+\aleph_1$. But $\delta,\aleph_1<\overline{\overline{\tau}}$. □

Now if we had been able to prove the covering lemma without the uncountability restriction then we could prove lemma 21.1 without the restriction $\tau\geq\omega_2$. Model 1 in appendix I shows the consistency of

(i) $0^{\#}$ does not exist;
(ii) $\underline{cf}(\omega_2^L)=\omega$;
(iii) $\omega_2^L=\aleph_1$.

Hence the uncountability assumption is essential.

<u>Corollary 21.2</u>. *If there is no ρ-model and β is a singular cardinal then β is singular in K.*
Proof: $\beta>\omega_2$. If β were regular in K then $cf(\beta)=\overline{\overline{\beta}}=\beta$, that is, β is regular. □

It is, of course, essential that β be a cardinal.

Lemma 21.3. *If there is no ρ-model and β is a singular cardinal then $(\beta^+)^K=\beta^+$.*
Proof: Otherwise $(\beta^+)^K<\beta^+$. Since $(\beta^+)^K$ is regular in K $cf((\beta^+)^K)=$ $=(\beta^+)^K$ by lemma 21.1. But $(\beta^+)^K<\beta^+ \Rightarrow (\beta^+)^K=\beta$. Thus $cf((\beta^+)^K)=\beta$: but β is singular. □

Corollary 21.4. *If there is no ρ-model and β is a singular cardinal then $\square_\beta$.*
Proof: $K\models\square_\beta$ by lemma 17.25. But then as $(\beta^+)^K=\beta^+$ $\square_\beta$. □

Of course lemma 21.3 is another restriction on forcing; you cannot generically collapse the successor of a singular cardinal without collapsing the cardinal (provided there is no ρ-model). The reader is left to formulate modifications of the above on the assumption that $0^\dagger$ does not exist.

Now we turn to the singular cardinals hypothesis. Recall that this states: if β is a singular cardinal and $2^{cf(\beta)}<\beta$ then $\beta^{cf(\beta)}=$ $=\beta^+$. This is abbreviated SCH. The background to this hypothesis has been explained in the introduction.

Before showing that SCH follows from the non-existence of $0^\dagger$ we shall show that it completely determines cardinal exponentiation.

Lemma 21.5. *Suppose β is singular. Then $2^\beta=(2^{<\beta})^{cf(\beta)}$ ($2^{<\beta}$ denotes $\sup_{\gamma<\beta} 2^\gamma$).*

Proof: Let $\gamma=cf(\beta)$, $\beta=\sup_{i<\gamma}\beta_i$, $\beta_i<\beta$ for all $i<\gamma$. Then

$$2^\beta=2^{(\sum_{i<\gamma}\beta_i)} \tag{21.1}$$
$$=\prod_{i<\gamma}2^{\beta_i}$$
$$\leq (2^{<\beta})^\gamma$$
$$\leq 2^{\beta\cdot\gamma}$$
$$=2^\beta. \quad \square$$

Lemma 21.6. *If SCH holds then*
$$2^\beta=2^{<\beta} \text{ if for some } \gamma<\beta\ 2^\gamma=2^{<\beta}$$
$$=(2^{<\beta})^+ \text{ otherwise.}$$
Proof: If $2^\gamma=2^{<\beta}$ then $2^\beta=2^{\gamma\cdot cf(\beta)}$ by lemma 21.5; but $\gamma\leq\gamma\cdot cf(\beta)<\beta$ so $2^\beta=2^{<\beta}$. Otherwise there is a sequence $\{\gamma_\alpha:\alpha<cf(\beta)\}$ cofinal in β with $\alpha<\alpha' \Rightarrow 2^{\gamma_\alpha}<2^{\gamma_{\alpha'}}$. So $cf(2^{<\beta})=cf(\beta)$. And

$$2^\beta=(2^{<\beta})^{cf(\beta)} \tag{21.2}$$
$$=(2^{<\beta})^{cf(2^{<\beta})}.$$

But $2^{cf(\beta)}<2^{<\beta}$ so by SCH $(2^{<\beta})^{cf(2^{<\beta})}=(2^{<\beta})^+$. □

Hence the value of 2^β for singular β is determined by 2^α for regular α. In particular, SCH implies that if GCH holds up to a singular cardinal then it holds at that cardinal. For example $(\forall n\in\omega)2^{\aleph_n}=\aleph_{n+1}\Rightarrow$

$\Rightarrow 2^{\aleph_\omega}=\aleph_{\omega+1}$. Of course, Silver's result quoted in the introduction gives this result outright for singular cardinals of uncountable cofinality.

Lemma 21.7. *If κ,λ are infinite cardinals and $2^\lambda \geq \kappa$ then $\kappa^\lambda=2^\lambda$.*
Proof: $\kappa^\lambda \leq (2^\lambda)^\lambda = 2^\lambda \leq \kappa^\lambda$. □

Lemma 21.8. *Suppose SCH holds. Suppose $2^\lambda<\kappa$; then $\kappa^\lambda=\kappa$ if $\lambda<cf(\kappa)$, $\kappa^\lambda=\kappa^+$ otherwise (κ,λ infinite).*
Proof: By induction on κ. Suppose $\kappa=\nu^+$. Then assume $\kappa>2^\lambda$; if $\nu>2^\lambda$ then $\nu^\lambda \leq \nu^+=\kappa$ (by induction hypothesis) or if $\nu=2^\lambda$ then $\nu^\lambda=2^\lambda=\nu<\kappa$. Either way $\nu^\lambda\leq\kappa$. Now

$$(\nu^+)^\lambda=\overline{\overline{\{f:f:\lambda\to\nu^+\}}}. \tag{21.3}$$

And $f:\lambda\to\nu^+$ implies $f:\lambda\to\gamma$, some $\gamma<\nu^+$. So

$$\begin{aligned}(\nu^+)^\lambda &= \overline{\overline{\bigcup_{\gamma<\nu^+}\{f:f:\lambda\to\gamma\}}} \\ &\leq \nu^+.\nu^\lambda \\ &\leq (\nu^+)^\lambda.\end{aligned} \tag{21.4}$$

So $\kappa^\lambda=\kappa.\nu^\lambda=\kappa$.

Now suppose κ is a limit cardinal. If $\lambda<cf(\kappa)$ then $\{f:f:\lambda\to\kappa\}= = \bigcup_{\gamma<\kappa}\{f:f:\lambda\to\gamma\}$, so $\kappa^\lambda=\sup_{\gamma<\kappa}\gamma^\lambda$. If $2^\lambda<\kappa$ and $\gamma>2^\lambda$ then $\gamma^\lambda\leq\gamma^+$ by induction hypothesis, so $\kappa^\lambda=\kappa$.

Suppose, then, that $\lambda\geq cf(\kappa)$. Let $\gamma=cf(\kappa)$, $\kappa=\sup_{i<\gamma}\kappa_i$ with each $\kappa_i<\kappa$. So

$$\begin{aligned}\kappa^\lambda &= (\sum_{i<\gamma}\kappa_i)^\lambda \leq (\prod_{i<\gamma}\kappa_i)^\lambda \\ &= \prod_{i<\gamma}\kappa_i^\lambda \\ &\leq (\sup_{\alpha<\kappa}\alpha^\lambda)^\gamma.\end{aligned} \tag{21.5}$$

But $\sup_{\alpha<\kappa}\alpha^\lambda=\kappa$, so $\kappa^\lambda=\kappa^{cf(\kappa)}$. $2^{cf(\kappa)}\leq 2^\lambda<\kappa$ so $\kappa^{cf(\kappa)}=\kappa^+$ by SCH. □

Lemmas 21.7 and 21.8 tell us that SCH completely determines cardinal exponentiation, given the values of 2^α for regular α. We turn to the proof of SCH.

Lemma 21.9. *Suppose β is a singular cardinal and M is an inner model with the property that for all $X\subseteq\beta$ there is $Y\subseteq\beta$ with $Y\in M$, $Y\supseteq X$ and $\overline{\overline{Y}}=\overline{\overline{X}}+\aleph_1$. Then every function $f:\gamma\to\beta$ ($\gamma<\beta$) is of the form $g'g$ where $g:\gamma\to\gamma^+$ and $g'\in M$.*
Proof: Let $Y\supseteq rng(f)$, $\overline{\overline{Y}}=\overline{\overline{rng(f)}}+\aleph_1$, $Y\in M$. Then $M\models\overline{\overline{Y}}\leq\gamma^+$. Let $g'\in M, g':\gamma^+\to Y$ onto. Let $g=g'^{-1}f$, where g'^{-1} is some inverse for g'. □

Lemma 21.10. *If there is no ρ-model then SCH holds.*

Proof: Take β singular with $\gamma=cf(\beta)$ and $2^{\gamma}<\beta$. By lemma 21.9

$$\beta^{cf(\beta)}\leq(\beta^{\delta})^{K}.(\gamma^{+})^{\gamma} \tag{21.6}$$

where $\delta=cf(\beta)^{+}$. $K\models\beta^{\delta}\leq 2^{\beta.\delta}=2^{\beta}=\beta^{+}$. And

$$(\gamma^{+})^{\gamma}\leq(2^{\gamma})^{\gamma}=2^{\gamma}<\beta. \tag{21.7}$$

So $\beta^{cf(\beta)}=\beta^{+}$. □

Lemma 21.11. *If $0^{\dagger}$ does not exist then SCH holds.*
Proof: Let L[U] be the minimal ρ-model. Suppose β is singular, $\gamma=cf(\beta)$ and $2^{\gamma}<\beta$. Let κ be the critical measure ordinal. If $\beta<\kappa$ then the hypothesis of lemma 21.9 holds for M=K and the proof is as in lemma 21.10. Similarly if there is no Prikry sequence for U and β is arbitrary, since $L[U]\models GCH$. Suppose $\beta\geq\kappa$ and C is a Prikry sequence on U with the hypothesis of lemma 21.9 holding over L[C]. So (21.6) can be proved with L[C] in place of K. But

$$L[C]\models\beta^{\delta}\leq 2^{\beta.\delta}=2^{\beta}=\beta^{+} \tag{21.8}$$

by exercise 2 of chapter 20; so $\beta^{cf(\beta)}=\beta^{+}$. □

MORE ABOUT #

A brief result motivates the following. First note that for any real a the covering lemma holds over L[a] unless there is a non-trivial $j:L[a]\to_e L[a]$. Exactly the usual proof works. Now suppose $0^{\#}$ exists but $0^{\#}\notin L[a]$. Then $L[a]\models$the covering lemma over L. Suppose there were no non-trivial $j:L[a]\to_e L[a]$. So the covering lemma holds over L[a]. Given $X\subseteq On$ uncountable there is $Y\in L[a]$ with $Y\supseteq X$ and $\overline{\overline{Y}}=\overline{\overline{X}}$. And there is $Z\in L$ with $Z\supseteq Y$ and $\overline{\overline{Z}}=\overline{\overline{Y}}$. So the covering lemma holds over L. But $0^{\#}$ exists, so this is impossible. So there is a non-trivial $j:L[a]\to_e L[a]$. This can be coded by a real, just as for L; we call this real $a^{\#}$. So: if $0^{\#}$ exists and $0^{\#}\notin L[a]$ then $a^{\#}$ exists. We can define a "mouse over a" and show generally that if there is a mouse not in K^{M} and a is a real in M then there is a mouse over a not in K^{M}. It would be nice directly to extract indiscernibles for a from those for K^{M}, but this we cannot do. We can show that there is a neat relationship between these two sets of indiscernibles. Unexpectedly this gives a result in descriptive set theory.

We defined $M^{\#}$ only for $M\models V=K$; but we shall let $M^{\#}$ denote $(K^{M})^{\#}$ for arbitrary inner models M.

We want also to define $a^{\#}$ for arbitrary $a\subseteq\omega$; this will not always $\in L[a]^{\#}$. The following are equivalent:

(i) there is a non-trivial $j:L[a]\to_e L[a]$;

(ii) there is a class C with the properties of lemma 12.9 for L[a];

(iii) there is an a-mouse

where by an a-mouse is meant an n-iterable premouse $\underline{J}_\alpha^{a,U}$ with critical point κ and $\omega\rho_N^{n+1} \leq \kappa$ ($\underline{N}=\underline{J}_\alpha^{a,U}$, some n).

Definition 21.12. *If any of these conditions is satisfied then* $a^\#$ *denotes* $\{n : \underline{J}^a_{\aleph_\omega} : \phi_n(\aleph_1 \ldots \aleph_{m(n)})\}$.

Lemma 21.13. *Suppose* $a^\#$ *does not exist; then the covering lemma holds over* $L[a]$.

Proof: Just as for L. □

Indeed, letting $K[a]=\bigcup\{M : \underline{M}$ is an a-mouse$\}\cup L[a]$, we get an inner model with a covering property conditional on there being no models of the form $L[a,U]$ where U is a normal measure in $L[a,U]$.

Lemma 21.14. *Suppose* $K^{L[a]} \neq K$. *Then* $a^\#$ *exists.*

Proof: Let $M=L[a]^\#$. Suppose $a^\#$ does not exist. Then the covering lemma holds over $L[a]$. Now $L[a] \models$ there are no ρ-models, because if $L[U] \subseteq L[a]$ were a ρ-model then $K \subseteq L[U] \subseteq L[a]$, so $K^{L[a]}=K \cap L[a]=K$. Hence $L[a] \models$ the covering lemma over $K^{L[a]}$. Thus the covering lemma really does hold over $K^{L[a]}$. Let $\langle\langle \underline{M}_\alpha\rangle_{\alpha \in On}, \langle \pi_{\alpha\beta}\rangle_{\alpha \leq \beta \in On}\rangle$ be the mouse iteration of $\underline{M}$ and let κ_α be the critical point of $\underline{M}_\alpha$. Then κ_α is regular in $K^{L[a]}$ for all α; but $\kappa_{\aleph_\omega}=\aleph_\omega$ is a singular cardinal, contradicting corollary 21.2. □

Corollary 21.15. *If* $0^\#$ *exists, a is a real but* $0^\# \notin L[a]$ *then* $a^\#$ *exists.*

Fix $a \subseteq \omega$, then, such that $K^{L[a]} \neq K$ and let $\underline{M}$ be $L[a]^\#$ with mouse iteration $\langle\langle \underline{M}_\alpha\rangle_{\alpha \in On}, \langle \pi_{\alpha\beta}\rangle_{\alpha \leq \beta \in On}\rangle$ and κ_α the critical point of $\underline{M}_\alpha$. Recall that $K^{L[a]} = \bigcup_{\alpha \in On} H^{M_\alpha}_{\kappa_\alpha}$. Let $\underline{N}=\underline{J}_\alpha^{aV}$ be an a-mouse with critical point τ, $\omega\alpha=\tau+\omega$. As in chapter 12, exercise 4, we may show that the weak iterated ultrapower of $L[a]$ by V is of the form $\langle L[a], \langle \pi'_{\alpha\beta}\rangle_{\alpha \leq \beta \in On}\rangle$. Let $\tau_\alpha = \pi'_{0\alpha}(\tau)$, $C'=\{\tau_\alpha : \alpha \in On\}$, $K'=K^{L[a]}$.

Lemma 21.16. $(\tau_\alpha^+)^{K'} = (\tau_\alpha^+)^{L[a]}$.

Proof: Let $\delta=(\tau_\alpha^+)^{L[a]}$, $\delta'=(\tau_\alpha^+)^{K'}$; suppose $\delta'<\delta$. Let $V_\alpha=\{X \in L[a] \cap P(\tau_\alpha) : \tau_\alpha \in \pi'_{\alpha\alpha+1}(X)\}$; so $V_0=V$ and $\langle J^a_\delta, V_\alpha\rangle$ is amenable, and so $V_\alpha \cap K' = V_\alpha \cap K'_{\delta'} \in L[a]$.

Claim $\langle K'_{\delta'}, V_\alpha \cap K'\rangle$ is amenable.

Proof: Suppose $\beta<\delta'$ and let $X \subseteq \tau_\alpha^2$ be such that $\{\{\gamma : \langle\gamma,\delta\rangle \in X\} : \zeta<\tau_\alpha\} = P(\tau_\alpha) \cap K'_\beta$, $X \in K'$. Now since $X\in K'$ $\pi'_{\alpha\alpha+1}(X)\in K'$; and letting $X_\zeta=\{\gamma:\langle\gamma,\zeta\rangle\in X\}$, $K'_\beta\cap V_\alpha=$ $=\{X_\zeta:\zeta<\tau_\alpha \wedge \langle\tau_\alpha,\zeta\rangle\in\pi'_{\alpha\alpha+1}(X)\}\in K'$. □(Claim)

Now the covering lemma holds over K' in L[a] so in L[a] $cf(\delta')=\overline{\overline{\delta'}}>\omega$. It follows that $V_\alpha\cap K'$ is countably complete in L[a]. Hence L[a] has an inner model with a measurable cardinal by lemma 16.11. So K=K'. Contradiction! □

Let $C=\{\kappa_\alpha:\alpha\in On\}$.

Let $\kappa=\kappa_0$ and set $W=P(K'_\kappa)\cap K'$. For $X\in W$ let $\tilde{X}$ denote $\bigcup_{i\in On}\pi_{0i}(X)$.

Lemma 21.17. $K'=\bigcup_{X\in W}\tilde{X}$.

Proof: Suppose $a\in K'_{\kappa_\alpha}$. Then there is $x\in rng(\pi_{0\alpha})$ such that $a\in x$, since $\pi_{0\alpha}:M\to_{\Sigma_0}M_\alpha$ cofinally. But then if $y=x\cap K'_\kappa$ then $a\in\pi_{0\alpha}(y)$ and $y\in W$. □

Lemma 21.18. $C=\bigcap_{X\in U}\tilde{X}$.

Proof: Suppose $X\in U$. Then for any α $\pi_{0\alpha}(X)\in U_\alpha$, so $\kappa_\alpha\in\pi_{0\alpha+1}(X)$, and so $\kappa_\alpha\in\tilde{X}$. So $C\subseteq\bigcap_{X\in U}\tilde{X}$.

Conversely suppose $\delta\notin C$. Let α be least with $\kappa_\alpha>\delta$; $\alpha=\beta+1$, say. Then there is $f\in W$ such that $\delta=\pi_{0\alpha}(f)(\kappa_{\beta_1}\ldots\kappa_{\beta_n})$, say, with $\beta_1<\ldots$ $\ldots<\beta_n\leq\beta$. Let $X=\{\xi:\forall\vec{\beta}<\xi\,\xi\neq f(\vec{\beta})\}$. If $X\notin U$ then $\{\xi:\exists\vec{\beta}<\xi\,\xi=f(\vec{\beta})\}\in U$ so for all α $\exists\vec{\beta}<\kappa_\alpha$ $\kappa_\alpha=\pi_{0\alpha}(f)(\vec{\beta})$. But this would imply $\kappa_\alpha\in rng(\pi_{\alpha\alpha+1})$ which is impossible. So $X\in U$: but $\delta\notin\pi_{0\alpha}(X)$. If $\beta<\alpha$ then $\delta\notin\pi_{0\beta}(X)$ as $\delta>\kappa_\beta$. But if $\beta>\alpha$ then $\delta\notin\pi_{0\beta}(X)$ as $\delta=\pi_{0\alpha}(f)(\kappa_{\beta_1}\ldots\kappa_{\beta_n})\Rightarrow\delta=\pi_{0\beta}(f)(\kappa_{\beta_1}\ldots\kappa_{\beta_n})$ So $\delta\notin X$. But $X\in U$ so $\delta\notin\bigcap_{X\in U}\tilde{X}$. □

Lemma 21.19. *There is i_0 such that whenever $i_0\leq i\leq j$ then $\pi'_{ij}(\tilde{X}\cap V_\nu)=\tilde{X}\cap V_{\pi_{ij}(\nu)}$ for all ν and all $X\in W$.*

Proof: Let $\Gamma=\{\eta:\eta=\tau_\eta \wedge cf(\eta)>\kappa\}$. Let $\eta\in\Gamma$.

Claim For all $x\in L[a]$ there is $\xi<\eta$ such that $\xi\leq i\leq j<\eta\Rightarrow\pi'_{ij}(x)=x$.

Proof: First of all, whenever α is a limit ordinal there is $\bar{\alpha}<\alpha$ such that $x\in rng(\pi^{\perp}_{\bar{\alpha}\alpha})$. So there is a stationary subset Δ of η such that for some $\bar{\alpha}$ $\alpha\in\Delta\Rightarrow x\in rng(\pi^{\perp}_{\bar{\alpha}\alpha})$. Let < be the canonical order of L[a]; then, letting $x_\alpha=\pi^{\perp\,-1}_{\bar{\alpha}\alpha}(x)$, if $\alpha<\alpha'$ $\alpha,\alpha'\in\Delta$ then since $\pi'_{\alpha\alpha'}(x)\geq x$ so $\pi'_{\alpha\alpha'}(\pi^{\perp}_{\bar{\alpha}\alpha}(x_\alpha))\geq\pi^{\perp}_{\bar{\alpha}\alpha'}(x_{\alpha'})$, i.e. $x_\alpha\geq x_{\alpha'}$. But since < is well-founded there is ξ such that $\alpha\geq\xi\Rightarrow x_\alpha=x_\xi$.

Now suppose $\xi\leq i\leq j<\eta$. Suppose $j<k<\eta$, $k\in\Delta$. Then

$$x\leq\pi'_{ij}(x)\leq\pi'_{ij}(\pi'_{\xi i}(x)) \quad (21.9)$$

$=\pi'_{\xi j}(x)$
$\leq\pi'_{jk}\pi'_{\xi j}(x)$
$=\pi'_{\xi k}(x)$
$=\pi'_{\xi k}\pi^{\perp}_{\alpha\xi}(x_\xi)$
$=\pi^{\perp}_{\alpha k}(x_\xi)$
$=\pi^{\perp}_{\alpha k}(x_k)$
$=x.$

So $\pi'_{ij}(x)=x$. □(Claim)

Given $X\in W$ let ξ_X be such that $\xi_X\leq i\leq j<\eta \Rightarrow \pi'_{ij}(\tilde{X}\cap V_\eta)=\tilde{X}\cap V_\eta$. Then since $\overline{\overline{W}}=\kappa<\mathrm{cf}(\eta), \sup_{X\in W}\xi_X<\eta$; call this ξ_η. Now Γ is stationary and the map $\xi:\Gamma\to On$ is regressive so there is $i_0, \Gamma'\subseteq\Gamma$ stationary such that $\xi_\eta=i_0$ for all $\eta\in\Gamma'$.

Suppose $i_0\leq i\leq j$, $X\in W, \nu<\infty$. Pick $\eta\in\Gamma'$ with $j,\nu<\eta$. Then

$$\begin{aligned}\pi'_{ij}(\tilde{X}\cap V_\nu)&=\pi'_{ij}((\tilde{X}\cap V_\eta)\cap V_\nu)\\&=(\tilde{X}\cap V_\eta)\cap V_{\pi'_{ij}(\nu)}\\&=\tilde{X}\cap V_{\pi'_{ij}(\nu)}.\end{aligned} \tag{21.10}$$

□

Lemma 21.20. *$\gamma\in C \leftrightarrow \pi'_{ij}(\gamma)\in C$ $(i_0\leq i\leq j)$.*

Proof: Suppose $\gamma\in C$. Then for all $X\in U$ $\gamma\in\tilde{X}$. Pick $\nu>\gamma$: then $\gamma\in\tilde{X}\cap V_\nu$ so by lemma 21.19 $\pi'_{ij}(\gamma)\in\tilde{X}\cap V_{\pi'_{ij}(\nu)}$, so $\pi'_{ij}(\gamma)\in\tilde{X}$. Hence $\pi'_{ij}(\gamma)\in\bigcap_{X\in U}\tilde{X}$. The reverse argument is similar. □

Corollary 21.21. *$C'\setminus\tau_{i_0}\subseteq C$.*

Proof: Given $i>i_0$ pick η regular with $\eta>i$. Then $\tau_\eta=\kappa_\eta=\eta$. $\pi'_{i\eta}(\tau_i)=\tau_\eta=$ $=\kappa_\eta\in C$, so by lemma 21.20 $\tau_i\in C$. □

This is not very surprising.

If γ is an ordinal then for some $f\in W$ $\gamma=f(\beta_1\ldots\beta_n)$ with $\beta_1\ldots\beta_n\in$ $\in C$. So by lemma 21.19 given $i_0\leq i\leq j$

$$\begin{aligned}\pi'_{ij}(\gamma)&=\pi'_{ij}(\tilde{f})(\pi'_{ij}(\vec{\beta}))\\&=\tilde{f}(\pi'_{ij}(\vec{\beta})).\end{aligned} \tag{21.11}$$

So π'_{ij} is determined by its action on C. The next lemma is much more surprising.

Lemma 21.22. *Suppose $\tau_{i_0}=\kappa_\alpha$. There is β such that*

(i) $\tau_{i_0+j}=\kappa_{\alpha+\beta j}$

(ii) $\pi'_{i_0+i,i_0+j}(\kappa_{\alpha+\beta i+\nu})=\kappa_{\alpha+\beta j+\nu}$.

Proof: Let $C_i=\{\gamma\in C:\tau_i\leq\gamma<\tau_{i+1}\}$. It suffices to show that $\pi'_{ij}"C_{i+h}=C_{j+h}$ $(i_0\leq i<j)$ since then, letting β be the order-type of C_{i_0}, $\pi'_{i_0j}"C_{i_0}=C_j$ so the order-type of each C_j is β $(j\geq i_0)$ giving (i). Furthermore (ii) follows as π'_{ij} is order-preserving.

If $\pi'_{ij}"C_i=C_j$ $(i_0\leq i\leq j)$ then $\pi'_{ij}"C_{i+h}=\pi'_{ij}\pi'_{ii+h}"C_i=\pi'_{ij+h}"C_i$ (as in chapter 12) $=C_{j+h}$. So it suffices to show $\pi'_{ij}"C_i=C_j$ $(i_0\leq i\leq j)$. This follows by an easy induction from $\pi'_{ii+1}"C_i=C_{i+1}$.

Suppose this failed, then, and $\gamma\in C_{i+1}\smallsetminus\mathrm{rng}(\pi'_{ii+1})$. Then there is $f:\tau_i\to\tau_{i+1}$, $f\in L[a]$ such that $\gamma=\pi'_{ii+1}(f)(\tau_i)$. We may assume f monotone since it is not constant on a set in V_i.

Let $f_n=\pi'_{ii+n}(f)$ $(n\leq\omega)$. $f_1(\tau_i)=\gamma\in C\cap\tau_{i+2}=\bigcap_{X\in U}\tilde{X}\cap\tau_{i+2}$ so

$$\{\nu<\tau_i:f(\nu)\in\tilde{X}\cap\tau_{i+1}\}\in V_i \quad \text{(all } X\in U) \tag{21.12}$$

Hence

$$\{\nu<\tau_{i+n}:f_n(\nu)\in\tilde{X}\cap\tau_{i+n+1}\}\in V_{i+n} \tag{21.13}$$

so $f_{n+1}(\tau_{i+n})\in\bigcap_{X\in U}\tilde{X}\cap\tau_{i+n+1}\subseteq C$. Let $\gamma_n=f_{n+1}(\tau_{i+n})$. Let $\delta_n=f_\omega(\tau_{i+n})$ (so $\delta_n=\pi'_{i+n+1,i+\omega}(\gamma_n)$). Let $\delta=\sup_{n\in\omega}\delta_n$. Each $\gamma_n\in C$, so each $\delta_n\in C$, so $\delta\in C$. Since $\sup_{n\in\omega}\tau_{i+n}=\tau_{i+\omega}$, $\delta=\sup f_\omega"\tau_{i+\omega}$. $\tau_{i+\omega}$ is regular in L[a] so $L[a]\models cf(\delta)=\tau_{i+\omega}$. But the covering lemma over K' holds in L[a]; and $\delta\in C$, so δ is inaccessible in K'. So $L[a]\models\bar{\bar{\delta}}=cf(\delta)=\tau_{i+\omega}$. But since $\gamma>\tau_i$, $\delta>\tau_{i+\omega}$. So $\delta>(\tau^+_{i+\omega})^{K'}$. But by lemma 21.16 $\delta>(\tau^+_{i+\omega})^{L[a]}$. Contradiction! □

What we have shown is that after a certain point the indiscernibles for L[a] arise from those for K' by simply taking every βth. It would be nice if we could find some significance for the ordinal β as a measure of complexity of a. It would also be nice to give an example in which $\beta>1$, but this involves a deep forcing construction. For example, if a is generic over L and $0^\#$ exists then $\beta=1$.

In appendix II this result is applied in descriptive set theory.

Exercises

1. Suppose all uncountable cardinals are singular. Show (in ZF) that there is a ρ-model.

2. Show that if there is a ρ-model then there is γ such that for all singular cardinals β with $\beta>\gamma$ $(\beta^+)^K<\beta^+$.

3. Formulate a covering lemma on the assumption that $0^{\dagger\dagger}$ (the smallest triple mouse) does not exist.

4. Suppose $a^{\#}$ exists, $a \subseteq \omega$, $K = K^{L[a]}$ but $L[a] \models$ there is no ρ-model. If we let $\underline{M}$ be the minimal ρ-model, show that the results lemma 21.16-lemma 21.22 still hold.

5. GCH $\rightarrow$ SCH

PART SIX: LARGER CORE MODELS

In this last part we briefly review what is known of core models beyond K. Much, it turns out, is known about models and mice and very little about covering lemmas and SCH.

In the final chapter we discuss the present state of knowledge about SCH. It was once thought that large cardinal axioms might settle the GCH. This has turned out not to be the case: powers of regular cardinals do as they please, irrespective of large cardinals. On the other hand it looks possible that large cardinals might partially settle SCH (partially because we only get relative consistency results from the cardinals). The division of large cardinals into two camps may mark a fundamental change in the nature of cardinality axioms; or it may say no more than that Prikry sequences and fine structure do not go together.

Exercise 3 of chapter 21 challenged the reader to formulate the covering lemma on the assumption that $0^{\dagger\dagger}$ does not exist. This could be generalised still further to any finite number of †. In particular failure of SCH gives an inner model with n measurable cardinals for any n.

Let us forget about SCH until chapter 24 and concentrate on increasing the number of measurable cardinals. How far can we go? A proper class of measurables might seem to be the limit; but actually that is a relatively weak assumption (relative, that is, to the cardinals some people use. For all we know, of course, all these cardinals could be moonshine and the existence of a ρ-model equivalent to that most powerful of large cardinal axioms, $0 \neq 0$).

Suppose that at some point κ there were two different normal measures. Indeed, suppose that U,V are normal measures on κ and that if $j:V\to_V M$ then $U\in M$. This implies $U\neq V$ by chapter 5 exercise 1. $M\models\kappa$ measurable, so $\{\gamma<\kappa:\gamma \text{ measurable}\}\in V$. Thus κ is a limit of measurable cardinals. Indeed by iterating V through all the ordinals we get an inner model with a proper class of measurables. In terms of consistency two such measures at a point involve a stronger assumption than a proper class of measurables.

Definition 22.1. *Suppose U,V are normal measures on* κ . *Let* $j:V\to_V M$. *Then* $U<V$ *if and only if* $U\in M$.

A still stronger assumption would be the existence of three measures U,V,W with $U<V<W$. And so on.

Lemma 22.2. *If* $U<V$, $j_1:V\to_U M_1$, $j_2:V\to_V M_2$ *then* $j_1(\kappa)<j_2(\kappa)$.

Proof: Let $j^*:M_2\to_U M^*$. Define $\sigma:j_1(\kappa)\to j^*(\kappa)$ by

$$\sigma(j_1(f)(\kappa))=j^*(f)(\kappa) \qquad (f:\kappa\to\kappa). \tag{22.1}$$

Then

$$j_1(f)(\kappa)=j_1(f')(\kappa) \Leftrightarrow \{\xi:f(\xi)=f'(\xi)\}\in U \tag{22.2}$$
$$\Leftrightarrow j^*(f)(\kappa)=j^*(f')(\kappa).$$

So σ is well-defined. Similarly σ is one-one and order-preserving. Finally $\kappa^\kappa\subseteq M_2$ so σ is surjective. So $j_1(\kappa)=j^*(\kappa)$.

By lemma 5.18 $M_2\models j^*(\kappa)<(2^\kappa)^+$, and $M_2\models j_2(\kappa)\geq(2^\kappa)^+$. So $j_2(\kappa)>$

$>j_1(\kappa)$. □

So $<$ is irreflexive and well-founded.

Lemma 22.3. *$<$ is transitive.*

Proof: Suppose $U<V<W$. Let $j:V\to_V M$. Suppose $U=j(\bar{U})(\kappa)$. Then

$$X\in U \Leftrightarrow X\in j(\bar{U})(\kappa) \qquad (X\subseteq\kappa) \tag{22.3}$$
$$\Leftrightarrow j(X)\cap\kappa\in j(\bar{U})(\kappa)$$
$$\Leftrightarrow \{\xi:X\cap\xi\in\bar{U}(\xi)\}\in V.$$

Let $j':V\to_W M'$. Then $V\in M'$. Without loss of generality $\bar{U}:\kappa\to H_\kappa$ so $\bar{U}\in M'$, hence $U\in M'$. That is, $U<W$. □

In general $<$ is not connected, but we shall restrict ourselves to models where it is. There is another difficulty: if $U<V$ and we form the model $L[U,V]$ then it is not necessarily the case that in $L[U,V]$ $U\cap L[U,V]<V\cap L[U,V]$. Indeed this never happens: we need, roughly speaking, to construct the model $L[\bar{U},V]$ with $\bar{U}$ as in the proof of lemma 22.3. This is the point of the following definition.

Definition 22.4. *$\mathcal{U}$ is a coherent sequence of measures provided:*

(i) $dom(\mathcal{U})=\Delta$, where $\Delta\subseteq On^2$ and $\langle\alpha,\beta\rangle\in\Delta \wedge \beta'<\beta \Rightarrow \langle\alpha,\beta'\rangle\in\Delta$ (call Δ $\Delta^{\mathcal{U}}$).

(ii) $\mathcal{U}(\alpha,\beta)$ is a normal measure on α (all $\langle\alpha,\beta\rangle\in\Delta$).

(iii) Suppose $j:V\to_{\mathcal{U}(\alpha,\beta)}M$. Then for all β', letting $\bar{\mathcal{U}}=j(\mathcal{U})$, $\langle\alpha,\beta'\rangle\in\Delta^{\bar{\mathcal{U}}} \Leftrightarrow$

$\Leftrightarrow \beta'<\beta$; and for $\beta'<\beta$ $\bar{\mathcal{U}}(\alpha,\beta')=\mathcal{U}(\alpha,\beta')$.

So $\langle\alpha,\beta'\rangle,\langle\alpha,\beta\rangle\in\Delta \wedge \beta'<\beta \Rightarrow \mathcal{U}(\alpha,\beta')<\mathcal{U}(\alpha,\beta)$.

We shall not linger over these models for we are in search of still more extensive generalisations. It can be shown that every normal measure in $L[\mathcal{U}]$ is of the form $\mathcal{U}(\alpha,\beta)$ (where $\mathcal{U}$ is coherent in $L[\mathcal{U}]$), so $<$ is indeed a well-order, and $L[\mathcal{U}]\models$GCH.

Definition 22.5. *$o^{\mathcal{U}}(\alpha)=\{\beta:\langle\alpha,\beta\rangle\in\Delta^{\mathcal{U}}\}$.*

Lemma 22.6. *Suppose $\mathcal{U}$ is coherent in $L[\mathcal{U}]$. Then $L[\mathcal{U}]\models o^{\mathcal{U}}(\alpha)\leq\alpha^{++}$.*

Proof: Suppose $\beta<o^{\mathcal{U}}(\alpha)$. Let $j:L[\mathcal{U}]\to_{\mathcal{U}(\alpha,\beta)}L[\bar{\mathcal{U}}]$. So for $\beta'<\beta$ $\mathcal{U}(\alpha,\beta')\in PP(\alpha)\cap L[\bar{\mathcal{U}}]$. Now given $X\subseteq P(\alpha)$, $X\in L[\bar{\mathcal{U}}]$ there is f such that $X=j(f)(\alpha)$ and $f(\gamma)\subseteq P(\gamma)$, all γ. So $L[\mathcal{U}]\models\overline{\overline{PP(\alpha)\cap L[\bar{\mathcal{U}}]}}\leq 2^\alpha$. Hence $L[\mathcal{U}]\models\bar{\bar{\beta}}\leq 2^\alpha$, and $L[\mathcal{U}]\models$GCH so $L[\mathcal{U}]\models\beta<\alpha^{++}$. Hence $o^{\mathcal{U}}(\alpha)\leq\alpha^{++}$. □

Now why is this? Recall that, although when given an embedding $j:V\to_e M$ we may always define $U=\{X:\kappa\in j(X)\}$, a normal measure on κ, yet if we let $j_U:V\to_U M'$ we shall not always have $j=j_U$. So the restriction on the number of normal measures at κ is not a limitation on the number of elementary embeddings with critical point κ. M' is the minimal model with a map j_U such that $j_U:V\to_U M'$: other models and maps may be too "big" to be coded by a normal measure. What we need is a more powerful mechanism for representation of embeddings.

It may be noted that if $j:V\to_U M$ with κ critical then $j(\kappa)<(2^\kappa)^+$. Since $U<V \Rightarrow j_U(\kappa)<j_U(\kappa)$ this is another form of the restriction in lemma 22.6, but it is a more useful starting point for generalisation. How could we possibly get $j(\kappa)\geq(2^\kappa)^+$?

For some θ we shall allow ordinals less than θ to be used to generate M; that is, instead of $M=\{j(f)(\kappa):f:\kappa\to V\}$ we shall allow $M=\{j(f)(\kappa,a):f:\kappa\times[\kappa]^n\to V \wedge a\in[\theta]^n\}$. Then the restriction becomes $j(\kappa)<(2^\kappa.\overline{\overline{\theta}})^+$ which can be made as large as we please by increasing θ.

The representation of the embedding follows naturally from this for if we are to get a unique "ultrapower" M then properties of $j(f)(\kappa,a)$ must be obtainable in V from the representation. Instead of

$$X\in U \Leftrightarrow \kappa\in j(X) \tag{22.4}$$

we shall require

$$X\in E_a \Leftrightarrow \langle\kappa,a\rangle\in j(X) \tag{22.5}$$

where $a\in[\theta]^n$, say, and $X\subseteq\kappa\times[\kappa]^n$.

Now for the definition: it should be compared with definition 5.11. In this case we simplify by ignoring the possibility of non well-foundedness.

Definition 22.7. *Suppose $E=\langle E_a:a\in[\theta]^{<\omega}\rangle$. E is a θ-extender on κ provided that there are j,M such that:*

(i) $j:V\to_e M$;

(ii) κ is the critical point of j and $j(\kappa)\geq\theta$;

(iii) $M=\{j(f)(\kappa,a)$: for some n $f:\kappa\times[\kappa]^n\to V$ and $a\in[\theta]^n\}$;

(iv) for $n\in\omega$, $a\in[\theta]^n$, $X\in P(\kappa\times[\kappa]^n)$ $X\in E_a \Leftrightarrow\langle\kappa,a\rangle\in j(X)$.

In chapter 5 we obtained an alternative characterisation that worked generally and had the additional advantage that when we were dealing with models of ZFC the model M would be well-founded. In this case things are less neat. There does not seem to be any reasonable equivalent for definition 22.7, even in ZFC. In exercise 1 a characterisation for all but the requirement that M be well-founded is given.

Note that if j',M' also satisfy the definition then the map

$\sigma:M\to M'$ defined by $\sigma(j(f)(\kappa,a))=j'(f)(\kappa,a)$ is an isomorphism. So j,M are uniquely determined by E.

Definition 22.8. *$j:V\to_E M$ means that j,M are as in definition 22.7.*

Next we consider the generalisation of coherent sequences. Here we encounter a difficulty. In applications of coherence as in definition 22.4 we know that if we take an ultrapower by $U(\alpha,\beta)$ we shall get a model $L[\overline{U}]$ where $\overline{U}(\alpha',\beta')=U(\alpha',\beta')$ whenever $\alpha'<\alpha$ or $\alpha=\alpha'$ and $\beta'<\beta$. The former did not need stating: it follows at once from the fact that α is the critical point. But with extenders things are different. We may have extender sequences E with $\alpha'<\alpha$ such that $\langle\alpha,0\rangle\in\Delta^E$ and for some $\beta>\alpha$ $\langle\alpha',\beta\rangle\in\Delta^E$. Then the extenders at α will upset those at α': are we to demand that extenders $E(\alpha',\beta)$ are fixed by $j:V\to_{E(\alpha,0)}M$? Actually the answer is simple (no!) but the resulting technical difficulties are substantial. Let us, therefore, restrict ourselves to sequences where the difficulty does not arise until chapter 24.

It is helpful to insist that $E(\alpha,\beta)$ be a β-extender on α. It is easy to change a β-extender into a β'-extender for many other β', so this is not very restrictive. E will no longer enumerate all the extenders in $L[E]$ but that was a lost cause anyway.

Definition 22.9. *E is a coherent sequence of extenders provided that*

(i) dom $(E)=\Delta$, where $\Delta\subseteq On^2$ and $\langle\alpha,\beta\rangle\in\Delta \wedge \beta'<\beta \Rightarrow \langle\alpha,\beta'\rangle\in\Delta$ (call Δ Δ^E);

(ii) $E(\alpha,\beta)$ is a β-extender on α (all $\langle\alpha,\beta\rangle\in\Delta$)

(iii) Suppose $j:V\to_{E(\alpha,\beta)}M$. Then for all β' $\langle\alpha,\beta'\rangle\in\Delta^{\overline{E}} \Leftrightarrow \beta'<\beta$; and for $\beta'<\beta$ $\overline{E}(\alpha,\beta')=E(\alpha,\beta')$ where $\overline{E}=j(E)$ (or $\bigcup_{i\in On} j(E\cap V_i)$ if E is a proper class);

(iv) $\alpha'<\alpha \wedge \langle\alpha,0\rangle\in\Delta \Rightarrow \exists\gamma<\alpha\langle\alpha',\gamma\rangle\notin\Delta$.

Definition 22.10. *$o^E(\alpha)=\{\beta:\langle\alpha,\beta\rangle\in\Delta^E\}$.*

The connection between normal measures and extenders is not hard to see. κ has an extender if and only if κ is measurable; we must go on saying measurable though, because extendible means something quite different.

Suppose U is a normal measure on κ. Provided $\theta\leq j(\kappa)$ U can be turned into a θ-extender on κ by setting $X\in E_a \Leftrightarrow \langle\kappa,a\rangle\in j(X)$. Since letting $j:V\to_U M$ $M=\{j(f)(\kappa):f:\kappa\to V\}$, certainly $M=\{j(f)(\kappa,a):$ for some n $f:\kappa\times[\kappa]^n\to V$ and $a\in[\theta]^n\}$. So E is a θ-extender. If U is a coherent sequence then if $j:V\to_{U(\kappa,\beta)}M$ certainly $j(\kappa)\geq\beta$ by lemma 22.2, so coherent sequences of measures can be turned into coherent sequences of extenders.

Conversely, given a θ-extender E, we define U_E by

$X\in U_E \Leftrightarrow X\times\{\phi\}\in E_\phi$. (22.6)

Suppose $j:V\to_E M$, $j':V\to_{U_E} M'$. In this case M and M' are not necessarily the same. They <u>will</u> be equal provided $M=\{j(f)(\kappa):f:\kappa\to V\}$ which will be true provided $\theta\subseteq\{j'(f)(\kappa):f:\kappa\to\kappa\}=j'(\kappa)$. If $\theta\leq\kappa+1$ this is obviously true, and for quite a way beyond $\kappa+1$. By chapter 5 exercise 3 $\kappa^+=(\kappa^+)^M$ so they will be equal provided $\theta\leq\kappa^++1$: and for quite a bit beyond.

On the other hand $j'(\kappa)<\kappa^{++}$ so $\theta=\kappa^{++}$ cannot possibly give equality (assuming GCH). In general the equivalence stops somewhere between κ^+ and κ^{++}.

The remainder of this chapter summarises without proof the equivalents of the main results in parts two and three for coherent sequences. It should be stressed that this is done just to show what can be done, and is not a full discussion.

<u>Definition 22.11</u>. *Let σ be the sentence asserting that E is a coherent sequence of extenders (see exercise 1). A standard model of $R^+ + \sigma$ is called a premouse.*

Remember the complexities in the discussion of $0^\dagger$ in chapter 13 caused by the fact that we had to keep track of which of the two normal measures we were iterating by. Naturally the complexities are much greater in the present case.

<u>Definition 22.12</u>. *Suppose $k=\langle\langle\kappa_\alpha,\nu_\alpha\rangle:\alpha<\xi\rangle$. Then $\langle\langle \underline{M}_\alpha\rangle_{\alpha\in\xi+1},\langle\pi_{\alpha\beta}\rangle_{\alpha\leq\beta\in\xi+1}\rangle$ is an iterated ultrapower of $\underline{M}$ with index k provided*

(i) $\underline{M}_0=\underline{M}$;

(ii) each $\underline{M}_\alpha$ is a premouse of the form $\langle M_\alpha,\in_{M_\alpha},E_\alpha,\vec{A}_\alpha\rangle$;

(iii) $\langle\langle \underline{M}_\alpha\rangle_{\alpha\in\xi+1},\langle\pi_{\alpha\beta}\rangle_{\alpha\leq\beta\in\xi+1}\rangle$ is a commutative system;

(iv) $\langle\kappa_\alpha,\nu_\alpha\rangle\in\Delta^{E_\alpha}$ and $\pi_{\alpha\alpha+1}:\underline{M}_\alpha\to_{E_\alpha(\kappa_\alpha,\nu_\alpha)}\underline{M}_{\alpha+1}$ $(\alpha<\xi)$;

(v) if λ is a limit then $\langle\underline{M}_\lambda,\langle\pi_{\alpha\lambda}\rangle_{\alpha<\lambda}\rangle$ is a direct limit of $\langle\langle\underline{M}_\alpha\rangle_{\alpha<\lambda},\langle\pi_{\alpha\beta}\rangle_{\alpha\leq\beta<\lambda}\rangle$.

Remember: $j:M\to_E M'$ must be redefined for non well-founded structures for this discussion to make sense.

Each $\underline{M}_\alpha$ is called an iterate of $\underline{M}$; the $\pi_{\alpha\beta}$ are called iteration maps.

<u>Definition 22.13</u>. *$\underline{M}$ is iterable provided each of its iterates is well-founded.*

Suppose $\underline{M}$ is an iterable premouse. For simplicity suppose N=O.

<u>Lemma 22.14</u>. *If $\underline{N}$ is an iterate of $\underline{M}$ with iteration map π and $\sigma:\underline{M}\to_{\Sigma_0}\underline{N}$ then for all $\xi\in On_M$, $\sigma(\xi)\geq\pi(\xi)$.*

(Compare lemma 8.19). It follows that π is the unique iteration map, although its index need not be unique.

<u>Definition 22.15</u>. *π_{MN} is the unique iteration map from $\underline{M}$ to $\underline{N}$.*

<u>Definition 22.16</u>. *Iterable premice $\underline{M},\underline{N}$ are comparable provided $\underline{M}=\underline{N}$ or for some α $\underline{M}=\underline{J}^N_\alpha$ or $\underline{N}=\underline{J}^M_\alpha$.*

<u>Lemma 22.17</u>. *If $\underline{M}$, $\underline{N}$ are comparable premice then $\underline{M},\underline{N}$ have comparable iterates, provided they are sets.*

(Compare lemma 8.18).

<u>Lemma 22.18</u>. *If $\underline{M},\underline{N}$ have a common iterate $\underline{Q}$ and $rng(\pi_{MQ})\subseteq rng(\pi_{NQ})$ then $\underline{N}$ is an iterate of $\underline{M}$.*

This result, which generalises an argument in lemma 10.16, is not true if clause (iv) in definition 22.9 is omitted.

<u>Definition 22.19</u>. *An iterated ultrapower $\langle\langle\underline{M}_\alpha\rangle_{\alpha\in\xi+1},\langle\pi_{\alpha\beta}\rangle_{\alpha\leq\beta\in\xi+1}\rangle$ with index $k=\langle\langle\kappa_\alpha,\nu_\alpha\rangle:\alpha<\xi\rangle$ is normal provided that $\alpha<\beta<\xi \Rightarrow \kappa_\alpha<\kappa_\beta$. Each $\underline{M}_\alpha$ is said to be a normal iterate of $\underline{M}_0$.*

<u>Lemma 22.20</u>. *Suppose that $\underline{M}$ is an iterable premouse and that $\underline{N}$ is an iterate of $\underline{N}$. Then $\underline{N}$ is a normal iterate of $\underline{M}$, and the normal index for the iteration is unique.*

<u>Lemma 22.21</u>. *If $X\prec_{\Sigma_0}\underline{M}$ and $P^M(\kappa)\subseteq X$, and if $\pi:\underline{N}=\underline{M}|X$ with $\underline{N}=\underline{J}^{\bar{E}}_\alpha$ and $\underline{M}=\underline{J}^E_\alpha$ then for all $\gamma<o^{\bar{E}}(\kappa)$ $\bar{E}(\kappa,\gamma)=E(\kappa,\gamma)$.*

This is a sort of condensation lemma. Next we state a criterion for iterability.

<u>Definition 22.22</u>. *An extender $E=\langle E_a:a\in[\theta]^{<\omega}\rangle$ is countably complete if and only whenever $\langle b_i:i\in\omega\rangle$ is a sequence with $b_i\in[\theta]^{<\omega}$ (all $i\in\omega$) and $\langle X_i:i\in\omega\rangle$ is such that $X_i\in E_{b_i}$, all $i\in\omega$, then there is τ and an order-preserving map $\delta:\bigcup_{i\in\omega} b_i\to On$ such that $\langle\tau,\delta(b_i)\rangle\in X_i$ for all $i\in\omega$.*

Lemma 22.23. *Suppose* $\underline{N}=\underline{J}^E_\alpha$ *is a premouse and for all* $\langle\kappa,\nu\rangle\in\Delta^E$ $E(\kappa,\nu)$ *is countably complete. Then* $\underline{N}$ *is iterable.*

As we have already hinted, the situation is not quite like it was for a single measure: it is not necessarily the case that $L[E]\models E(\alpha,\beta)$ is countably complete.

Finally a word about generalised mice. Iterable premice have already been introduced. There is a bit of a problem in getting further. Our fine-structure enabled us to extend a map of N^p to H to some $\pi:\underline{N}\to_{\Sigma_1}\underline{M}$ with $H=M^{p'}$, $\tilde{\pi}\supseteq\pi$ and $p'=\tilde{\pi}(p)$ on condition that $\underline{N}$ be p-sound. Since mice $\underline{N}$ with $\omega\rho_{\underline{N}}\geq\kappa$ (κ critical) always have this property for some p the definition of mice was fairly simple. But if $\underline{N}=\underline{J}^E_\alpha$ is an acceptable iterable premouse we may have κ,κ' with $\omega\alpha>\kappa>\omega\rho_N>\kappa'$, $o(\kappa),o(\kappa')>0$. But then how are we to define $\eta:N\to^1_{E(\kappa',0)}N'$ say ?

Actually generalised fine-structure does give an extension of embeddings result without the soundness restriction. And applying this to mice with extenders it turns out that mouse iteration, like ordinary iteration, can be "normalised", so that the problem did not really matter; we can always iterate below the projectum before we iterate above it. Our worries, though, are by no means over.

As with simple mice, the main result that we need is that iterable premice are acceptable. Consulting the proof of lemma 11.24 the reader will see that this is essentially a question of indiscernibles; so we must get a theory of indiscernibles for generalised mice. Just as when $\underline{N}$ is a premouse with a single measure and $\underline{M}=\underline{J}^N_\beta$, $\omega\beta\in N$, and $\underline{M}$ is a mouse then $\underline{M}$ has indiscernibles of order-type a multiple of ω^ω, so in the general case $\underline{M}$ has a smoothness property giving nice indiscernibles. The reason why we supressed the ordinal arithmetic in the proof of lemma 7.17 was to give something of the flavour of this more general case.

Our first major obstacle is the fact that the crucial corollary 11.21 can no longer be proved. It is possible that $\omega\rho_N=\kappa$, $o(\kappa)>0$ but $H^N_\kappa\neq H^{N^+}_\kappa$. This may occur when $o(\kappa)$ is a limit ordinal. Indeed, if $\underline{N}$ is such that $\underline{N}=\underline{J}^E_{\kappa+1}$, $o^E(\kappa)=\omega$ but for all $\lambda<\kappa$ $o^E(\lambda)$ is finite then $\rho^m_{J^E_\kappa}=\kappa$, all $m\in\omega$, but $P(\omega)\cap N\neq P(\omega)\cap J^E_\kappa$. The proof of this is not hard and is left to the reader.

So there can be a new subset of γ at a level without the projectum falling to γ. This, we hope the reader will have perceived from the arguments in this book, contradicts the whole purpose of fine-structure. We must replace J^N_κ by a structure H with $P(\kappa)\cap N=\Sigma_\omega(H)\cap\cap P(\kappa)$, in which our usual results hold. H is called a hydra; we may have to repeat the process any finite number of times. The cardinal principle is that the hierarchy may only grow at a speed that enables the projectum to act as an index of formation of new subsets of given ordinals; if it cannot do this the hierarchy must be slowed down by dividing up levels.

Also irritating, though of less interest, is the case $\underline{N}=\underline{J}^{E}_{\alpha+1}$ where for some κ $o^{E}(\kappa)>\omega\alpha$. Then the usual rule that $N=\mathrm{rud}_{E|N}(J^{N}_{\alpha})$ breaks down because some extenders are not available in N.

All these problems can be solved and we can prove

Lemma 22.24. *Iterable premice are acceptable.*

But all the applications are cluttered up with cases about hydras and so forth; life is never quite the same again.

Exercises

1. Formulate definition 22.7 without the requirement that M be well-founded. Show that with this definition $E=\langle E_a:a\in[\theta]^{<\omega}\rangle$ is a θ-extender on κ if and only if

(i) $E_a\subseteq P(\kappa\times[\kappa]^n)$ where $a\in[\theta]^n$.

Let $a\in[\theta]^n$, $b\in[\theta]^m$, $a\subseteq b$. Then given $u\in[\kappa]^m$, u^{ab} denotes $\{u_{h_1}\ldots u_{h_n}\}$, where $u=\{u_1\ldots u_m\}$, $u_1<\ldots<u_m$, $b=\{b_1\ldots b_m\}$, $b_1<\ldots<b_m$, $a=\{b_{h_1}\ldots b_{h_n}\}$. For $X\subseteq\kappa\times[\kappa]^n$ X^{ab} denotes $\{\langle\xi,u\rangle\in\kappa\times[\kappa]^m:\langle\xi,u^{ab}\rangle\in X\}$.

(ii) $X\in E_a \Leftrightarrow X^{ab}\in E_b$.

(iii) E_a is a non-principal ultrafilter on $P(\kappa\times[\kappa]^n)$ $(a\in[\theta]^n)$

i.e. (a) for all $\gamma<\kappa$, $u\in[\kappa]^n$ $\{\langle\gamma,u\rangle\}\notin E_a$;

(b) $X,Y\in E_a \Leftrightarrow X\cap Y\in E_a$;

(c) $X\in E_a \Leftrightarrow (\kappa\times[\kappa]^n\setminus X)\in E_a$.

(iv) Suppose $f:\kappa\times[\kappa]^n\to\kappa$ and $\{\langle\xi,u\rangle\in\kappa\times[\kappa]^n:f(\xi,u)<\xi\}\in E_a$. Then for some $\zeta<\kappa$ $\{\langle\xi,u\rangle\in\kappa\times[\kappa]^n:f(\xi,u)=\zeta\}\in E_a$.

(v) Let $f:\kappa\times[\kappa]^n\to\kappa$. Suppose $\{\langle\xi,u\rangle\in\kappa\times[\kappa]^n:f(\xi,u)<\max(u)+1\}\in E_a$. Then there is $\nu<\max(a)+1$ such that, letting $m=n$ if $\nu\in a$, $n+1$ otherwise, defining $f^{ab}(\xi,u)=f(\xi,u^{ab})$, and supposing that $a\cup\{\nu\}=$ $=\{a'_1\ldots a'_m\}$ with $a'_i=\nu$, $u=\{u_1\ldots u_m\}$ $(a'_1<\ldots<a'_m,\ u_1<\ldots<u_m)$ then

$\{\langle\xi,u\rangle\in\kappa\times[\kappa]^m:f^{a\,a\cup\{\nu\}}(\xi,u)=u_i\}\in E_{a\cup\{\nu\}}$.

2. Show that there is a sentence σ such that $L[E]\models\sigma \Leftrightarrow E$ is coherent.

3. If E is coherent in $L[E]$ show that $L[E]$ has a Δ^1_3 well-order of $P(\omega)$

4. Show that if $o^{E}(\kappa)=n+1$ then κ is a limit of κ' with $o^{E}(\kappa')=n$. Can this be extended to infinite n? What if $o^{E}(\kappa)=\mathrm{On}$?

Definition 23.1. *A cardinal κ is β-strong provided that there is an embedding $j:V\to_e M$ with κ critical such that $P(\beta)\subseteq M$.*

Definition 23.2. *A cardinal κ is strong provided that it is β-strong for all β*

For $\beta\leq\kappa$ κ is β-strong if and only if κ is measurable. For $\beta>\kappa$ it turns out that β-strong is connected with the existence of extenders at κ. This chapter will show that models of the form $L[E]$ are inner models for β-strong cardinals; that $L[E]\models$GCH (this follows from lemma 22.24, of course, but we shall try to supply some details in this case); and then we shall solve the equation $x/L[E]=0^{\#}/L$.

But first consider iterability of $L[E]$. If U is a coherent sequence of measures in $L[U]$ then $L[U]$ is always iterable, but this is not true for extenders.

If $\underline{N}=\underline{J}^E_\alpha$ is an iterable premouse let $\langle k_\alpha\rangle$ enumerate Δ^E in ascending order $(\langle\kappa,\nu\rangle<\langle\kappa',\nu'\rangle \Leftrightarrow \kappa<\kappa' \vee (\kappa=\kappa' \wedge \nu<\nu'))$. Let $\langle\langle \underline{N}_\alpha\rangle_{\alpha<\theta},\langle\pi_{\alpha\beta}\rangle_{\alpha\leq\beta<\theta}\rangle$ be the iteration of $\underline{N}$ with index $\langle k'_\alpha\rangle$ where $k'_\alpha=\pi_{0\alpha}(k_\alpha)$. Call $\underline{N}_\theta$ $\underline{N}^+$. Let $\underline{N}^0=\underline{N}$, $\underline{N}^{\delta+1}=(\underline{N}^\delta)^+$ and let $\underline{N}^\lambda$ be the direct limit of $\langle\langle\underline{N}^\alpha\rangle_{\alpha<\lambda},\langle\pi_{N^\alpha N^\beta}\rangle_{\alpha\leq\beta<\lambda}\rangle$, ($\lambda$ a limit). Each $\underline{N}^\alpha$ is an iterate of $\underline{N}$ definable in $L[E]$; and each iterate of $\underline{N}$ is embeddable in some $\underline{N}^\alpha$. So if $L[E]\models$"V is iterable" then $L[E]$ is iterable.

Lemma 23.3. *Suppose E is coherent in $L[E]$. The following are equivalent*

(i) $L[E]$ is iterable;

(ii) For all $\langle\kappa,\nu\rangle\in\Delta^E$, if $j:L[E]\to_{E(\kappa,\nu)}M$ then M is well-founded;

(iii) Every $E(\kappa,\nu)$ is countably complete in $L[E]$.

Proof: (i)$\Rightarrow$(ii) is trivial.

Suppose (ii). Let $X_i\in E_{a_i}$, all $i\in\omega$, where $E=E(\kappa,\nu)$, $\langle X_i:i\in\omega\rangle$ and $\langle a_i:i\in\omega\rangle\in L[E]$. Let $j:L[E]\to_E L[\bar{E}]$. Then for all $i\in\omega$ $\langle\kappa,a_i\rangle\in j(X_i)$. Let $a'_i=j(a_i)$. Let R be the set of all $\langle f,n\rangle$ such that $n\in\omega$, $f:\bigcup_{i<n} a'_i\to On$ is order-preserving, and $\langle\kappa,f"a'_i\rangle\in j(X_i)$. Given $\langle f,n\rangle$, $\langle g,m\rangle\in R$ $\langle f,n\rangle<_R\langle g,m\rangle \Leftrightarrow n>m$ and $g=f|\bigcup_{i<m} a'_i$. Letting $f_n=(j|\bigcup_{i<n} a_i)^{-1}$ $\langle f_{n+1},n+1\rangle<_R\langle f_n,n\rangle$; so $\langle R,<_R\rangle$is not well-founded in $L[E]$ and hence not in $L[\overline{E}]$. Say $\langle\langle g_n,n\rangle : n\in\omega\rangle$ is a descending $<_R$ sequence in $L[\overline{E}]$. Let $g=\bigcup_{n\in\omega} g_n$. Then $g: \bigcup_{i\in\omega} a_i^! \to On$ is order-preserving and for all $i\in\omega$ $\langle\kappa,g''a_i^!\rangle\in j(X_i)$. So

$$(\exists\tau<j(\kappa))(\exists g: \bigcup_{i\in\omega} j(a_i)\to On \text{ order-preserving})(\forall i\in\omega)(\langle\tau,g''j(a_i)\rangle\in j(X_i)). \quad (23.1)$$

Applying j^{-1}

$$(\exists\tau<\kappa)(\exists g: \bigcup_{i\in\omega} a_i\to On \text{ order-preserving})(\forall i\in\omega)(\langle\tau,g''a_i\rangle\in X_i). \quad (23.2)$$

(iii)$\Rightarrow$(i) by lemma 22.23 and the remark preceding this lemma.□

Suppose that $L[E]$ is an iterable premouse.

Lemma 23.4. *$L[E]\models GCH$*

Proof: Work in $L[E]$. Given an infinite cardinal α we must show $2^\alpha=\alpha^+$. We use induction on infinite α.

Claim 1 For all $a\subseteq\alpha$ there is an iterable premouse $\underline{N}=J_\beta^{\overline{E}}$ such that

(a) $a,\alpha\in N$, $o^{\overline{E}}(\alpha)=0$, $\underline{N}\models ZF^-$;

(b) every x in N is $\underline{N}$-definable from $\alpha\cup\{p\}$ (some $p\subseteq\alpha$, $p\in N$)

(c) for $\tau<\alpha$

(i) $o^E(\tau)<\alpha^+ \Rightarrow o^{\overline{E}}(\tau)=o^E(\tau)$;

(ii) $o^E(\tau)\geq\alpha^+ \Rightarrow o^{\overline{E}}(\tau)\geq(\alpha^+)^N$;

(iii) $\nu<o^{\overline{E}}(\tau) \Rightarrow E(\tau,\nu)=\overline{E}(\tau,\nu)$.

Proof: We may assume that $o^E(\alpha)=0$, since otherwise it is sufficient to prove the result in $L[E']$ where $\pi: L[E]\to_{E(\alpha,0)} L[E']$. Pick a regular $\gamma>\alpha$ with $a\in J_\gamma^E$; we may assume also that $\bigcup_{\delta<\alpha} P(\delta)\subseteq J_\gamma^E$. Let $p\subseteq\alpha$ code a, $\bigcup_{\delta<\alpha} P(\delta)$ (this is possible as $\overline{\overline{\bigcup_{\delta<\alpha} P(\delta)}}=\alpha$). Let $X\prec J_\gamma^E$ be least with $\alpha\cup\{p\}\subseteq X$. Let $\sigma:\underline{N}\cong\underline{J}_\gamma^E|X$. The reader is left to check (a)-(c) (lemma 22.21 is needed for (c)(iii)) □(Claim 1)

Claim 2 Suppose $\underline{N},\underline{M}$ are as in claim 1 and $\max(\alpha^{+N},\sup_{\tau<\alpha} o^{E^N}(\tau))= =\max(\alpha^{+M},\sup_{\tau<\alpha} o^{E^M}(\tau))$. Then $\underline{N}=\underline{M}$.

Proof: Let $\underline{Q},\underline{Q}'$ be comparable iterates of $\underline{N},\underline{M}$. Suppose $Q\in Q'$. We may assume $\pi_{NQ}|\alpha+1=\pi_{MQ'}|\alpha+1=id|\alpha+1$. Let p_N be as in claim 1(b). Since $\pi_{NQ}|\alpha\cup\{\alpha,a\}=id|\alpha\cup\{\alpha,a\}$, letting $X\prec Q$ be least such that $\alpha\cup\{\alpha,a\}\subseteq X$, $\underline{N}\cong\underline{Q}|X$. Hence $\underline{N}\in Q'$. But $\overline{\overline{N}}^{Q'}\leq\alpha$, so there is a map of α to α^{+M} in Q', hence in M. Contradiction!

Similarly $Q'\notin Q$, so $Q=Q'$. Hence letting $X\prec Q$ be as above $\underline{N}\cong\underline{Q}|X\cong\underline{M}$. So $\underline{N}=\underline{M}$. □(Claim 2)

Now given $a\subseteq\alpha$ there is $\underline{N}$ as in claim 1 with $a\in N$. And $\max(\alpha^{+N},\sup_{\tau<\alpha} o^{E^N}(\tau))<\alpha^+$. So $\overline{\overline{P(\alpha)}}\leq\alpha^+$. □

Lemma 23.5. *Suppose* $\kappa<(\alpha^+)^{L[E]}\leq\nu<o^E(\kappa)$ *and* $\pi:L[E]\to_{E(\kappa,\nu)}L[\bar{E}]$. *Then* $P(\alpha)\cap L[E]\subseteq L[\bar{E}]$.

Proof: (in L[E]) Suppose $a\subseteq\alpha$. Then as in lemma 23.4 pick γ regular with $\gamma>o^E(\kappa)$ and $a\in J^E_\gamma$ and $\bigcup_{\delta<\alpha}P(\delta)\subseteq J^E_\gamma$. With $p,\underline{N},\sigma$ as before $a\in N$, and letting $\underline{N}=\underline{J}^{E'}_\beta$, $o^{E'}(\kappa)<\alpha^+$. Set $\bar{\nu}=o^{E'}(\kappa)$. So $\bar{\nu}<\nu$, and we may define $\bar{\pi}:J^{\bar{E}}_\gamma\to_{E(\kappa,\bar{\nu})}J^{E''}_\gamma$. Then N and $M=J^{E''}_\gamma$ have comparable iterates Ω,Ω' with $\pi_{N\Omega}|\bar{\nu}=\pi_{M\Omega'}|\bar{\nu}=id|\bar{\nu}$. If $\Omega'\in\Omega$ then $\pi_{M\Omega'}\bar{\pi}\pi\sigma:\underline{N}\to_{\Sigma_0}\underline{\Omega}$ non cofinally; so $\Omega\subseteq\Omega'$. Hence $a\in M$.

Now for $\gamma<\alpha$, supposing that $a=\bar{\pi}(f)(\kappa,b)$, $b\in[\bar{\nu}]^n$, $f\in J^{\bar{E}}_\gamma$

$$\gamma\in a \Leftrightarrow \gamma\in\bar{\pi}(f)(\kappa,b) \tag{23.2}$$
$$\Leftrightarrow \{\langle\xi,u\rangle:u_i\in f(\xi,u^{bb\cup\{\gamma\}})\}\in E(\kappa,\bar{\nu})_{b\cup\{\gamma\}}$$

with notation as in chapter 22 exercise 1. Hence $a\in J^{\bar{E}}_\gamma$. □

Corollary 23.6. *Suppose* $\kappa<(\alpha^+)^{L[E]}<o^E(\kappa)$. *Then* $L[E]\models\kappa$ *is* α*-strong.*

Corollary 23.7. *Suppose* $o^E(\kappa)=On$. *Then* $L[E]\models\kappa$ *is strong.*

Our aim is to get converses to these corollaries.

Define inductively a sequence E_κ as follows: Suppose $\tau,\nu<\kappa$ and $E_\kappa(\tau,\nu)$ is defined for all $\langle\tau',\nu'\rangle$ with $\tau'<\tau$ or $\tau'=\tau$ and $\nu'<\nu$ (where by defined we mean that either $\langle\tau',\nu'\rangle\in\Delta^E\kappa$ and E_κ has been specified or it has been decided that $\langle\tau',\nu'\rangle\notin\Delta^E\kappa$). If there is a coherent sequence E^* of countably complete extenders with $E^*(\tau',\nu')=E(\tau',\nu')$ (or both undefined) for $\tau'<\tau$ or $\tau'=\tau$ and $\nu'<\nu$, and if $\langle\tau,\nu\rangle\in\Delta^{E^*}$, then let $\langle\tau,\nu\rangle\in\Delta^E\kappa$ and $E_\kappa(\tau,\nu)=E^*(\tau,\nu)$ for some such E^*. Otherwise $\langle\tau,\nu\rangle\notin\Delta^E\kappa$.

Note that this construction gives priority to existing measurables: we cannot have a τ'-extender on τ and an extender on τ' if $\tau<\tau'$: we add the τ'-extender on τ rather than the extender on τ.

Lemma 23.8. *Suppose* κ *is* β*-strong,* $\beta\geq 2^\kappa$ *is a cardinal. Suppose* $o^E\kappa(\tau)\neq\kappa$ *(all* $\tau<\kappa$*). Then there is a coherent* E *with* $E(\tau,\nu)=E_\kappa(\tau,\nu)$ *(all* $\tau<\kappa$*) and* $o^E(\kappa)>\beta^+$.

Proof: Suppose not. Let $j:V\to_e M$ with κ critical and $P(\beta)\subseteq M$. Let $E=j(E_\kappa)$; then $E(\tau,\nu)=E_\kappa(\tau,\nu)$, all $\tau<\kappa$. Suppose $\theta=o^E(\kappa)\leq\beta^+$. Define a θ-extender on κ by

$$X\in E_a \Leftrightarrow \langle\kappa,a\rangle\in j(X) \tag{23.3}$$

where $X\subseteq\kappa\times[\kappa]^n$, $a\in[\theta]^n$. Let $\Delta^{E^+}=\Delta^E\cup\{\langle\kappa,\theta\rangle\}$, $E^+(\tau,\nu)=E(\tau,\nu)$ if $\tau<\kappa$ or $\tau=\kappa$ and $\nu<\theta$; $E^+(\kappa,\theta)=E$.

Claim E^+ is coherent and E is countably closed.

Proof: Let $\sigma:V\to_E M'$. Then there is a map $\sigma':M'\to_{\Sigma_0}M$ defined by $\sigma'(\sigma(f)(\kappa,a))=j(f)(\kappa,a)$; so M' is well-founded (as M is).

Let $E'=\sigma(E^+)$.

Now $\sigma'(o^{E'}(\kappa))=o^E(\kappa)=\theta$ so $\theta\geq o^{E'}(\kappa)$. On the other hand $E(\kappa,\nu)\in\mathrm{rng}(\sigma')$, all $\nu<\theta$, so $\theta=o^{E^+}(\kappa)$. Also $\sigma'(E'(\kappa,\nu))=E'(\kappa,\nu)$ $(\nu<\theta)$ so E^+ is coherent. E is countably complete by lemma 23.3. □(Claim)

If $\theta=\beta^+$ we are done. So suppose $\theta<\beta^+$. Then $E\in M$, since it can be coded as a subset of β $(\beta\geq 2^\kappa)$. So

$M\models E$ can be properly extended by a θ-extender on κ, where $\theta=o^E(\kappa)$. (23.4)

Applying j^{-1}

E_κ can be properly extended by a θ'-extender on κ', where $\theta'=o^{E_\kappa}(\kappa')$ (some $\kappa'<\kappa$) (23.5)

But this contradicts the definition of E_κ. □

<u>Corollary 23.9</u>. *Suppose κ is β-strong, $\beta\geq 2^\kappa$ is a cardinal. Then there is an inner model $L[E]$ which is an iterable premouse with $o^E(\kappa)>\beta^+$.*

Proof: By lemma 23.8 this is true provided $\forall\tau<\kappa\, o^{E_\kappa}(\tau)<\kappa$. So suppose $o^{E_\kappa}(\tau)=\kappa$. Let $j:V\to_e M$ with $P(\beta)\subseteq M$, κ critical. So $j(\kappa)>\beta^+$. Hence if $E'=j(E_\kappa)$, then $o^{E'}(\tau)>\beta^+$.

Now let $\langle\langle \underline{M}_\alpha\rangle_{\alpha\in On},\langle\pi_{\alpha\beta}\rangle_{\alpha\leq\beta\in On}\rangle$ be the iterated ultrapower of M by $E'(\tau,0)$. Let $M'=M_\kappa$, $E''=\pi_{0\kappa}(E')$. As $\kappa>\tau$, κ regular, $\pi_{0\kappa}(\tau)=\kappa$. Since $M\models o^{E'}(\tau)>\beta^+$, $M'\models o^{E''}(\kappa)>\pi_{0\kappa}(\beta^+)=\beta^+$. Thus $L[E'']$ is the desired model.□

If κ is a strong cardinal then corollary 23.9 gives, for each β, an $L[E]$ with $o^E(\kappa)>\beta^+$. To get $L[E]$ with $o^E(\kappa)=On$ requires a bit of work. We may assume $\forall\tau<\kappa\, o^E(\tau)<\kappa$: for if $o^E(\tau)=\kappa$ let $U=\{X\in P(\kappa):\kappa\in j(X)\}$ and let $\langle\langle \underline{N}_\alpha\rangle_{\alpha\in On},\langle\pi_{\alpha\beta}\rangle_{\alpha\leq\beta\in On}\rangle$ be the iterated ultrapower of $L[E]$ by U. Letting $E_\alpha=\pi_{0\alpha}(E_\kappa)$ we have $o^{E_\alpha}(\tau)=\pi_{0\alpha}(\kappa)$. Hence letting $N=\bigcup_{\alpha\in On} H^{N_\alpha}_{\kappa_\alpha}$ $N\models ZFC$ (just as in earlier arguments) and if $E=\bigcup_{\alpha\in On} E_\alpha$ then $o^E(\tau)=On$. Just as in corollary 23.9 we deduce that there is an inner model with $o^E(\kappa)=On$. We shall return to this argument in a bit.

<u>Lemma 23.10</u>. *Suppose (i) E,E' are coherent;*

(ii) $\tau<\kappa\Rightarrow E(\tau,\nu)\simeq E'(\tau,\nu)$ all ν and $o^E(\tau)<\kappa$;

(iii) $\nu<\kappa^{++}\Rightarrow E(\kappa,\nu)\simeq E'(\kappa,\nu)\wedge o^E(\kappa)=o^{E'}(\kappa)$;

(iv) $\tau>\kappa\Rightarrow o^E(\tau)=o^{E'}(\tau)=0$.

Then $L[E']\cap E=L[E]\cap E'$ (hence $L[E]=L[E']$).

Proof: Take δ regular, $\delta>o^E(\kappa)$. Take $X\prec V_\delta$ with $P(\kappa)\cap(J^E_\alpha\cup J^{E'}_\alpha)\cup\cup\{J^E_\alpha,J^{E'}_\alpha\}\subseteq X$ (where $E\cap L[E']\subseteq J^{E'}_\alpha$, similarly for E',E) and $\overline{\overline{X}}=\kappa^+$. Let $\pi:N\tilde{=}X$, N transitive. Let $J^E_\alpha=\pi(J^{\bar E}_{\bar\alpha})$, $J^{E'}_\alpha=\pi(J^{\bar E'}_{\bar\alpha})$. Now since $P(\kappa)\cap(J^E_\alpha\cup J^{E'}_\alpha)\subseteq X$ and $\pi|J^{\bar E}_{\bar\alpha}:\underline{J}^{\bar E}_{\bar\alpha}\to_e \underline{J}^E_\alpha$, $P(\kappa)\cap J^E_\alpha\subseteq J^{\bar E}_{\bar\alpha}$: similarly for $E',\bar E'$. Thus for

$\xi<\bar{\alpha}\ \bar{E}(\kappa,\xi)=E(\kappa,\xi)$. Hence $\underline{J}_{\alpha}^{\bar{E}}=\underline{J}_{\alpha}^{\bar{E}'}$ (as $\bar{\alpha}<\kappa^{++}$) so $\underline{J}_{\alpha}^{E}=\underline{J}_{\alpha}^{E'}$, and

$$\begin{aligned} E\cap L[E']=E\cap J_{\alpha}^{E'} &=\pi(\bar{E}\cap J_{\bar{\alpha}}^{\bar{E}'}) \\ &=\pi(\bar{E}'\cap J_{\bar{\alpha}}^{\bar{E}}) \\ &=E'\cap J_{\alpha}^{E} \\ &=E'\cap L[E]. \qquad \square \end{aligned} \tag{23.5}$$

Lemma 23.11. *Suppose* κ *is strong. Then there is an inner model* $L[E]$ *that is an iterable premouse with* $o^{E}(\kappa)=On$.

Proof: We may, as we said, assume $o^{E}\kappa(\tau)<\kappa$ for all $\tau<\kappa$. For each $\beta\geq 2^{\kappa}$ that is a cardinal let E^{β} be as in lemma 23.8. Let B= $=\{f:\ f:\kappa^{++}\to V$ and $\forall\gamma<\kappa^{++}f(\gamma)$ is a γ-extender on $\kappa\}$. Then B is a set. For $f\in B$ let $b_f=\{\beta>\kappa^{++}:(\forall\gamma<\kappa^{++})E^{\beta}(\kappa,\gamma)=f(\gamma)\}$. Since B is a set and $\bigcup_{f\in B} b_f$ is a proper class, some b_f must be a proper class. But then by lemma 23.10 $\beta,\beta'\in b_f$ and $\beta<\beta'$ implies $E^{\beta}\cap L[E^{\beta'}|\Delta^{E^{\beta}}]=(E^{\beta'}|\Delta^{E^{\beta}})\cap L[E^{\beta}]$. Then letting $E=\bigcup_{\beta\in b_f}E^{\beta}$ gives the required E. $\square$

PISTOLS

Definition 23.12. *A pistol is a structure* $\underline{N}=\langle N,\in,E,U\rangle$ *such that*

(a) E *is coherent in* $\underline{N}$;

(b) $\underline{N}\models R^{+}$;

(c) U *is a normal measure on* κ *(say) in* $\underline{N}$;

(d) *for some* τ $o^{E}(\tau)=\kappa$;

(e) $On_{N}=\kappa+\omega$;

(f) $\underline{N}$ *is iterable (i.e. every iterate by an extender in* E *or by* U *or by a combination of the two is transitive).*

Pistols are like sharps and daggers. It can be shown that if $\underline{M},\underline{N}$ are pistols then $\underline{M}$, $\underline{N}$ have comparable iterates, $\underline{Q},\underline{Q}'$ say. $\underline{Q},\underline{Q}'$ will be pistols. But $Q\in Q'$ and $Q'\in Q$ contradict (e) so $\underline{Q}=\underline{Q}'$.

Take some pistol $\underline{N}$ and let $X=h_N(\phi)$. Then let $\pi:\underline{M}\cong\underline{N}|X$, with M transitive. Since X is cofinal, $\pi:\underline{M}\to_{\Sigma_1}\underline{N}$. But if $\underline{N}'$ were some other pistol then since $\underline{N}$, $\underline{N}'$ have a common iterate we may deduce that the same $\underline{M}$ is obtained. $\underline{M}$ is the unique pistol that may be embedded in any pistol by a Σ_1-preserving embedding (in fact every pistol is an iterate of $\underline{M}$, but this is not easy to prove. Pistols are acceptable and for any pistol $\underline{N}$ $\rho_N=1$, $p_N=\phi$ just as for sharps).

Definition 23.13. *Let* τ *be as in chapter 13. Then* $0^{\P}=A_{M}^{\tau}$.

So $0^{\P}\subseteq\omega$. (This is not really consistent with $0^{\#}$ and $0^{\dagger}$, which were only interconstructible with the corresponding master code. But after all, that only complicated our descriptive set theory, and we are free to make such definitions as we please).

As with $0^{\#}$, $0^{¶}$ is Δ^1_3 and if $a=0^{¶}$, $a\in M$, $\underline{M}$ an inner model, then $\underline{M}\models a=0^{¶}$.

The following standard lemma can be proved

Lemma 23.14. *The following are equivalent*

(i) $0^{¶}$ *exists;*

(ii) There is an inner model $L[E]$*, an iterable premouse, with* $o^E(\kappa)=On$ *and a class C satisfying:*

(a) *C contains all uncountable cardinals above* κ*;*

(b) $C\cap\theta$ *has order-type* θ *(all cardinals* $\theta>\kappa$*);*

(c) *if* θ *is regular and greater than* κ *then* $C\cap\theta$ *is closed unbounded in* θ

(d) *C are generating* Σ_ω *indiscernibles for* $\langle L[E],\gamma\rangle_{\gamma<\kappa}$*;*

(iii) There is an inner model $L[E]$ *with* $o^E(\tau)=On$ *and non-trivial* $j:L[E]\to_e L[E]$ *with critical point above* κ.

Proof: (i)⇒(ii). Let $\underline{N}$ be the minimal pistol, $\underline{N}=\langle N,\in,E,U\rangle$; let the iterated ultrapower of $\underline{N}$ by U be $\langle\langle \underline{N}_\alpha\rangle_{\alpha\in On},\langle\pi_{\alpha\beta}\rangle_{\alpha\leq\beta\in On}\rangle$. Suppose $o^E(\tau)=\kappa$. Let $\underline{N}_\alpha=\langle N_\alpha,\in,E_\alpha,U_\alpha\rangle$ and let $On_{N_\alpha}=\kappa_\alpha+\omega$. Then if we let $E=\bigcup_{\alpha\in On} E_\alpha$ $L[E]\models o^E(\kappa)=On$. Let $C=\{\kappa_\alpha:\alpha\in On\}$. Only (d) presents any difficulty, and that is proved just as in chapter 12.

(ii)⇒(iii) Any order-preserving map of C to C generates an embedding.

(iii)⇒(i) τ-minimality and τ-maximality both follow from

Claim Suppose $\pi:L[\bar{E}]\to_e L[E]$, $o^{\bar{E}}(\tau)=o^E(\tau)=On$, $\pi|\tau=id|\tau$. Then $L[\bar{E}]=L[E]$.

Proof: For $\tau'<\tau$ $\bar{E}(\tau',\nu)=E(\tau',\nu)$ is clear. Since $\pi(\tau)=\tau$, $\pi|\tau^+=id|\tau^+$ so $P(\tau)\cap L[E]\subseteq L[\bar{E}]$. Hence $\bar{E}(\tau,\nu)=E(\tau,\nu)$ (all ν). □(Claim)

Hence there is U a normal measure on some $\kappa>\tau$ with $\langle J^E_{\kappa^+},U\rangle$ amenable and such that L[E] is iterable by U. From this we may as usual construct a pistol. □

Actually in any model $L[E]$ with $o^E(\kappa)=On$ E is wholly determined by $E(\tau,\nu)$ with $\tau<\kappa$.

Pistols are special cases of more general mice (just as daggers are special cases of double mice), and these mice will be discussed in the next chapter. With this more powerful apparatus we see that the analogy with daggers is very close, and that we could write "every $L[E]$ with $o^E(\kappa)=On$" in (ii) and (iii) of lemma 23.14.

Exercise

1. Show that if there is a strong cardinal and A is a set then $V\neq L[A]$.

24: THE GREAT BLUE YONDER

This final chapter considers two questions. Firstly, what comes "after" strong cardinals; and secondly, what about SCH?

BEYOND STRONG CARDINALS

Suppose we were to lift the assumption on coherent sequences that $\kappa<\kappa', o(\kappa')>0 \Rightarrow o(\kappa)<\kappa'$. Then we should get premice of which the pistols in chapter 23 were primitive forms. We should have to change the definition of coherence, although this is less of a nuisance than it might be because we can (at least for our immediate purposes) safely assume that $\kappa<\kappa'\wedge\kappa'<o(\kappa) \Rightarrow o(\kappa')\leq o(\kappa)$.

The main new feature we have to deal with is the "virtual extender" that arises as follows: let $\underline{N}=J^{EU}_{\kappa+1}$ be a pistol. Let $\pi:\underline{N}\to_U\underline{N}'$ and let $\kappa<\nu<\pi(\kappa)$. Then if $E=E^{N'}(\tau,\nu)$ we may define $\pi':\underline{N}'\to_E\underline{M}'$ in the usual way; but it is more natural to deal with the virtual extension $\pi'':\underline{N}\to^{\pi}{}_E\underline{M}''$. Now obviously there is no hope of arranging all iterations "normally", that is, with increasing indices. In fact iterations are nested (the depth of $\pi:\underline{N}\to^{\sigma}_E\underline{M}$ being one greater than that of σ) and each nest index is normal. This complicates the argument appallingly.

The other restriction we might try to lift was that if E is a θ-extender on κ and $\pi:V\to_E M$ then $\theta\leq\pi(\kappa)$. This is a weaker restriction than the one just discussed but it is essential to the definition of extender: for if $x=\pi(f)(\kappa,a)$ and $f:\kappa\times[\kappa]^n\to V$ then $\pi(f):\pi(\kappa)\times[\pi(\kappa)]^n\to V$ so any a beyond $\pi(\kappa)$ could not be in dom(f). Consider what happens if there is a θ-extender with $\pi(\kappa)=\theta$. Actually this is trivially attained: we need to add that E must be coherent with $o^E(\kappa)>\theta$ and $\pi(\kappa)=\theta$ where $\pi:L[E]\to_{E(\kappa,\theta)}L[\overline{E}]$. $E(\kappa,\theta)$ is called a $\pi(\kappa)$-extender.

In $L[\overline{E}]$ $o^{\overline{E}}(\kappa)=\pi(\kappa)$: and $o^{\overline{E}}(\pi(\kappa))=\pi(o^E(\kappa))>0$ so certainly $0^\P$ exists. Actually the assumption is much stronger than $0^\P$.

Definition 24.1. *κ is superstrong provided there is $j:V\to_e M$ with κ critical such that $P(\gamma)\subseteq M$ for all $\gamma<j(\kappa)$.*

It can be shown that sequences E with $\pi(\kappa)$-extenders do for superstrong cardinals what sequences E with $o^E(\kappa)$=On do for strong cardinals. That is, if E has a $\pi(\kappa)$-extender then $L[E]\models\kappa$ superstrong and if κ is superstrong then there is an E with a $\pi(\kappa)$-extender. It follows that although a superstrong cardinal gives an inner model with a strong cardinal, yet by chapter 23 exercise 1 there could be a model with a superstrong but no strong cardinal.

This is the limit of the use we may make if extenders as so far defined. However, other considerations motivate the following definition.

Definition 24.2. *κ is β-supercompact provided there is $j:V\to_e M$ with κ critical and $M^\beta\subseteq M$.*

κ is supercompact provided κ is β-supercompact for all β.

β-supercompact implies β-strong, of course. No extender can code a supercompact embedding. For suppose $j:V\to_E M$ and $M^\kappa\subseteq M$. Then $j''\kappa^+$ is cofinal in $(j(\kappa)^+)^M$. For given $j(f)(\kappa,a)<(j(\kappa)^+)^M$, $f:\kappa\times[\kappa]^n\to\kappa^+$ without loss of generality and letting $\gamma=\sup f''\kappa\times[\kappa]^n$ $j(f)(\kappa,a)<j(\gamma)$. But since $M^\kappa\subseteq M$ $j''\kappa^+\in M$. So $M\models cf(j(\kappa)^+)\leq\kappa^+$. But $(\kappa^+)^M<j(\kappa)<(j(\kappa)^+)^M$ and $M\models j(\kappa)^+$ is regular. Contradiction!

A final note on generalisations: κ is n-superstrong if (letting $j^0(\kappa)=\kappa$, $j^{n+1}(\kappa)=j(j^n(\kappa))$) there is $j:V\to_e M$ with κ critical and $P(j^n(\kappa))\subseteq M$. And κ is ω-superstrong provided there is $j:V\to_e M$ with with κ critical and $P(\lambda)\subseteq M$ where $\lambda=\sup_{n\in\omega} j^n(\kappa)$. But

Lemma 24.3. *κ is not ω-superstrong.*

Proof:

Claim Suppose $\lambda\geq\omega$, λ a cardinal, $2^\lambda=\lambda^{\aleph_0}$. There exists a function $f:\lambda^\omega\to\lambda$ such that whenever $A\subseteq\lambda$, $\overline{\overline{A}}=\lambda$ and $\gamma<\lambda$ then there is $s\in A^\omega$ with $f(s)=\gamma$.

Proof: Let $\{\langle A_\alpha,\gamma_\alpha\rangle:\alpha<2^\lambda\}$ enumerate $\{x\subseteq\lambda:\overline{\overline{x}}=\lambda\}\times\lambda$. Define s_α for $\alpha<2^\lambda$ inductively so that $s_\alpha\in A_\alpha^\omega$, $s_\alpha\neq s_\beta$ (all $\beta<\alpha$). Because $\overline{\overline{A}}_\alpha=\lambda$, $\overline{\overline{A_\alpha^\omega}}=\lambda^{\aleph_0}=2^\lambda$ so there is always such an s_α.

Given $s\in\lambda^\omega$ let $f(s)=\gamma_\alpha$ if $s=s_\alpha$, 0 otherwise. Given $A\subseteq\lambda$, $\overline{\overline{A}}=\lambda$ and $\gamma<\lambda$ suppose $A=A_\alpha$, $\gamma=\gamma_\alpha$. Then $f(s_\alpha)=\gamma$. □(Claim)

Now suppose $j:V\to_e M$ with $\lambda=\sup_{n\in\omega} j^n(\kappa)$, κ critical and $P(\lambda)\subseteq M$. $j(\lambda)=\lambda$, of course. Let $G=j''\lambda$. So $G\in M$. $2^\lambda=\lambda^{\aleph_0}$: for $\lambda=\sup\{j^n(\kappa):n\in\omega\}$. Hence

$$2^\lambda=2^{(\sum_{n\in\omega} j^n(\kappa))}=\prod_{n\in\omega}2^{j^n(\kappa)} \tag{24.1}$$

$$\leq\prod_{n\in\omega}\lambda \quad (j^n(\kappa)\text{ is measurable in }M\text{, hence a strong limit})$$

$$=\lambda^\omega$$

$$\leq 2^{\lambda.\omega}$$

$=2^\lambda$.

Let $f:\lambda^\omega\to\lambda$ be as in the claim. $j(f)$ must have the same property. So there is $s\in G^\omega$ with $(j(f))(s)=\kappa$.

Let $t(n)=j^{-1}(s(n))$. So $s=j(t)$. Thus $\kappa=(j(f))(j(t))=j(f(t))$ so $\kappa\in rng(j)$. This is impossible. □

Corollary 24.4. *There is no non-trivial* $j:V\to_e V$.

So there is a point where the sequence of large cardinal properties becomes inconsistent.

SCH

Given the existence of a supercompact cardinal the SCH may fail. Model 4 in appendix I explains why. Indeed SCH may fail in very simple forms: it is possible (given a suitable large cardinal) that $2^{\aleph_n}=\aleph_{n+1}$ for all $n<\omega$ but $2^{\aleph_\omega}=\aleph_{\omega+2}$. ZFC+-SCH is a large cardinal axiom: at least as strong as $O^\dagger$ by lemma 21.11. At least as strong as $O^{\dagger\ n\ \text{times}}$ for any finite n. Beyond this difficulties arise.

We may no longer take K as the union of all mice in our generalised sense of mice. Consider just the case of double mice. Let $\underline{M}=\underline{J}_\alpha^{UV}$. Are we to insist (supposing U a normal measure on κ, V on λ) that $\omega\rho_{\underline{M}}^n\le\kappa$? Or that $\omega\rho_{\underline{M}}^n\le\lambda$? We skirted the definition of critical because in daggers $\underline{M}$ $\rho_M=1$.

If we are too strict and make $\omega\rho_{\underline{M}}^n\le\kappa$ then we shall get too small a core model - a single embedding of K will never give us two normal measures! If any old mouse with $\omega\rho_{\underline{M}}^n\le\lambda$ is allowed in then chaos ensues. What we do is to fix a single measure U and only allow mice $\underline{J}_\alpha^{UV}$: then allow all <u>such</u> mice with $\omega\rho_{\underline{M}}^n\le\lambda$. U is picked minimally (in some sense).

In general, just as we added a maximal sequence to get $L[E]$ in chapter 23, now we add one to get $K[E]$. If $\pi:K[E]\to_e K[E]$ is non-trivial then as in lemma 23.8 use π to get a new extender. The resulting sequence E^+ will be coherent in $K[E^+]$ (otherwise it yields a mouse) but cannot coherently extend E(nothing can). So what has gone wrong? We must have violated the coherence condition by getting $\kappa<\kappa'$, $o(\kappa)=\kappa'$ and $o(\kappa')>0$. But then $O^\P$ exists! So perhaps -SCH implies $O^\P$ exists?

Perhaps. But not by the proof just given, for we have reckoned without our Prikry sequences. The covering lemma was proved (on the assumption $-O^\dagger$) either over K or over L[U] or over L[C]. The more measures, the more Prikry sequences we need. And once we need an infinite number, we shall no longer be able to disregard finite initial segments of C: this was essential to the proof of lemma 20.15. By careful choice of sequences we can get a covering lemma on the assumption that there is no inner model with a measurable that is a limit of measurables. Compared with the cardinals of chapter 23, that is not very far (such a cardinal need not even be of order 2).

The proofs that SCH may fail rely on an abundance of Prikry sequences given by large cardinals: and the question is; at what point do the Prikry sequences get out of control? If $0^{\P}$ does not exist can they be chosen neatly to preserve SCH? Alternatively is a strong cardinal sufficient for forcing proofs of independance of SCH?

A related question concerns $\square_\lambda$. If $0^\dagger$ does not exist then $\square_\lambda$ holds for singular λ. If there is a supercompact below λ it fails. As all that we need here is $K \models \square_\beta$ and $(\beta^+)^K = \beta^+$ for singular cardinals β this may be easier than the full covering lemma. A similar situation exists with the question whether all uncountable cardinals can be singular; see chapter 21 exercise 1.

Unless the universe is so obliging as to exclude large cardinals that proliferate Prikry sequences overmuch, the best we can hope for is a precise boundary cardinal - an X cardinal - such that

(a) if SCH fails then there is an inner model with an X-cardinal;

(b) Consis(ZFC+there is an X-cardinal) $\Rightarrow$ Consis(ZFC+-SCH).

Exercise

1. Show that if L[E] is as in chapter 23 then L[E]$\models$"there is no κ that is κ^+-supercompact".

APPENDIX I: SOME GENERIC MODELS

All of the following descriptions of models are of the vaguest kind, and are only intended as guides to the literature. The reader should consult Jech ([25]) for any unfamiliar definitions; also for full proofs.

Before any models are given, observe one general point: forcing does not generate new large cardinal properties. Otherwise we should have relative consistency proofs ruled out by Gödel's second incompleteness theorem. More precisely, suppose that M is a generic extension of N; then $K^M=K^N$. For otherwise there is a mouse in K^M but not in K^N, $N^\#$. Suppose M=N[G] where G is a generic ultrafilter on B. If $\lambda=\overline{\overline{B}}$ (in N) then B satisfies the $(\lambda^+)^N$-chain condition, so for all N-cardinals $\gamma>\lambda$ γ is a cardinal in M. Let $\langle\langle \underline{N}_\alpha\rangle_{\alpha\in On},\langle\pi_{\alpha\beta}\rangle_{\alpha\leq\beta\in On}\rangle$ be the mouse iteration of $N^\#$. If κ_α is the critical point of $\overline{\underline{N}}_\alpha$ and $\theta>\lambda$ is regular in M then $\kappa_\theta=\theta$. Now $\overline{\overline{\kappa_{\theta+1}}}^M=\theta$ so $\kappa_{\theta+1}$ is not a cardinal in M. But the κ_α are indiscernibles in K^N and $\kappa_\theta=\theta$ is a cardinal in N so $\kappa_{\theta+1}$ must be. Contradiction!

For example, if M is a generic extension of L then $0^\#\notin M$. The above proof is easily adapted to show:

(i) if M is a generic extension of K then M contains no ρ-model

(ii) if M is a generic extension of L[U] (a ρ-model) then $0^\dagger\notin M$.

Hence SCH holds in every generic extension of L[U]. A further restriction on large cardinals in generic extensions is proved in appendix II. Now for the models.

<u>Model 1</u> $cf(\omega_2^L)=\omega$ but $\overline{\overline{(\omega_2^L)}}=\omega_1$.

This model was discovered by Namba ([46])

A <u>tree</u> is a set T such that:

(i) $t\in T \Rightarrow t\in\omega_2^{<\omega}$;

(ii) $t\in T \wedge s=t|n \Rightarrow s\in T$.

T is called <u>perfect</u> provided it is non-empty and every $t\in T$ has $\aleph_2$ extensions $s\supseteq t$ in T. The set P of conditions is the set of perfect trees: the ordering is inclusion.

Let M be some model of ZFC+V=L. Let M[G] be a generic extension in which we are to define a cofinal subset of $(\omega_2)^M$ of order-type ω. It is quite easily shown that there is a unique infinite path that goes through each tree in G, that is, an ω-sequence $f:\omega\to(\omega_2)^M$ such that $\forall T\in G\exists s\in T\exists n s=f|n$. Also by genericity $f:\omega\to(\omega_2)^M$ is cofinal. So $cf(\omega_2^M)=\omega$.

The crux of the proof is the argument to show that ω_1 is preserved in the extension.

Model 2 $\kappa\to(\alpha)^{<\omega}$ but $K\not\models\kappa\to(\alpha)^{<\omega}$.

This model was discovered by Jensen([28]).

By a result of Silver if $\kappa\to(\alpha)^{<\omega}$ and M is an inner model in which α is countable then $M\models\kappa\to(\alpha)^{<\omega}$. Now suppose $\kappa\to(\omega_1)^{<\omega}$: so $0^{\#}$ exists. Let P be the conditions in L for collapsing ω_1 to ω; so P is countable, and hence there is a generic extension L[G] of L with $(\omega_1^L)=\omega$ in L[G]. Since $\kappa\to(\omega_1)^{<\omega}$ certainly $\kappa\to(\omega_1^L)^{<\omega}$, and since ω_1^L is countable in M=L[G], $M\models\kappa\to(\omega_1^L)^{<\omega}$. But $K^M=L$ and $L\not\models\kappa\to(\omega_1)^{<\omega}$ (since $0^{\#}\notin L$). So M is the model required.

Model 3 κ is measurable in an inner model M, $H_\kappa^M=H_\kappa$ and $cf(\kappa)=\omega$.

This model was discovered by Prikry([48]).

Take a model M with κ measurable. We are going to add a Prikry sequence to M. The set of conditions P is $\{\langle s,A\rangle : s\in[\kappa]^{<\omega} \wedge A\in U\}$ where U is the normal measure on κ. $\langle s,A\rangle<\langle t,B\rangle$ means:

(a) $t=s\cap\alpha$, some α;

(b) $A\subseteq B$;

(c) $s\setminus t\subseteq B$.

Let G be a generic filter on P. Let $S=\{s:\langle s,A\rangle\in G$, some $A\}$. If $\langle s,A\rangle$, $\langle t,B\rangle\in G$ then either $s\subseteq t$ or $t\subseteq s$ so S is of order-type ω. $S\subseteq\kappa$, obviously, and by genericity S is cofinal. So $cf(\kappa)=\omega$.

S is indeed a Prikry sequence. For suppose $X\in U$. Then $\{\langle s,A\rangle : A\subseteq X\}$ is dense in P: for given any $\langle t,B\rangle\in P$ $\langle t,B\cap X\rangle<\langle t,B\rangle$. So there is $\langle s,A\rangle\in G$ with $A\subseteq X$. Suppose $\langle t,B\rangle\in G$, $\langle t,B\rangle<\langle s,A\rangle$. Then $t\setminus s\in A\subseteq X$. Hence $S\setminus s\subseteq X$, so S is a Prikry sequence for U.

The difficult part of the proof is $H_\kappa=H_\kappa^M$. This cannot be proved by κ-closure since P adds a new ω-sequence. It follows in fact from a technical lemma: if σ is a sentence of the forcing language and $\langle s_0,A_0\rangle\in P$ then there is $A\subseteq A_0$, $A\in U$ such that $\langle s_0,A\rangle\Vdash\sigma$ or $\langle s_0,A\rangle\Vdash\neg\sigma$.

A converse also holds: Mathias ([40]) showed that any Prikry sequence C for U yields a generic ultrafilter $G=\{\langle s,A\rangle : s\subseteq C \wedge C\subseteq A\}$ on P.

Model 4 κ measurable and $2^\kappa>\kappa^+$.

This model was discovered by Silver (see Menas [41]).

Since GCH cannot fail for the first time at a measurable it is necessary not only to disrupt $P(\kappa)$ but also $P(\alpha)$ for many $\alpha<\kappa$. By iterated forcing add α^{++} new subsets to α for all inaccessible $\alpha \leq \kappa$. The resulting model has $2^{\kappa}=\kappa^{++}$, of course, but κ may not be measurable. It turns out that provided that κ was supercompact in the ground model we can obtain an elementary embedding of M[G] into an inner model with κ critical, so that κ remains measurable.

Now take a Prikry extension of M[G], M[G][C] and obtain a model where $cf(\kappa)=\omega$ and $2^{\kappa}>\kappa^{+}$ and $H_{\kappa}^{M[G]}=H_{\kappa}^{M[G][C]}$. But since κ is measurable in M[G] it is a strong limit in M[G], so for all $\gamma<\kappa$ $(2^{\gamma})^{M[G]}<\kappa$, so this holds too in M[G][C]. So SCH fails in M[G][C].

APPENDIX II: ABSOLUTENESS RESULTS

This appendix gathers together some absoluteness properties for ZFC models that are not true in general for models of R. These are used to prove the Σ^1_3-absoluteness of K (under certain conditions) and to give a negative result prophesied in chapter 16. The proofs in this area seem to me dreadfully indirect, but no doubt descriptive set theorists think the same about fine-structure.

The absoluteness theorem in its most general form - lemma 4 below - is well-known, but I have not found it as stated here in the literature, so I have had to include a proof. It is a trivial modification of the proof sketched in Jech ([25]), based in turn on the proof of Barwise and Fisher (see [1]).

All absoluteness results in this appendix stem from the absoluteness of well-foundedness; this is proved by using a well-known consequence of the replacement axioms to turn the apparently Π_1 property of well-foundedness into a Σ_1 one.

Lemma 1 *Suppose M is an inner model of ZF and $E \in M$ is a partial order. Then E is well-founded if and only if $M \models E$ is well-founded.*

Proof: If E is well-founded then $M \models E$ is well-founded. On the other hand, suppose $M \models E$ is well-founded. Recall that $\rho_E(x)$ is defined inductively on well-founded E by $\rho_E(x) = \sup\{\rho_E(y)+1 : yEx\}$. So in M there is a function f whose range is a subset of On such that $xEy \Rightarrow f(x)<f(y)$. But then f really does have this property: so E is well-founded. □

The foundation of the absoluteness theorem is the following lemma:

Lemma 2 *Suppose M is an inner model of ZFC and $\Theta \in M$ is a set of $\forall\exists$-sentences, countable in M, in a language $L_{E,\vec{A},\vec{c}}$ with no function symbols. Suppose further that Θ has a model $\underline{N}$ for which E^N is well-founded. Then Θ has a model $N' \in M$ for which $E^{N'}$ is well-founded.*

Note: By an $\forall\exists$ sentence we mean one which is of the form $\forall x_1 \ldots x_m \exists y_1 \ldots y_n \psi(x_1 \ldots x_m, y_1 \ldots y_n)$ where ψ is quantifier-free.

Proof: If Θ has a finite model we are done, so suppose it does not. Let $\langle \sigma_k : k \in \omega \rangle \in M$ enumerate Θ . We may assume $L_{E,\vec{A},\vec{c}}$ is countable: if $\langle \underline{c}_k : k \in \omega \rangle$ enumerates its constant symbols we may also assume that all the constants of σ_k lie among $\underline{c}_0 \ldots \underline{c}_{k-1}$. Let P be the set of triples $\langle \underline{B}, f, k\rangle$ such that, letting $\underline{B}^*$ be some model of Θ with $\overline{\overline{B^*}}<\kappa$, E^{B^*} well-founded and $\rho_E B^*: B^* \to \kappa$

(i) $k \in \omega$;

(ii) $B \subseteq \kappa$, $\underline{B}$ is a finite model of the type of $L_{E,\vec{A},\underline{c}_0 \ldots \underline{c}_{k-1}}$;

(iii) $f: B \to \kappa$ is such that $bEb' \Rightarrow f(b) < f(b')$.

A partial order < of P is defined by $\langle \underline{B}, f, k\rangle \prec \langle \underline{B}', f', k'\rangle$ provided:

(i) $k > k'$;

(ii) $\underline{B}'$ is a substructure of the reduct of $\underline{B}$ to $L_{E,\vec{A},\underline{c}_0 \ldots \underline{c}_{k'-1}}$;

(iii) $f' \subseteq f$;

(iv) for all $h \leq k$, supposing σ_h is $\forall x_1 \ldots \forall x_m \exists y_1 \ldots \exists y_n \psi(\vec{x}, \vec{y})$ and $a_1 \ldots a_m \in B'$ then there are $b_1 \ldots b_n \in B$ such that $\underline{B} \models \psi(a_1 \ldots a_m, b_1 \ldots b_n)$.

Then since $\langle \sigma_k : k \in \omega\rangle \in M$, $\langle P, <\rangle$ is in M.

We want to show that $\langle P, <\rangle$ is not well-founded. To do this we define inductively an infinite descending chain in < ,$\langle\langle \underline{B}_n, f_n, n\rangle : n \in \omega\rangle$ as follows: let $\underline{B}$ be some B^* as above with E^B well-founded. We may assume $B \subseteq \kappa$. Pick some $b \in B$ arbitrarily: let $B_0 = \{b\}$, $\underline{B}_0$=the unique substructure of the reduct of $\underline{B}$ to $L_{\vec{E},A}$ with domain B_0, $f_0 = \{\langle b, \rho_E B(b)\rangle\}$.

Now suppose $\underline{B}_n, f_n$ have been picked with $\langle \underline{B}_n, f_n, n\rangle \in P$ and $f_n = \rho_E B | B_n$. Suppose $h \leq n+1$; let σ_h be $\forall x_1 \ldots x_m \exists y_1 \ldots \exists y_n \psi_h(x, y)$ with ψ_h quantifier free. $\underline{B} \models \sigma_h$ so there is a finite set X_h such that for all $a_1 \ldots a_m \in B_n$ there are $b_1 \ldots b_n \in X_h$ $\underline{B} \models \psi_h(a_1 \ldots a_m, b_1 \ldots b_n)$. Let $B_{n+1} = B_n \cup X_0 \cup \ldots \cup X_n \cup \{c_n^B\}$. Let $\underline{B}_{n+1}$ be the unique substructure of the reduct of $\underline{B}$ to $L_{E,\vec{A},\underline{c}_0 \ldots \underline{c}_n}$ with domain B_{n+1}. Then since ψ_h is quantifier free and involves only constants from among $\underline{c}_0 \ldots \underline{c}_n$ ($h \leq n+1$) $\forall a_1 \ldots \forall a_m \in B_n \exists b_1 \ldots b_n \in B_{n+1}$ $\underline{B}_{n+1} \models \psi_h(a_1 \ldots a_m, b_1 \ldots b_n)$. Let $f_{n+1} = \rho_E B | B_{n+1}$. So $\langle \underline{B}_{n+1}, f_{n+1}, n+1\rangle < \langle \underline{B}_n, f_n, n\rangle$. So $\langle P, <\rangle$ is not well-founded; by lemma 1, then, it is not well-founded in M.

So there is, in M, an infinite descending <-chain: call it $\langle\langle \underline{C}_n, g_n, k_n\rangle : n \in \omega\rangle$. Let $C = \bigcup_{n \in \omega} C_n$. Define E^C by $xE^C y \Leftrightarrow \exists n(x, y \in C_n \wedge xE^{C_n} y)$; similarly for $\vec{A}$. Define c_k^C to be $c_k^{C_h}$ where $k_h > k$. This defines $\underline{C}$. Let $g = \bigcup_{n \in \omega} g_n$; then $g: C \to On$ such that $cE^C c' \Rightarrow g(c) < g(c')$, so E^C is well-founded. Finally we must show that $\underline{C} \models \Theta$. Suppose $a_1 \ldots a_m \in C_h$: we may also assume $k_h > k$. $\langle \underline{C}_{h+1}, g_{h+1}, k_{h+1}\rangle < \langle \underline{C}_h, g_h, k_h\rangle$ so $k < k_{h+1}$ and there are $b_1 \ldots b_n \in C_{h+1}$ such that $\underline{C}_{h+1} \models \psi_h(a_1 \ldots a_m, b_1 \ldots b_n)$. Hence $\underline{C} \models \psi_h(a_1 \ldots a_m, b_1 \ldots b_n)$. □

Note that we could have stipulated that N' be countable in M. The apparently essential restriction to $\forall\exists$-sentences can be removed by the following lemma.

<u>Lemma 3</u> *Let σ be a sentence of a language L'. Then there is $k\in\omega$ and an $\forall\exists$-sentence σ' of $L'_{\underline{R}_1,\ldots,\underline{R}_k}$ such that:*

(i) if $\underline{A}\models\sigma$ then there are $R_1\ldots R_k$ such that $\langle\underline{A},R_1\ldots R_k\rangle\models\sigma'$;

(ii) if $\langle\underline{A},R_1\ldots R_k\rangle\models\sigma'$ then $\underline{A}\models\sigma$.

Note: σ' is called the Skolem normal form of σ.

Proof: Suppose σ is in prenex normal form, $Q_1x_1\ldots Q_kx_k\psi(x_1\ldots x_k)$ where Q_i is $\forall$ or $\exists$. Define sentences σ_h:

σ_1 is $\forall x_1\ldots x_{k-1}(Q_kx_k\psi(x_1\ldots x_k) \Leftrightarrow R_1(x_1\ldots x_{k-1}))$;

σ_h is $\forall x_1\ldots x_{k-h}(Q_{k-h+1}x_{k-h+1}R_{h-1}(x_1\ldots x_{k-h+1}) \Leftrightarrow R_h(x_1\ldots x_{k-h}))$

$(1<h\leq k)$. So in particular σ_k is $Q_1x_1R_{k-1}(x_1) \Leftrightarrow R_k$. Let σ' be $R_k\wedge\sigma_1\wedge\ldots\wedge\sigma_k$. It remains to prove that σ' is $\forall\exists$. We need only consider σ_h: this is of the form $\forall x_1\ldots x_{h-1}(QxRx \Leftrightarrow Sx)$. This can be written as a conjunct of an $\forall$ and an $\forall\exists$ sentence.

Now for the absoluteness theorem: in essence it is due to Levy:

<u>Lemma 4</u> *Suppose $\underline{M}$ is an inner model of ZFC and suppose $a_1\ldots a_n$ are hereditarily countable in M. Let $\phi(x_1\ldots x_n)$ be a Σ_1 formula with at most $x_1\ldots x_n$ free. Suppose $\phi(a_1\ldots a_n)$. Then $M\models\phi(a_1\ldots a_n)$.*

Proof: We may assume n=1. Let $x=x_1$. Let $y=TC(\{x\})$ and let $f:\omega\to y$ onto with $f\in M$. Consider a theory Θ in the language $L_{E,\underline{f},\underline{y},\langle\underline{n}:n\in\omega\rangle}$ (in standard set-theoretic formulae E is to stand for $\in$):

(i) extensionality;

(ii) y is transitive;

(iii) f is a function with domain ω and range y;

$(iv)_{n,m}$ $\underline{f}(\underline{n})=\underline{f}(\underline{m})$ (if $f(n)=f(m)$);

$(v)_{n,m}$ $\underline{f}(\underline{n})\neq\underline{f}(\underline{m})$ (if $f(n)\neq f(m)$);

$(vi)_{n,m}$ $\underline{f}(\underline{n})\in\underline{f}(\underline{m})$ (if $f(n)\in f(m)$);

$(vii)_{n,m}$ $\underline{f}(\underline{n})\notin\underline{f}(\underline{m})$ (if $f(n)\notin f(m)$);

$(viii)_n$ $\underline{n+1}=\underline{n}+1$; $\underline{0}=0$;

(ix) $\phi(\underline{f}(\underline{k}))$ (where $x=f(k)$).

Then $\langle H_{\omega_1},\in,f,y,n\rangle_{n\in\omega}$ is a well-founded model of Θ. Also $\Theta\in M$. Now the only axiom of Θ that is not necessarily $\forall\exists$ is (ix). But there are $R_1\ldots R_k$ and (ix)' which is $\forall\exists$ as in lemma 3: so $\langle H_{\omega_1},\in,R_1\ldots R_k,f,y,n\rangle$ is a well-founded model of $\Theta'=(i)-(viii)_n+(ix)'$. By lemma 2, then, there is a well-founded model of Θ' in M, call it $\underline{B}=\langle B,E,S_1\ldots S_k,g,z,m_n\rangle_{n\in\omega}$. Thus $\langle B,E,g,z,m_n\rangle\models\Theta$. By (i) $\langle B,E\rangle$ is extensional, so there is a transitive C with $\pi:\langle C,\in\rangle\cong\langle B,E\rangle$. Let $g'=\pi^{-1}(g)$, $z'=\pi^{-1}(z)$, $m'_n=\pi^{-1}(m_n)$. By induction using (viii) $m'_n=n$. Since C is transitive z' is transitive (by (ii)) and g' is a function of ω onto z' (by (iii)). Let $\sigma:y\to z'$ be defined by $\sigma(f(n))=g'(n)$. Then (iv)-(vii) show that σ is well-defined and $\sigma:\langle y,\in\rangle\cong\langle z',\in\rangle$

so z=y' and f=g'. In particular g'(k)=x, so $\langle C,\in\rangle\models\phi(x)$. But ϕ is Σ_1 so $M\models\phi(x)$. □

By the remark after lemma 2 this could be extended to $H_{\omega_1}^M M\models\phi(x)$. For example $L_{\omega_1^L}\prec_{\Sigma_1} V$. Hereditary countability is essential: if $\omega_1^L\neq\omega_1$ then "ω_1^L is countable" but $L\not\models$"ω_1^L is countable".

Now we quote a standard result of descriptive set theory, due in essence to Kleene:

Lemma 5 *If ϕ is a Σ^1_2 formula then there is a Σ_1 formula ψ such that $\phi(a_1\ldots a_n)\Leftrightarrow H_{\omega_1}\models\psi(a_1\ldots a_n)$.*

For a proof see Jech [25] section 41. Hence

Lemma 6 *If ϕ is a Σ^1_2 formula and $a\in M\cap P(\omega)$, M an inner model of ZFC, then $M\models\phi(a)\Leftrightarrow\phi(a)$.*

Proof: There is a Σ_1 formula ϕ' such that $\phi(a)\Leftrightarrow H_{\omega_1}\models\phi'(a)$. Suppose $\phi(a)$; then $H_{\omega_1}\models\phi'(a)$. So by lemma 4 $H_{\omega_1}^M M\models\phi'(a)$, so $M\models\phi(a)$. If, conversely, $M\models\phi(a)$ then $H_{\omega_1}^M M\models\phi'(a)$ and $H_{\omega_1}^M M\subseteq H_{\omega_1}$ so $H_{\omega_1}\models\phi'(a)$; thus $\phi(a)$. □

In particular any true Σ^1_2 sentence is true in M. If a is a Σ^1_2 singleton - i.e. there is a Σ^1_2 formula ϕ such that $x=a\Leftrightarrow\phi(x)$ - then a is constructible, because $\exists x\phi(x)\Rightarrow\exists x\in L\phi(x)$. On the other hand $0^\#$ is a Π^1_2 singleton that is not constructible. "$0^\#$ exists" is a Σ^1_3 sentence, so lemma 6 is the best possible. Indeed we do not need $0^\#$: $\exists x\subseteq\omega(x\notin L)$ is Σ^1_3.

Suppose our inner model was the core model K. The $0^\#$ example would cause no difficulty, of course. K cannot be Σ^1_3 absolute, for it could be L. To say that K is not L is to say that $0^\#$ exists; in fact we must not let K be of the form L[a] for any $a\subseteq\omega$; this is effected if we insist that the reals be closed under #. $0^\dagger$ is a Π^1_2 singleton, so "$0^\dagger$ exists" is Σ^1_3 (but fails in K). So we must also stipulate that $0^\dagger$ should not exist. Given these restrictions we can show that K is Σ^1_3-absolute. Since $x\in K\cap P(\omega)$ is Σ^1_3(($\exists$M)(M is a real mouse and $x\in M$); we know that being a real mouse can be coded as a Π^1_2 property of reals) $P(\omega)\neq K\cap P(\omega)$ is Σ^1_4, so this is the best we can hope for.

We shall make considerable use of the discussion of $a^\#$ in chapter 21. The main result we need is:

<u>Lemma 7</u> *If ϕ is a Π^1_2 formula of one free variable and $a\subseteq\omega$ such that $\phi(a)$, $a^{\#}$ exists and $L[a]\models$there are no ρ-models then $\exists a\in K\phi(a)$.*

Proof: Suppose first of all that $K^{L[a]}\neq K$ and let $\underline{M}$ be $L[a]^{\#}$ with mouse iteration $\langle\langle\underline{M}_\alpha\rangle_{\alpha\in On},\langle\pi_{\alpha\beta}\rangle_{\alpha\leq\beta\in On}\rangle$. Let κ_α be the critical point of $\underline{M}_\alpha$. Let $\underline{N}=\underline{J}^{aV}_{\tau+1}$ be the minimal core a-mouse and let $\langle\langle L[a]\rangle,\langle\pi'_{\alpha\beta}\rangle_{\alpha\leq\beta\in On}\rangle$ be the weak iterated ultrapower of L[a] by V. Let $\tau_\alpha=\pi'_{0\alpha}(\tau)$. So $C'=\{\tau_\alpha:\alpha\in On\}$ are the canonical indiscernibles for L[a]. By lemma 21.22 there are α,β such that, letting $\kappa_\alpha=\tau_{i_0}$,

(a) $\tau_{i_0+j}=\kappa_{\alpha+\beta j}$ (all j);

(b) $\pi'_{i_0+i,i_0+j}(\kappa_{\alpha+\beta i+\nu})=\kappa_{\alpha+\beta j+\nu}$ (all $i\leq j$, all ν).

Suppose $\mu<\nu\in On$. $t(\mu,\nu)$ is defined to be $\langle 0,\nu\rangle$if $\nu<\alpha$; if $\nu=\alpha+\beta j+\gamma$ $(\gamma<\beta)$ and $\mu<\alpha+\beta j$ then $t(\mu,\nu)$ is $\langle 1,\gamma\rangle$; otherwise if $\nu=\alpha+\beta j+\gamma$ $(\gamma<\beta)$ then $t(\mu,\nu)$ is $\langle 2,\gamma\rangle$. Given $j_0<\ldots<j_{m-1}\in On$ $t^*_{\alpha\beta}(j_0\ldots j_{m-1})$ denotes $\langle t(0,j_0)\ldots t(j_{m-2},j_{m-1})\rangle$. Suppose $t=\langle t_0\ldots t_{m-1}\rangle\in(3\times On)^m$: t is called (α,β)-apt provided:

(i) $t_i=\langle 0,\gamma\rangle\Rightarrow\forall j<i\exists\gamma'<\gamma\, t_j=\langle 0,\gamma'\rangle$;

(ii) $t_i=\langle 0,\gamma\rangle\Rightarrow\gamma<\alpha$;

(iii) $t_i=\langle 1,\gamma\rangle$ or $t_i=\langle 2,\gamma\rangle\Rightarrow\gamma<\beta$;

(iv) $t_i=\langle 2,\gamma\rangle\Rightarrow i>0, t_{i-1}=\langle 1,\gamma'\rangle$ or $\langle 2,\gamma'\rangle$ and $\gamma'<\gamma$.

So $t^*_{\alpha\beta}(j_0\ldots j_{m-1})$ $(j_0<\ldots<j_{m-1}\in On)$ is (α,β)-apt. Say $\langle j_0\ldots j_{m-1}\rangle\sim_{\alpha\beta}$ $\sim_{\alpha\beta}\langle j'_0\ldots j'_{m-1}\rangle$ provided $t^*_{\alpha\beta}(j_0\ldots j_{m-1})=t^*_{\alpha\beta}(j'_0\ldots j'_{m-1})$.

Suppose $\langle t_0\ldots t_{m-1}\rangle$ is (α,β)-apt. Define inductively k_h,γ_h,β_h by:

(i) if $t_h=\langle 0,\gamma\rangle$ then k_h,γ_h are undefined and $\beta_h=\gamma$;

(ii) if $t_h=\langle 1,\gamma\rangle$ then $k_h=k_{h-1}+1$, or 0 if k_{h-1} is undefined, $\gamma_h=\gamma$ and $\beta_h=\alpha+\beta k_h+\gamma_h$;

(iii) if $t_h=\langle 2,\gamma\rangle$ then $k_h=k_{h-1}$ (which must be defined), $\gamma_h=\gamma$ and $\beta_h=\alpha+\beta k_h+\gamma_h$.

Now fix some m-tuple $\langle j_0\ldots j_{m-1}\rangle$ with $t^*_{\alpha\beta}(j_0\ldots j_{m-1})=\langle t_0\ldots t_{m-1}\rangle$ Suppose $j_h=\alpha+\beta\mu_h+\nu_h$ $(\nu_h<\beta)$ (if $j_h<\alpha$ then μ_h,ν_h are undefined). Let $\tilde{k}=i_0+k_{m-1}+1$, $\tilde{\mu}=i_0+\mu_{m-1}+1$. We are to define a map $\sigma:L[a]\to_e L[a]$. Work by induction on h. σ_0 is $id|L[a]$. Suppose $\sigma_h:L[a]\to_e L[a]$. If k_h is undefined let $\sigma_{h+1}=id|L[a]$. If $k_h=k_{h-1}$ then let $\sigma_{h+1}=\sigma_h$. Suppose $k_h=k_{h-1}$. If k_{h-1} is undefined then $k_h=0$; let $\sigma_{h+1}:L[a]\to_e L[a]$be the unique map such that $\sigma_{h+1}\pi'_{i_0 i_0+1}=\pi'_{i_0 i_0+\mu_h+1}$ and $\sigma_{h+1}(\tau_{i_0})=\tau_{i_0+\mu_h}$. If $k_h=k_{h-1}+1$ let $\sigma_{h+1}:L[a]\to_e L[a]$ be the unique map such that $\sigma_{h+1}\pi'_{i_0+k_h,i_0+k_h+1}=\pi'_{i_0+\mu_{h-1}+1,i_0+\mu_h+1}\sigma_h$ and $\sigma_{h+1}(\tau_{i_0+k_h})=\tau_{i_0+\mu_h}$. Let $\sigma=\sigma_m$.

<u>Claim 1</u> If $X\in W$ and $\nu\in On$ then $\sigma_h(\tilde{X}\cap V_\nu)=\tilde{X}\cap V_{\sigma_h(\nu)}$.

Proof: Let $\nu<\nu'=\pi'_{i_0 i_0+k_{h-1}+1}(\bar{\nu})$. Then

$$\sigma_h(\tilde{X}\cap V_\nu)=\sigma_h((\tilde{X}\cap V_{\nu'})\cap V_\nu)$$

$=\sigma_h\pi'_{i_0 i_0+k_{h-1}+1}(\tilde{X}\cap V_{\bar{\nu}})\cap V_{\sigma_h(\nu)}$ (by lemma 21.19)

$=\pi'_{i_0 i_0+\mu_{h-1}+1}(\tilde{X}\cap V_{\bar{\nu}})\cap V_{\sigma_h(\nu)}$

$=(\tilde{X}\cap V_{\pi'_{i_0 i_0+\mu_{h-1}+1}(\bar{\nu})})\cap V_{\sigma_h(\nu)}$ (by lemma 21.19)

$=\tilde{X}\cap V_{\sigma_h(\nu)}$. □(Claim 1)

Claim 2 $\gamma\in C \Leftrightarrow \sigma_h(\gamma)\in C$.

Proof: If $\gamma\in C$ then for all $X\in U$ $\gamma\in\tilde{X}$ ($\underline{M}=\underline{J}^U_\delta$). Pick $\nu>\gamma$: so $\gamma\in\tilde{X}\cap V_\nu$, hence $\sigma_h(\gamma)\in X\cap V_{\sigma_h(\nu)}\subseteq\tilde{X}$. So $\sigma_h(\gamma)\in C$. The converse is similar.

□(Claim 2)

Claim 3 $C\cap[\tau_{i_0+\mu_h},\tau_{i_0+\mu_h+1})\subseteq\mathrm{rng}(\sigma_{h+1})$.

Proof: Suppose $\xi\in C\cap[\tau_{i_0+\mu_h},\tau_{i_0+\mu_h+1})$; then $\xi=\pi'_{i_0 i_0+\mu_h}(\bar{\xi})$ for some $\bar{\xi}$ in $C\cap[\tau_{i_0},\tau_{i_0+1})$. There is a term t and a $k\in\omega$, $\vec{\gamma}<\tau_{i_0}$ such that $L[a]\models\bar{\xi}=t(\vec{\gamma},\tau_{i_0},\tau_{i_0+1}\ldots\tau_{i_0+k})$. Hence $L[a]\models\xi=t(\vec{\gamma},\tau_{i_0+\mu_h},\tau_{i_0+\mu_h+1}\ldots$ $\ldots\tau_{i_0+\mu_h+k})$ (for $\pi'_{i_0 i_0+\mu_h}(\tau_{i_0+k})=\pi'_{i_0 i_0+\mu_h}\pi'_{i_0 i_0+k}(\tau_{i_0})=$ $=\pi'_{i_0 i_0+\mu_h+k}(\tau_{i_0})=\tau_{i_0+\mu_h+k}$). But for $k>0$ $\tau_{i_0+\mu_h+k}=\pi'_{i_0 i_0+\mu_h+1}(\tau_{i_0+k-1})$ $=\sigma_{h+1}\pi'_{i_0 i_0+k_h+1}(\tau_{i_0+k-1})\in\mathrm{rng}(\sigma_{h+1})$. And $\tau_{i_0+\mu_h}\in\mathrm{rng}(\sigma_{h+1})$. So $\xi\in\mathrm{rng}(\sigma_{h+1})$. □(Claim 3)

It follows at once that $\sigma_{h+1}(\kappa_{\beta_h})=\kappa_{j_h}$. And so $\sigma(\kappa_{\beta_h})=\kappa_{j_h}$ for all $h<m$. Hence

$$L[a]\models\phi(\tilde{f}(\kappa_{\beta_0}\ldots\kappa_{\beta_{m-1}}))\Leftrightarrow\phi(\tilde{f}(\kappa_{j_0}\ldots\kappa_{j_{m-1}}))$$

(using claim 1 again). But $\beta_1\ldots\beta_{m-1}$ depended only on $t^*_{\alpha\beta}(j_0\ldots j_{m-1})$. So $\langle j_0\ldots j_{m-1}\rangle\sim_{\alpha\beta}\langle j'_0\ldots j'_{m-1}\rangle\Rightarrow L[a]\models\phi(\tilde{f}(\kappa_{j_0}\ldots\kappa_{j_{m-1}}))\Leftrightarrow$ $\Leftrightarrow\phi(\tilde{f}(\kappa_{j'_0}\ldots\kappa_{j'_{m-1}}))$.

Consider the statement $\psi_{M\alpha\beta}(a)$ that says:

$a\subseteq\omega$ and, letting $\bar{\kappa}$ be the critical point of $\underline{M}_{\alpha+\beta\omega}$ $\underline{J}^a_{\bar{\kappa}}\models\phi(a)$; and $\gamma_0<\ldots<\gamma_{m-1}<\alpha+\beta\omega$, $\beta_0<\ldots<\beta_{m-1}<\alpha+\beta\omega$, $f:\bar{\kappa}^m\to\bar{\kappa}$, $f\in\mathrm{rng}(\pi_{0\alpha+\beta\omega})$, $\langle\gamma_0\ldots\gamma_{m-1}\rangle\sim_{\alpha\beta}\langle\beta_0\ldots\beta_{m-1}\rangle$ imply $J^a_{\bar{\kappa}}\models\phi(f(\kappa_{\gamma_0}\ldots\kappa_{\gamma_{m-1}}))\Leftrightarrow\phi(f(\kappa_{\beta_0}\ldots$ $\ldots\kappa_{\beta_{m-1}}))$ for all formulae ϕ.

Claim 4 $\psi_{M\alpha\beta}(a)$ is true.

Proof: We need only show $\underline{J}^a_{\bar{\kappa}}\models\phi(a)$. We know $\phi(a)$, so $L[a]\models\phi(a)$ by lemma 6. But $\bar{\kappa}=\tau_{i_0+\omega}$ and $\{\tau_\alpha:\alpha\in On\}$ are the canonical indiscernibles for $L[a]$ so $\underline{J}^a_{\bar{\kappa}}\prec L[a]$. □(Claim 4)

$\psi_{M\alpha\beta}$ is not a Σ_1 property of M,α,β,a because "$\underline{M}$ is a mouse"is not Σ_1 (otherwise there would be mice in L! It is Π_2). And M,α,β are not necessarily countable in K, so lemma 4 cannot be applied directly. But lemma 2 can be adapted to prove:

Claim 5 There are $\bar{M},\bar{\alpha},\bar{\beta},\bar{a}\in K$, countable in K, such that $\psi'_{\bar{M},\bar{\alpha},\bar{\beta}}(\bar{a})$ and there is $\sigma:\bar{M}\to_e M$, where ψ' differs from ψ in only asserting that there is a transitive $\alpha+\beta\omega$-th mouse iterate of M.

Proof: A simple modification of the definition of $\langle P,<\rangle$; details are left to the reader. □(Claim 5)

But then $\bar{M}$ is a mouse. Let $\langle\langle \underline{\bar{M}}_\alpha\rangle_{\alpha\in On},\langle\bar{\pi}_{\alpha\beta}\rangle_{\alpha<\beta\in On}\rangle$ be its mouse iteration. Let $\bar{\kappa}_\delta$ be the critical point of $\underline{\bar{M}}_\delta$ and let $\bar{\kappa}=\bar{\kappa}_{\alpha+\beta\omega}$. So $J^{\bar{a}}_{\bar{\kappa}}\models\phi(\bar{a})$. The crucial step is the next one, which shows that $J^{\bar{a}}_{\bar{\kappa}}\prec L[\bar{a}]$.

From now on $\tilde{X}$ is to denote $\bigcup_{i\in On}\bar{\pi}_{0i}(X)$. Consider the language L', which is L augmented by a constant $\underline{a}$ and by constants $\underline{\nu}$ for each ordinal ν. For any ordinal ν there are $\gamma_0<\ldots<\gamma_{m-1}\in On$ and $f\in\bar{M}$, $f:\bar{\kappa}_0^m\to\bar{\kappa}_0$ such that $\nu=\tilde{f}(\kappa_{\gamma_0}\ldots\kappa_{\gamma_{m-1}})$.

Define a set of sentences S of the language L' as follows: suppose ϕ is a formula of L_a with n free variables. Suppose $\nu_i\in On$, $\nu_i=f_i(\kappa_{\gamma_0}\ldots\kappa_{\gamma_{m-1}})$ $(i<n)$. Then $\phi(\underline{\nu}_0\ldots\underline{\nu}_{n-1})$ is to be in S if and only if there are $\delta_0<\ldots<\delta_{m-1}<\bar{\alpha}+\bar{\beta}\omega$ such that $\langle\gamma_0\ldots\gamma_{m-1}\rangle\sim_{\overline{\alpha\beta}}\langle\delta_0\ldots\delta_{m-1}\rangle$ and, letting $\bar{\nu}_i=\tilde{f}_i(\kappa_{\delta_0}\ldots\kappa_{\delta_{m-1}})$, $J^{\bar{a}}_{\bar{\kappa}}\models\phi(\bar{\nu}_0\ldots\bar{\nu}_{n-1})$. The choice of δ_i are irrelevant; and S is consistent.

S has definable Skolem functions: that is, given a formula χ of L' there is a formula χ' that defines a function (i.e. $S\vdash\chi'(\vec{x},y)\wedge\chi'(\vec{x},z)\Rightarrow y=z$), call it h_χ, such that $S\vdash\exists y\chi(\vec{x},y)\Rightarrow$ $\Rightarrow\chi(\vec{x},h_\chi(\vec{x}))$. ($h_\chi(\vec{x})$ may be the L[a]-least y such that $\chi(\vec{x},y)$ when there is one, for example). So we may build a term model of S with domain $\{[\text{"}h_\chi(\vec{\nu})\text{"}]:\vec{\nu}\in On\}$, where [] denotes the equivalence class of a term t under $t\sim t'\Leftrightarrow S\vdash t=t'$. Call this model Ω.

Suppose "$\exists x\in On\chi(x)$"$\in S$, where χ is a formula of L'. Let $\chi(x)\Leftrightarrow\chi'(x,\underline{\tilde{f}_1(\kappa_{\gamma_0}\ldots\kappa_{\gamma_{m-1}})}\ldots\underline{\tilde{f}_{n-1}(\kappa_{\gamma_0}\ldots\kappa_{\gamma_{m-1}})})$ where χ' is a formula of L_a. Take $\beta_0<\ldots<\beta_{m-1}$ of $t^*_{\overline{\alpha\beta}}(\gamma_0\ldots\gamma_{m-1})$ as in the indiscernibility argument above. So $\beta_0\ldots\beta_{m-1}<\bar{\alpha}+\bar{\beta}\omega$. Suppose $J^{\bar{a}}_{\bar{\kappa}}\models\chi'(\tilde{g}(\kappa_{\beta'_0}\ldots\kappa_{\beta'_{p-1}}),\vec{\tilde{f}}(\kappa_{\beta_0}\ldots\kappa_{\beta_{m-1}}))$ with $\{\beta'_0\ldots\beta'_{p-1}\}\supseteq\{\beta_0\ldots\beta_{m-1}\}$. It is easily seen that there are $\{\gamma'_0\ldots\gamma'_{p-1}\}\supseteq\{\gamma_0\ldots\gamma_{m-1}\}$ such that $\langle\gamma'_0\ldots\gamma'_{p-1}\rangle\sim_{\overline{\alpha\beta}}\langle\gamma'_0\ldots\gamma'_{p-1}\rangle$. So $S\vdash\chi(\underline{\tilde{g}(\kappa_{\gamma'_0}\ldots\kappa_{\gamma'_{p-1}})})$. That is, there is is an ordinal ν such that $S\vdash\chi(\underline{\nu})$. In particular for any Skolem term $t(\vec{\underline{\nu}})$ of S there is $\gamma\in On$ such that $S\vdash\text{rank}(t(\vec{\underline{\nu}}))=\underline{\gamma}$. So Ω is well-founded. An indiscernibility argument shows that for $x\in\Omega$ $\{y:\Omega\models y\in x\}$ is a set, so there is a transitive class Ω' isomorphic to Ω. So $\bar{a}$ is the Ω'-interpretation of $\underline{a}$. Also, since clearly for $\vec{\nu}<\bar{\kappa}$ $J^{\bar{a}}_{\bar{\kappa}}\models\chi(\vec{\nu})\Leftrightarrow S\vdash\chi(\vec{\underline{\nu}})$, $J^{\bar{a}}_{\kappa}\prec\Omega'$. So $\Omega'\models V=L[\bar{a}]$, i.e. $\Omega'=L[\bar{a}]$, and $\Omega'\models\phi(\bar{a})$. By lemma 6 $\phi(\bar{a})$.

If $K=K^{L[a]}$ then M is to be the minimal ρ-model: as in chapter 21 the modifications are left to the reader. □

Lemma 8 *If $0^\dagger$ does not exist and ϕ is a Π^1_2 formula of one free variable, $a\subseteq\omega$ $\phi(a)$ and $a^\#$ exists then $\exists a\in K\phi(a)$.*

Proof: Suppose not, so that by lemma 7 $L[a]\models$ there is a ρ-model. Let $L[a]\models M$ is a ρ-model with critical point κ. There is an embedding $j:L[a]\to_e L[a]$ with critical point above κ^+: but then $j|M$ gives $0^\dagger$. □

Lemma 9 *Suppose that the reals are closed under # but that $0^\dagger$ does not exist. Then any true Σ^1_3 sentence is true in K.*

Proof: Suppose $\phi(a)$ for some $a\subseteq\omega$ and ϕ is Σ^1_2. Since $a^\#$ exists there is $a\in K$ such that $\phi(a)$. But then $K\models\phi(a)$ by lemma 6. So $K\models\exists a\phi(a)$. □

Lemmas 7-9 are due to Jensen ([28]), who also proved the following, unpublished, result.

Lemma 10 *Consis(ZFC+there is a measurable cardinal) $\Rightarrow$ Consis(ZFC+ there is $j:K\to_e K$ with κ critical but no ρ-model with κ critical).*

Proof: By generically collapsing cardinals if necessary we may assume that there is a countable κ that is the critical point of some ρ-model, and, letting the ρ-model be $L[U]$, that $(\kappa^+)^{L[U]}$ is countable. Let $j:L[U]\to_U L[U']$ and let $\kappa'=j(\kappa)$. Let $\tilde\kappa=(\kappa'^+)^{L[U']}$.

Claim 1 $\tilde\kappa$ is countable.

Proof: If $j(f)(\kappa)<\tilde\kappa$ we may assume $f:\kappa\to(\kappa^+)^{L[U]}$. Now $L[U]\models(\kappa^+)^K=\kappa^+$; and $(\kappa^+)^{L[U]}$ is countable. So $\tilde\kappa$ is countable. □(Claim 1)

Now let $\phi(U,j)$ be the formula asserting that U is a normal measure in $J^U_{\hat\kappa}$, where $\hat\kappa=(\kappa^+)^K$, $K_{\hat\kappa}\subseteq J^U_{\hat\kappa}$ and $j:J^U_{\hat\kappa}\to_U J^{U'}_{\tilde\kappa}$. Then $\phi(U,j|J^U_{\hat\kappa})$ holds. Since $\tilde\kappa$ is countable there is a generic f such that $\tilde\kappa$ is countable in $M=L[U'][f]$.

Claim 2 κ is not the critical point of any ρ-model in M.

Proof: Suppose it were: then by uniqueness $L[U]\subseteq M$. If γ is a cardinal of $L[U']$, $\gamma>\tilde\kappa$ then γ is a cardinal of M, hence of $L[U]$. So there is δ such that $j(\delta)=\delta$ and for all α $\aleph^{L[U]}_{\delta+\alpha}=\aleph^{L[U']}_{\delta+\alpha}$. Let $\gamma=(\aleph_{\delta+\kappa^+})^{L[U]}$. γ is a strong limit of cofinality greater than κ in $L[U]$ so $j(\gamma)=\gamma$. But

$$\begin{aligned}j(\gamma)=(\aleph_{\delta+j(\kappa^+)})^{L[U']}&=(\aleph_{\delta+j(\kappa^+)})^{L[U]}\\&>(\aleph_{\delta+j(\kappa)})^{L[U]}\\&>(\aleph_{\delta+\kappa^+})^{L[U]}\\&=\gamma.\end{aligned}$$

Contradiction! □(Claim 2)

Now by lemma 4 there are $\bar U,\bar j$ in M such that $\phi(\bar U,\bar j)$.

Claim 3 $\langle J^{\bar U}_{\hat\kappa},\bar U\rangle$ is iterable.

Proof: Its first iterate is $J^{U'}_{\tilde\kappa}$ which is transitive. Therefore its $(1+\alpha)$-th iterate is the α-th iterate of $J^{U'}_{\tilde\kappa}$, which is iterable.

□(Claim 3)

Now let $\langle\langle J^{U_\alpha}_{\rho_\alpha}\rangle_{\alpha\in On}, \langle \pi_{\alpha\beta}\rangle_{\alpha\le\beta\in On}\rangle$ be the iteration of $J^{\bar{U}}_{\rho}$. For $f\in J^{\bar{U}}_{\rho}$, $f:\kappa^n\to K_\kappa$ let $\tilde{f}$ denote $\bigcup_{i\in On} \pi_{0i}(f)$. Then $\tilde{f}:On^n\to K$ is a function. Given $x\in K$, suppose $x\in K_{\kappa_\alpha}$ (where $\kappa_\alpha=\pi_{0\alpha}(\kappa)$). Then there are $\alpha_0<\ldots<\alpha_{m-1}<\alpha$ such that for some $f\in J^{\bar{U}}_{\rho}$, $f:\kappa^n\to K_\kappa$ $x=\tilde{f}(\kappa_{\alpha_0}\ldots\kappa_{\alpha_{m-1}})$.

For any formula ϕ , $\alpha_0<\ldots<\alpha_{m-1}<\beta, \alpha'_0<\ldots<\alpha'_{m-1}<\beta$ $J^{U_\beta}_{\rho_\beta}\models\phi(\kappa_{\alpha_0}\ldots\kappa_{\alpha_{m-1}},\pi_{0\beta}(x)) \Leftrightarrow \phi(\kappa_{\alpha'_0}\ldots\kappa_{\alpha'_{m-1}})$. So $K_{\kappa_{\alpha_{m-1}}}\models\phi(\tilde{f}(\kappa_{\alpha_0}\ldots\kappa_{\alpha_{m-2}})) \Leftrightarrow K_{\kappa_{\alpha'_{m-1}}}\models\phi(\tilde{f}(\kappa_{\alpha'_0}\ldots\kappa_{\alpha'_{m-2}}))$. In particular $\alpha<\beta \Rightarrow K_{\kappa_\alpha}\prec K_{\kappa_\beta}$, so for all α $K_{\kappa_\alpha}\prec K$. Hence for $\alpha_0<\ldots<\alpha_{m-1}$, $\alpha'_0<\ldots<\alpha'_{m-1}$ $K\models\phi(\vec{\tilde{f}}(\kappa_{\alpha_0}\ldots\kappa_{\alpha_{m-1}})) \Leftrightarrow \phi(\vec{\tilde{f}}(\kappa_{\alpha'_0}\ldots\kappa_{\alpha'_{m-1}}))$. So we may define a map $\tau:K\to K$ by $\tau(\tilde{f}(\kappa_{\alpha_0}\ldots\kappa_{\alpha_{m-1}}))=\tilde{f}(\kappa_{\alpha_0+1}\ldots\kappa_{\alpha_{m-1}+1})$. τ is well-defined, $\tau:K\to_e K$, $\tau|\kappa=id|\kappa$ and $\tau(\kappa)=\kappa'>\kappa$. Hence M is the model we are looking for. □

One can even add the stipulation that κ be uncountable, but then the proof becomes considerably more complex. On the other hand Jensen has proved that if $j:K\to_e K$ with critical point κ and $\kappa\ge\omega_2$ then there is a ρ-model with κ critical. It is, so far as I know, an open question whether it is consistent that there be $j:K\to_e K$ with countable critical point but no ρ-model with ω_1 critical (by chapter 16 exercise 2 there would be a ρ-model with ω_2 critical). I suspect that it is.

HISTORICAL NOTES

In these notes we give brief attributions of results where we know them. References in square brackets are to the bibliography that follows. These notes refer only to the body of the text; the introduction and the appendices give references for results as they go along.

Part 1: Fine Structure

1: The theory R is the BST of Gandy [17]. Rudimentary functions were developed by Jensen [26] and Gandy [17]. The presentation in, e.g., [26] is generally more elegant than that using the basis functions: but as we need these functions later on we introduce them.

These basis functions are a modification of those used by Gödel to define L in [19]; the present form and the proof of lemma 1.5 are derived from Barwise [1], although the proof there is complicated by a different convention about ordered n-tuples.

2: This chapter is a restatement of §2 of [26] allowing firstly for relative constructibility and secondly for the weak theory.

This presentation of fine structure would be rather perverse were there not already elementary expositions in [26] and [6]. Apart from the use we must later make of non well-founded structures it is interesting to see what can be done internally.

3: The notion of acceptability was foreshadowed in Silver's proof [53] of GCH in L[U]. Otherwise this chapter, together with chapter 4, contains the substance of fine structure theory from [26]; the relativised form here first appeared in [10]. The presentation of lemma 3.3 in [10] contains an error.

The definition of V^p and the proof of corollary 3.19 are due to Solovay [58].

4: The extension of embeddings theorems were first proved in the downward case in [26] and upward in [8]. Lemma 4.27 and corollary 4.28 are, of course, from Gödel [19].

Part 2: Normal Measures.

5: Ulam [60] gave a definition of measurability based on classical measure theory. Keisler and Tarski [33] found the definition used here. Corollary 5.7 was already known to Ulam but lemma 5.8 remained open for some time; it was proved in [20]. Definition 5.9 is Ulam's original formulation.

Claim 1 in lemma 5.16 is known as Los' theorem [37]. Corollary 5.20 was proved by Scott [51]; the more general result mentioned after lemma 5.8 is proved by Kunen [36].

Premice are defined in [10], where in addition they must be transitive. In this book transitivity is not insisted on until we deal with iterable premice.

6: Gaifman invented iterated ultrapowers: they were developed by Kunen [35], who proved all the results in this chapter.

7: Kunen [35] is the source of lemmas 7.1-7.7. The remainder of the chapter is due to Solovay [58]. The notion of remarkability in exercise 2 is from Silver [52].

8: Most of this chapter is from Kunen [35]. Lemma 8.3 was used in [35] to prove corollary 7.3 - see exercise 1. The criteria for iterability are also in [35]. The proof of lemma 8.9 used here was discovered by Jensen [31]. Comparability was studied in [35] . Lemma 8.19 appeared first, to our knowledge, in [31]. It is not used in [10].

Part 3: Mice.

9: Real mice were used by Silver in [54] to obtain a Δ^1_3 well-order of the reals in L[U]. A version of this proof will be given in chapter 13. Silver's mice all had $n(\underline{N})=0$. The material here, with the main definition 9.25, derives from [10]. Our approach there depended more on well-foundedness; hence in the more abstract framework of the present work the definition has been complicated.

The ordering $<_*$ was discovered by Solovay [58].

10: Solovay proved the preservation results [58]. Core mice were introduced in [10]. Lemma 10.27 was proved by Solovay; the rest of the chapter is from [10].

11: The strategy of the case $\omega\rho_N^{n(N)+1}<\kappa$ was invented by Silver [54]. He worked entirely with elementary substructures: the fine structural analysis was presented by Solovay in [58]; this is also the source of the argument for the case $\omega\rho_N^{n(N)+1}=\kappa$.

12: The implication (i) was proved by Gaifman [16]. Silver proved the same result using combinatorial techniques [52]. Solovay extracted $0^\#$ from lemma 12.9 and proved that it is a Π^1_2 singleton [57], and hence lemma 12.22 and 12.23. (ii) was presumably noticed by everyone. Kunen obtained $0^\#$ from the existence of a non-trivial $j:L\to_e L$; his proof is easily adapted to give (iii).

13: All the results in the first part were proved by Kunen [35]. $0^\dagger$ was invented by Solovay, who as already remarked also proved that $0^\#$ is Δ^1_3. Silver [54] proved lemma 13.24; his proof was essentially the one given here. The full theory of double mice was worked out by Hodgetts in [23].

Part 4: The Core Model.

14: The results in this chapter are all from [10].

15: The first part of this chapter is taken from [28]. Sharplike mice are there called critical but this seems confusing. The examples are folklore. Koepke [34] showed that $L^{\#\infty}$ is really very like L indeed. The results on ρ-models are in [10].

16: The indiscernibles lemma was proved by Jensen [28], generalising a result of Mitchell (see chapter 17). The final argument is due to Rowbottom [49]. The embeddings lemma was proved in [11] on the assumption that the image of the critical point is regular. This suffices for the covering lemma. Jensen then used the covering lemma to prove lemma 16.21. The present direct proof was discovered later by Jensen ([29]).

17: Definition 17.2 is due to Erdős and Rado [15]. Lemma 17.4 was proved by Erdős and Hajnal [14]. The definition of α-Erdős is in Baumgartner [2]. Lemma 17.9 was proved by Schmerl [50]. Corollary 17.14 was proved by Mitchell [43]; the generalisation to lemma 17.13 is in [28]. Silver [52] proved corollary 17.16 directly.

Lemmas 17.19 and 17.21 are in [28], as is the conjecture. $\diamond$ and $\square_\kappa$ were proved in K by Welch [61] : many corollaries are examined in [7].

The results of [28] have appeared in [12].

Part 5: The Covering Lemma

18: The upward mapping theorem for mice was proved in [11]. The proof here incorporates simplifications discovered by Silver and Magidor - see [24]. These do not shorten the work on viciousness but they do enable us to make considerable abbreviations in chapter 19.

19: The result for L was proved in [8]. Lemma 19.26 was first proved in [11]; the Silver-Magidor simplification of vicious sequences simplifies the argument.

20: The proof of lemma 20.16 was given in [11]. Jensen discovered an error in the original proof and repaired it in [30]. Prikry sequences were introduced by Prikry in [48]. Lemma 20.17 is due to Dehornoy [5].

Our self-imposed "no forcing" rule is a particularly severe restriction in this chapter and the reader should consult appendix I for the real definition of Prikry sequence. The form used in this chapter is an equivalent proved by Mathias [40].

21: The corollaries about singular cardinals are straightforward generalisations of results in [8]. A detailed account of the development of SCH will be found in the introduction. Lemma 21.6 was proved by Bukovsky [3]. Lemma 21.8 embodies many results of which the oldest - the successor case - is due to Hausdorff [22]. Lemma 21.10 is the main result of [11]; for singular cardinals of uncountable cofinality Silver had proved a weak form without any assumption about large cardinals [55]. Lemma 21.11 is also from [11], where it is proved by the simpler expedient of appealing to the fact that $L[C]\models GCH$.

$a^{\#}$ was invented by Solovay [57]. The main result of this section, lemma 21.22, is an adaptation by Jensen [28] of a result of Paris [47]. Jensen also proved lemma 21.14 [27]. The model referred to at the end in which $\beta>1$ was discovered by Solovay.

Part 6: Larger Core Models.

22: The material on coherent sequences of normal measures was discovered by Mitchell [42]. Mitchell worked out the theory of mice based on coherent sequences [45] and also produced a generalisation of measure - called a hypermeasure - that is equivalent to the notion of extender - see [44].

Extenders were defined, and all the results on them stated here proved, in [31].

23: Mitchell proved the results up to lemma 23.11 in [44] using hypermeasures. The proofs here are taken from [31], as is the theory of $0^{¶}$. Lemma 23.8 adapts an argument of Solovay [59].

24: Superstrong cardinals were considered in [9] together with virtual extenders. Supercompact cardinals were defined by Reinhardt and Solovay. Lemma 24.3 and its corollary are due to Kunen [36]. The covering lemma on the assumption of no measurable limit of measurables was discovered by Mitchell.

BIBLIOGRAPHY

1. BARWISE K.J. Admissible Sets and Structures. Springer-Verlag, Berlin Heidelberg and New York, 1975.

2. BAUMGARTNER J.E. Ineffability properties of cardinals. In "Logic, Foundations of Mathematics and Computability Theory" (R.Butts and J.Hintikka eds.) D.Reidel, Dordrecht-Holland, 1977.

3. BUKOVSKY L. The Continuum Problem and the Powers of Alephs. Comment. Math. Univ. Carolinae 6(1965) 181-197.

4. COHEN P.J. The Independence of the Continuum Hypothesis. Proc. Nat. Acad. Sci. U.S.A. 50(1963), 1143-1148; 51(1964), 105-110.

5. DEHORNOY P. Iterated Ultrapowers and Prikry Forcing. Ann. Math. Log. 15(1978) 109-160.

6. DEVLIN K.J. Aspects of Constructibility. Lecture Notes in Mathematics 354, Springer-Verlag, Berlin and New York 1973.

7. DEVLIN K.J. The axiom of constructibility: a guide for the Mathematician, Lecture Notes in Mathematics 617, Springer-Verlag Berlin and New York, 1977.

8. DEVLIN K.J. & JENSEN R.B. Marginalia to a Theorem of Silver. ⊨ISILC Logic Conf. Lecture Notes in Mathematics 499, Springer-Verlag, Berlin and New York 1975.

9. DODD A.J. Superstrong Cardinals. Handwritten Notes, Oxford 1980.

10. DODD A.J. & JENSEN R.B. The Core Model. Ann. Math. Log. 20(1981) 43-75.

11. DODD A.J. & JENSEN. R.B. The Covering Lemma for K. Ann. Math. Log. To appear.

12. DONDER D., JENSEN R.B. & KOPPELBURG B.J. Some Applications of The Core Model. Freie Universität Berlin Preprint 115, 1981.

13. EASTON W.B. Powers of Regular Cardinals. Ann. Math. Log.

1(1970) 139-178.

14. ERDÖS P. & HAJNAL A. Some remarks concerning our paper "On the Structure of Set Mappings". Acta Math. Acad. Sci. Hungar. 13(1962) 223-226.

15. ERDÖS P. & RADO R. A Partition Calculus in Set Theory. Bull. Amer. Math. Soc. 62(1956) 427-489.

16. GAIFMAN H. Elementary Embeddings of Models of Set Theory and certain subtheories. In "Axiomatic Set Theory", Proc. Symp. Pure. Math. 13 II (T.Jech ed.) 33-101 Amer. Math. Soc. Providence, Rhode Island, 1974.

17. GANDY R.O. Set-theoretic Functions for Elementary Syntax. In "Axiomatic Set Theory", Proc. Symp. Pure. Math. 13 II (T.Jech ed.) 103-126 Amer. Math. Soc. Providence, Rhode Island, 1974.

18. GÖDEL K. The Consistency of the Axiom of Choice and of the Generalised Continuum Hypothesis. Proc. Nat. Acad. U.S.A. 24(1938) 556-557.

19. GÖDEL K. The Consistency of the Axiom of Choice and of the Generalised Continuum Hypothesis. Ann. Math. Studies 3(1940).

20. HANF W.P. Incompactness in Languages with Infinitely Long Expressions. Fund. Math. 53(1964) 309-324.

21. HANF W.P. & SCOTT D.S. Classifying Inaccessible Cardinals. Notices Amer. Math. Soc. 8(1961) 445.

22. HAUSDORFF F. Der Potenzbegriff in der Mengenlehre. Jahresber. Deutsch Math.-Verein. 13(1904) 569-571.

23. HODGETTS A. D.Phil. Thesis. Oxford 1980.

24. HOLZMANN R. Jensen's Covering Lemma - an Elementary Proof. Handwritten Notes, Jerusalem 1980.

25. JECH T. Set Theory. Academic Press, New York, 1978.

26. JENSEN R.B. The Fine Structure of the Constructible Hierarchy. Ann. Math. Log. 4(1972) 229-308.

27. JENSEN R.B. Pendant to a Theorem of Paris. Handwritten Notes, Bonn, 1977.

28. JENSEN R.B. Applications of K. Handwritten Notes, Bonn 1977.

29. JENSEN R.B. Embeddings of K. Handwritten Notes, Bonn 1977.

30. JENSEN R.B. A Correction to "The Core Model". Handwritten Notes, Freiburg 1980.

31. JENSEN R.B. & DODD A.J. The Core Model for Extenders. Handwritten Notes, Oxford 1978-80.

32. JENSEN R.B. & KARP C. Primitive Recursive Set Functions. In "Axiomatic Set Theory", Proc. Symp. Pure Math. 13 I (D.Scott ed.) 143-167 Amer. Math. Soc. Providence, Rhode Island, 1971.

33. KEISLER H.J. & TARSKI A. From Accessible to Inaccessible Cardinals. Fund. Math. 53(1964) 225-308.

34. KOEPKE P. $L^{\#}$. Dissertation, Bonn.

35. KUNEN K. Some Applications of Iterated Ultrapowers in Set Theory. Ann. Math. Log. 1(1970) 179-227.

36. KUNEN K. Elementary Embeddings and Infinitary Combinatorics. J. Symbolic Logic 36(1971) 407-413.

37. LOS J. Quelques Remarques, Théorèmes et Problèmes sur les Classes Définissables d'Algèbres. In "Mathematical Interpretation of Formal Systems (T.Skolem et al. eds.) 98-113 North Holland Publ. Co. Amsterdam 1970.

38. MAGIDOR M. On the Singular Cardinals Problem. I Israel J. Math. 28(1977) 1-31; II Ann. Math. 106(1977) 517-547.

39. MAREK W. & SREBRNY M. Gaps in the Constructible Universe. Ann. Math. Log. 6(1974) 359-394.

40. MATHIAS A. On Sets Generic in the sense of Prikry. J. Austral. Math. Soc. 15(1973) 409-414.

41. MENAS T.K. Consistency results concerning Supercompactness. Trans. Amer. Math. Soc. 223(1976) 61-91.

42. MITCHELL W.J. Sets constructible from Sequences of Ultrafilters J. Symbolic Logic 39(1974) 57-66.

43. MITCHELL W.J. Ramsey Cardinals and Constructibility. J. Symbolic Logic 44(1979) 260-266.

44. MITCHELL W.J. Hypermeasurable Cardinals. In "Logic Colloquium 1978" (M.Boffa et al. eds) 303-316. North Holland Publ. Co. Amsterdam 1979.

45. MITCHELL W.J. The Core Model for Sequences of Measures. Typewritten Notes. Penn. State 1980.

46. NAMBA K. Independence Proof of (ω,ω_α)-distributive Law in Complete Boolean Algebras. Comment. Math. Univ. St. Pauli 19(1970) 1-12.

47. PARIS J.B. Patterns of Indiscernibles. Bull. London Math. Soc. 6(1974) 183-188.

48. PRIKRY K.L. Changing Measurable into Accessible Cardinals. Diss. Math. 68(1970) 5-52.

49. ROWBOTTOM F. Some Strong Axioms of Infinity incompatible with the Axiom of Constructibility. Ann. Math. Log. 3(1971) 1-44.

50. SCHMERL J.H. On κ-like structures which embed Stationary and Closed Unbounded Subsets. Ann. Math. Logic 11(1976) 289-314.

51. SCOTT D.S. Measurable Cardinals and Constructible Sets. Bull. Acad. Polon. Sci. 9(1961) 521-524.

52. SILVER J.H. Some Applications of Model Theory in Set Theory. Ann. Math. Logic 3(1971) 45-110.

53. SILVER J.H. The Consistency of the GCH with the existence of a Measurable Cardinal. In "Axiomatic Set Theory" Proc. Symp. Pure Math. 13 I (D.Scott ed.) 391-396 Amer. Math. Soc. Providence, Rhode Island, 1971.

54. SILVER J.H. Measurable Cardinals and Δ^1_3 well orderings. Ann. Math. 94(1971) 414-446.

55. SILVER J.H. On the Singular Cardinals Problem. Proc. Int. Congr. Math. Vancouver 1974 265-268.

56. SOLOVAY R.M. Independence Results in the Theory of Cardinals. (Abstract). Notices Amer. Math. Soc. 10(1963) 595.

57. SOLOVAY R.M. A non-constructible Δ^1_3 set of Integers. Trans. Amer. Math. Soc. 127(1967) 50-75.

58. SOLOVAY R.M. The Fine Structure of L[μ]. Handwritten Notes Berkeley 1972.

59. SOLOVAY R.M., REINHARDT W.N. & KANAMORI A. Strong Axioms of Infinity and Elementary Embeddings. Ann. Math. Logic 13(1978) 73-116.

60. ULAM S. Zur Masstheorie in der algemeinen Mengenlehre. Fund. Math. 16(1930) 140-150.

61. WELCH P. D.Phil thesis. Oxford 1979.

INDEX OF DEFINITIONS

Note: n.Xm denotes exercise m in chapter n.

We have not listed terms that arise from defined terms by our subscripting convention: ρ_N for example. Note, in connection with definition 15.6, that K^M arises from the convention whereas K_M is defined in its own right.